Iron Catalysis in Organic Chemistry

Edited by
Bernd Plietker

Related Titles

Cornils, B., Herrmann, W. A., Muhler, M., Wong, C.-H. (eds.)

Catalysis from A to Z

A Concise Encyclopedia

2007

ISBN: 978-3-527-31438-6

Tietze, L. F., Brasche, G., Gericke, K. M.

Domino Reactions in Organic Synthesis

2006

SBN: 978-3-527-29060-4

Yudin, A. K. (ed.)

Aziridines and Epoxides in Organic Synthesis

2006

ISBN: 978-3-527-31213-9

Cornils, B., Herrmann, W. A., Horvath, I. T., Leitner, W., Mecking, S., Olivier-Bourbigou, H., Vogt, D. (eds.)

Multiphase Homogeneous Catalysis

2005

ISBN: 978-3-527-30721-0

Christoffers, J., Baro, A. (eds.)

Quaternary Stereocenters

Challenges and Solutions for Organic Synthesis

2005

ISBN: 978-3-527-31107-1

Dyker, G. (ed.)

Handbook of C-H Transformations

Applications in Organic Synthesis

2005

ISBN: 978-3-527-31074-6

Knochel, P. (ed.)

Handbook of Functionalized Organometallics

Applications in Synthesis

2005

ISBN: 978-3-527-31131-6

Iron Catalysis in Organic Chemistry

Reactions and Applications

Edited by
Bernd Plietker

WILEY-VCH Verlag GmbH & Co. KGaA

The Editor

Prof. Dr. Bernd Plietker
Institut für Organische Chemie
Universität Stuttgart
Pfaffenwaldring 55
70569 Stuttgart
Germany

Library of Congress Card No.: applied for

British Library Cataloguing-in-Publication Data
A catalogue record for this book is available from the British Library.

Bibliographic information published by the Deutsche Nationalbibliothek
Die Deutsche Nationalbibliothek lists this publication in the Deutsche Nationalbibliografie; detailed bibliographic data are available in the Internet at http://dnb.d-nb.de.

Typesetting Thomson Digital, Noida, India
Printing Strauss GmbH, Mörlenbach
Binding Litges & Dopf GmbH, Heppenheim
Cover Design Adam-Design, Weinheim

Printed in the Federal Republic of Germany
Printed on acid-free paper

ISBN: 978-3-527-31927-5

Contents

Iron Catalysis in Organic Chemistry. Edited by Bernd Plietker

ISBN: 978-3-527-31927-5

Preface

Sustainability has emerged as one of the keywords in discussions in the fields of politics, society and science within the past 20 years. The need for the production of high-quality products with minimum waste and energy demands is a key challenge in today's environment. This is even more salient when the increase in the world population and the decrease in fossil fuel resources are considered. In the field of chemistry, the concept of sustainability is clearly defined by the use of low-waste chemical transformations plus the use of catalysts in order to decrease the amount of energy needed for a process. Although most catalytic reactions fulfill the criteria for a sustainable transformation on the macroscopic scale, often the high price of catalysts, which are mostly transition metal based and their ligands paired with an inherent toxicity contradict these criteria on a microscopic scale.

This contradiction has spurred interest in developing transformations that make use of *sustainable catalysis*. Organocatalysis and biocatalysis both fulfill the criteria of sustainable catalysis: The catalysts are cheap, readily accessible and non-toxic. In the field *of sustainable metal catalysis*, iron-catalyzed transformations have evolved as powerful tools for performing organic synthesis. This development is somewhat surprising if one considers that the earliest iron catalysis dates back to the 1960s, a point in time where late transition metal catalysis using palladium, ruthenium or rhodium was still in its infancy. For some reason, which to me is one of the mysteries in metal catalysis, iron complexes never attracted the same interest in catalysis as their higher homologues in Group VIII metals, e.g. Ru, Os, Rh, Ir, Pd and Pt. This development is even more astonishing if one considers the rich organometallic chemistry of iron. It would appear that the current discussion about sustainability (energy resources, non-toxic reagents, catalysts and green solvents, etc.) has led to resurrection of *iron catalysis in organic synthesis* as a way to generate sustainable metal catalysis.

Due this recent revival, there is the need for an authoritative review of this important chemistry. It is the purpose of this book not only to introduce the chemistry community to the most recent achievements in the field of catalysis, but also to create a deeper understanding of the underlying fundamentals in the organometallic chemistry of iron complexes.

Iron Catalysis in Organic Chemistry. Edited by Bernd Plietker

ISBN: 978-3-527-31927-5

Consequently, the first chapter of this multi-authored book introduces the reader to the most general aspects of organoiron chemistry. The stability of complexes, together with prominent examples of stoichiometric iron-mediated organic transformations, are presented in a concise way, providing a first insight into and references to the leading review articles.

Iron complexes also play a dominant role in biological systems. The second chapter focuses on aspects connected with heme and non-heme iron catalysts in biological and biomimetic transformations. Most biological and biomimetic catalysts are employed in oxidation chemistry. Hence the reader can compare these systems with artificial catalytic oxidations, e.g. Gif chemistry and allylic and heteroatom oxidations, which are summarized in Chapter 3. Catalytic reductions in the presence of iron complexes are the synthetic counterpart to the oxidations and are reviewed in Chapter 4. Chapters 5–7 deal with different aspects of substitutions catalyzed by iron complexes. The cross-coupling of aryl or alkenyl halides with Grignard reagents in the presence of catalytic amounts of iron salts, which are reviewed in Chapter 5, has experienced almost explosive progress within the past 10 years. The current state of research is discussed with special emphasis on factors influencing the reactivity, e.g. solvent and temperature. Chapter 6 completes the aromatic substitution section by reviewing iron-catalyzed electrophilic substitutions. Chapter 7 focuses on nucleophilic substitutions either by using iron salts as Lewis acidic catalysts that facilitate the substitution of a leaving group by coordination or by employing low-valent ferrates as nucleophiles. Iron salts play a dominant role in catalytic addition and conjugate additions to carbonyl groups and these aspects are concisely presented in Chapter 8. The canon of iron-catalyzed transformation is rounded up by Chapter 9, which summarizes the current state of research in cycloaddition and ring expansion reactions.

I hope this book, *Iron Catalysis in Organic Chemistry. Reactions and Applications,* will stimulate further developments in this field and be of value to chemists both in academia and in industry.

Stuttgart, April 2008 *Bernd Plietker*

List of Contributors

Angelika Baro
Universität Stuttgart
Institut für Organische Chemie
Pfaffenwaldring 55
70569 Stuttgart
Germany

Ingmar Bauer
Technische Universität Dresden
Fachbereich Chemie
Bergstrasse 66
01069 Dresden
Germany

Matthias Beller
Leibniz-Institut für Katalyse e.V. an der
Universität Rostock
Albert-Einstein-Strasse 29a
18059 Rostock
Germany

Carsten Bolm
RWTH Aachen
Institut für Organische Chemie
Landoltweg 1
52056 Aachen
Germany

Martin Bröring
Philipps-Universität Marburg
Fachbereich Chemie
Hans-Meerwein-Strasse
35032 Marburg
Germany

Jens Christoffers
Carl von Ossietzky Universität
Oldenburg
Institut für Reine und Angewandte
Chemie
Carl-von-Ossietzky-Strasse 9–11
26111 Oldenburg
Germany

Stephan Enthaler
Leibniz-Institut für Katalyse eV und
Universität Rostock
Albert-Einstein-Strasse 29a
18059 Rostock
Germany

Herbert Frey
Carl von Ossietzky Universität
Oldenburg
Institut für Reine und Angewandte
Chemie
Carl-von-Ossietzky-Strasse 9–11
26111 Oldenburg
Germany

Iron Catalysis in Organic Chemistry. Edited by Bernd Plietker

ISBN: 978-3-527-31927-5

Olga García Mancheño
RWTH Aachen
Institut für Organische Chemie
Landoltweg 1
52056 Aachen
Germany

Gerhard Hilt
Philipps-Universität Marburg
Fachbereich Chemie
Hans-Meerwein-Strasse
35043 Marburg
Germany

Judith Janikowski
Philipps-Universität Marburg
Fachbereich Chemie
Hans-Meerwein-Strasse
35043 Marburg
Germany

Irina Jovel
Leibniz-Institut für Katalyse e.V. an der
Universität Rostock
Albert-Einstein-Strasse 29a
18059 Rostock
Germany

Kathrin Junge
Leibniz-Institut für Katalyse eV und
Universität Rostock
Albert-Einstein-Strasse 29a
18059 Rostock
Germany

Jette Kischel
Leibniz-Institut für Katalyse e.V. an der
Universität Rostock
Albert-Einstein-Strasse 29a
18059 Rostock
Germany

Hans-Joachim Knölker
Technische Universität Dresden
Fachbereich Chemie
Bergstrasse 66
01069 Dresden
Germany

Sabine Laschat
Universität Stuttgart
Institut für Organische Chemie
Pfaffenwaldring 55
70569 Stuttgart
Germany

Andreas Leitner
Karl-Dillinger Strasse 14
67071 Ludwigshafen
Germany

Agathe Christine Mayer
RWTH Aachen
Institut für Organische Chemie
Landoltweg 1
52056 Aachen
Germany

Kristin Mertins
Leibniz-Institut für Katalyse e.V. an der
Universität Rostock
Albert-Einstein-Strasse 29a
18059 Rostock
Germany

Jens Müller
Technische Universität Dortmund
Anorganische Chemie
Otto-Hahn-Strasse 6
44227 Dortmund
Germany

Bernd Plietker
Universität Stuttgart
Institut für Organische Chemie
Pfaffenwaldring 55
70569 Stuttgart
Germany

Volker Rabe
Universität Stuttgart
Institut für Organische Chemie
Pfaffenwaldring 55
70569 Stuttgart
Germany

Anna Rosiak
Evonik Degussa GmbH
Rodenbacher Chaussee 4
63457 Hanau-Wolfgang
Germany

Alexander Zapf
Leibniz-Institut für Katalyse e.V. an der
Universität Rostock
Albert-Einstein-Strasse 29a
18059 Rostock
Germany

1 Iron Complexes in Organic Chemistry

Ingmar Bauer and Hans-Joachim Knölker

1.1 Introduction

Catalysis is an important field in both academic and industrial research because it leads to more efficient reactions in terms of energy consumption and waste production. The common feature of these processes is a catalytically active species which forms reactive intermediates by coordination of an organic ligand and thus decreases the activation energy. Formation of the product should occur with regeneration of the catalytically active species. The efficiency of the catalyst can be described by its turnover number, providing a measure of how many catalytic cycles are passed by one molecule of catalyst.

For efficient regeneration, the catalyst should form only labile intermediates with the substrate. This concept can be realized using transition metal complexes because metal–ligand bonds are generally weaker than covalent bonds. The transition metals often exist in different oxidation states with only moderate differences in their oxidation potentials, thus offering the possibility of switching reversibly between the different oxidation states by redox reactions.

Many transition metals have been applied as catalysts for organic reactions [1]. So far, iron has not played a dominant role in catalytic processes. Organoiron chemistry was started by the discovery of pentacarbonyliron in 1891, independently by Mond [2] and Berthelot [3]. A further milestone was the report of ferrocene in 1951 [4]. Iron catalysis came into focus by the Reppe synthesis [5]. Kochi and coworkers published in 1971 their results on the iron-catalyzed cross-coupling of Grignard reagents with organic halides [6]. However, cross-coupling reactions became popular by using the late transition metals nickel and palladium. More recently, the increasing number of reactions using catalytic amounts of iron complexes indicates a renaissance of this metal in catalysis. This chapter describes applications of iron complexes in organic chemistry and thus paves the way for an understanding of iron catalysis.

Iron Catalysis in Organic Chemistry. Edited by Bernd Plietker

ISBN: 978-3-527-31927-5

1.2
General Aspects of Iron Complex Chemistry

1.2.1
Electronic Configuration, Oxidation States, Structures

In complexes iron has an electronic configuration of $[Ar]4s^03d^8$. The most common oxidation states for iron are +2 and +3. Moreover, the oxidation states +6, 0, −1 and −2 are of importance. In contrast to osmium, iron never reaches its potential full oxidation state of +8 as a group VIII element. In air, most iron(II) compounds are readily oxidized to their iron(III) analogs, which represent the most stable and widespread iron species. For iron(II) complexes ($[Ar]4s^03d^6$) a coordination number of six with an octahedral ligand sphere is preferred. Iron(III) ($[Ar]4s^03d^5$) can coordinate three to eight ligands and often exhibits an octahedral coordination. Iron(III) generally is a harder Lewis acid than iron(II) and thus binds to hard Lewis bases. Iron(0) mostly coordinates five or six ligands with trigonal bipyramidal and octahedral geometry. Iron(–II) is tetrahedrally coordinated. Iron in low oxidation states is most interesting for organometallic chemistry and in particular for iron-catalyzed reactions because they can form more reactive complexes than their iron(II) and iron(III) counterparts. Therefore, iron(0) and iron(–II) compounds are favored for iron catalysis. Iron carbonyl complexes are of special interest due to their high stability with an iron(0) center capable of coordinating complex organic ligands, which represents the basis for organoiron chemistry.

1.2.2
Fundamental Reactions

The following fundamental reactions play a key role in organo-transition metal chemistry: halogen–metal exchange, ligand exchange, insertion, haptotropic migration, transmetallation, oxidative addition, reductive elimination, β-hydride elimination and demetallation. Generally, several of these reactions proceed sequentially to form a catalytic cycle. No stable product should be generated, as this would interrupt the catalytic cycle by preventing the subsequent step.

Oxidative addition generally increases the oxidation state of the metal by two units and, based on the common oxidation states of iron, leads from iron(0) to iron(II) or iron(–II) to iron(0). The former represents the most widespread system for iron catalysis in organic synthesis but the latter also has enormous potential for applications (see Section 1.4).

Oxidative additions are frequently observed with transition metal d^8 systems such as iron(0), osmium(0), cobalt(I), rhodium(I), iridium(I), nickel(II), palladium(II) and platinum(II). The reactivity of d^8 systems towards oxidative addition increases from right to left in the periodic table and from top to down within a triad. The concerted mechanism is most important and resembles a concerted cycloaddition in organic chemistry (Scheme 1.1). The reactivity of metal complexes is influenced by their

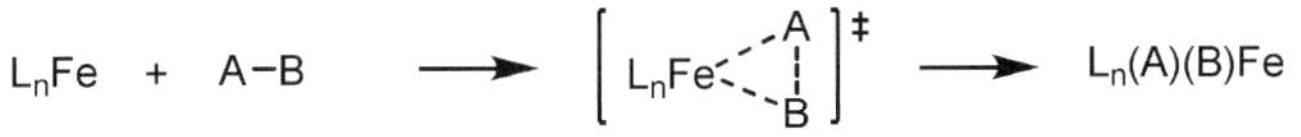

$$L_nFe + A{-}B \longrightarrow [L_nFe(A)(B)]^{\ddagger} \longrightarrow L_n(A)(B)Fe$$

Scheme 1.1

ligand sphere. Thus, strong σ-donor ligands and more poor π-acceptor ligands favor the oxidative addition due to increased electron density at the metal.

Reaction of the nucleophilic Collman's reagent ($Na_2Fe[CO]_4$) with two alkyl halides affords ketones via successive oxidative additions (Scheme 1.2) [7]. However, no catalytic cycle is achieved because the reaction conditions applied do not lead to regeneration of the reagent.

The reductive elimination eventually releases the newly formed organic product in a concerted mechanism. In the course of this process, the electron count is reduced by two. Iron has a great tendency for coordinative saturation, which in general does not favor processes such as ligand dissociation and reductive elimination. This aspect represents a potential limiting factor for catalytic reactions using iron.

Another important reaction typically proceeding in transition metal complexes is the insertion reaction. Carbon monoxide readily undergoes this process. Therefore, the insertion reaction is extremely important in organoiron chemistry for carbonylation of alkyl groups to aldehydes, ketones (compare Scheme 1.2) or carboxylic acid derivatives. Industrially important catalytic processes based on insertion reactions are hydroformylation and alkene polymerization.

Many metal-mediated reactions do not release the organic product by elimination but generate a stable transition metal complex, which prohibits a catalytic cycle. This is generally observed for diene–iron complexes which provide the free ligand only after removal of the metal by demetallation. Demetallation can be achieved using harsh oxidative conditions, which destroy the metal complex to give inert iron oxides. However, such conditions may lead to the destruction of sensitive organic ligands. In these cases, milder demetallation procedures are required to obtain the free ligand. For example, the demetallation has been a limiting factor for application of the

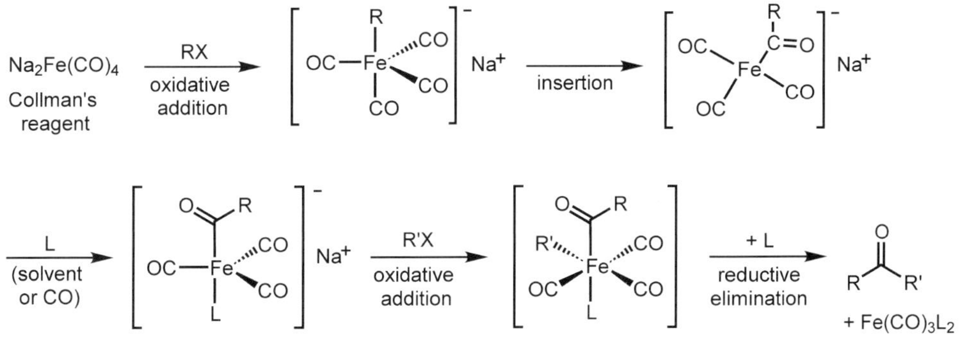

$Na_2Fe(CO)_4$ (Collman's reagent) —RX, oxidative addition→ $[RFe(CO)_4]^-\,Na^+$ —insertion→ $[(RC{=}O)Fe(CO)_3]^-\,Na^+$ —L (solvent or CO)→ $[(RCO)Fe(CO)_3L]^-\,Na^+$ —R'X, oxidative addition→ $(RCO)R'Fe(CO)_3L$ —+ L, reductive elimination→ $RC(O)R'$ + $Fe(CO)_3L_2$

Scheme 1.2

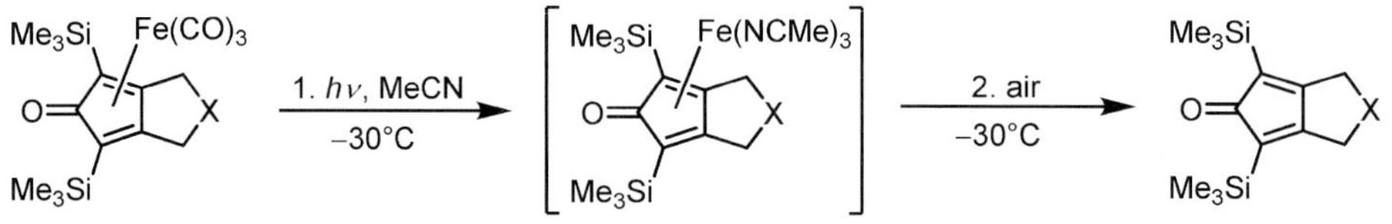

$X = CH_2, (CH_2)_2, C(COOMe)_2, S, O, NR$

Scheme 1.3

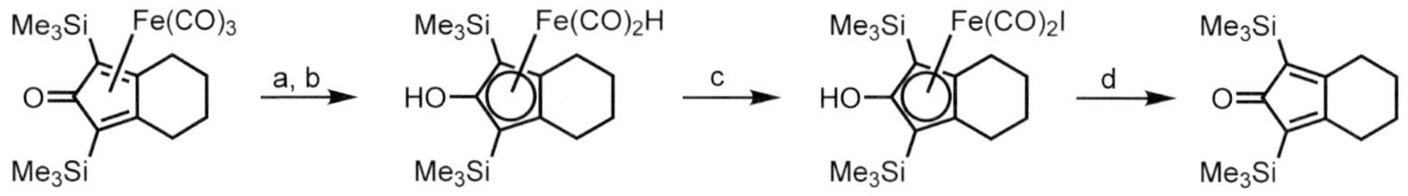

a) 1M NaOH/THF (1/2); b) H_3PO_4; c) $C_5H_{11}I$; d) air, daylight, Et_2O/THF, $Na_2S_2O_3$, Celite®

Scheme 1.4

iron-mediated [2 + 2 + 1]-cycloaddition. The demetallation of tricarbonyl(η^4-cyclopentadienone)iron complexes using trimethylamine *N*-oxide provides low yields. Photolytically induced ligand exchange of carbon monoxide by the poor π-accepting acetonitrile leads to intermediate very labile tri(acetonitrile)iron complexes. Demetallation by bubbling of air through the solution at low temperature affords the free ligands in high yields (Scheme 1.3) [8].

A further novel method for demetallation provides even higher yields. Hieber-type reaction of the tricarbonyl(η^4-cyclopentadienone)iron complex with sodium hydroxide to the corresponding hydride complex followed by ligand exchange with iodopentane affords an intermediate iodoiron complex, which is readily demetallated in the presence of air and daylight at room temperature (Scheme 1.4) [9]. Combining steps a–c in a one-pot procedure without isolation of the intermediate hydride complex gave yields of up to 98%.

These examples demonstrate that ligand exchange of carbon monoxide by poor π-acceptor ligands provides, due to decreased back-bonding, labile complexes which can be demetallated under mild reaction conditions, providing the corresponding free ligands in high yields.

1.3 Organoiron Complexes and Their Applications

In order to understand catalytic systems based on iron, the chemistry of organoiron complexes is briefly described and their reactivity is demonstrated. A comprehensive summary of the applications of iron compounds in organic synthesis has been given by Pearson [7].

1.3.1 Binary Carbonyl–Iron Complexes

Iron forms three stable homoleptic complexes with carbon monoxide, pentacarbonyliron ($Fe[CO]_5$), nonacarbonyldiiron ($Fe_2[CO]_9$) and dodecacarbonyltriiron ($Fe_3[CO]_{12}$) (Figure 1.1).

Pentacarbonyliron is a stable 18-electron complex of trigonal-bipyramidal geometry and represents the primary source of most organoiron complexes. Nonacarbonyldiiron is prepared in a photolytic reaction from pentacarbonyliron. Dodecacarbonyltriiron can be obtained from nonacarbonyldiiron by a thermal reaction. Both, nonacarbonyldiiron and dodecacarbonyltriiron, contain metal–metal bonds. They are slowly degraded to give pyrophoric iron and therefore should be handled with care.

Iron carbonyls have been used in stoichiometric and catalytic amounts for a variety of transformations in organic synthesis. For example, the isomerization of 1,4-dienes to 1,3-dienes by formation of tricarbonyl(η^4-1,3-diene)iron complexes and subsequent oxidative demetallation has been applied to the synthesis of 12-prostaglandin PGC_2 [10]. The photochemically induced double bond isomerization of allyl alcohols to aldehydes [11] and allylamines to enamines [12, 13] can be carried out with catalytic amounts of iron carbonyls (see Section 1.4.3).

Iron carbonyls also mediate the cycloaddition reaction of allyl equivalents and dienes. In the presence of nonacarbonyldiiron α,α'-dihaloketones and 1,3-dienes provide cycloheptenes (Scheme 1.5) [14, 15]. Two initial dehalogenation steps afford a reactive oxoallyliron complex which undergoes a thermally allowed concerted [4 + 3]-cycloaddition with 1,3-dienes. The 1,3-diene system can be incorporated in cyclic or heterocyclic systems (furans, cyclopentadienes and, less frequently, pyrroles). Noyori and coworkers applied this strategy to natural product synthesis, e.g. α-thujaplicin and β-thujaplicin [14, 16].

The reducing ability of iron(0) complexes has been exploited for functional group interconversion, for example reduction of aromatic nitro compounds to amines by dodecacarbonyltriiron [17].

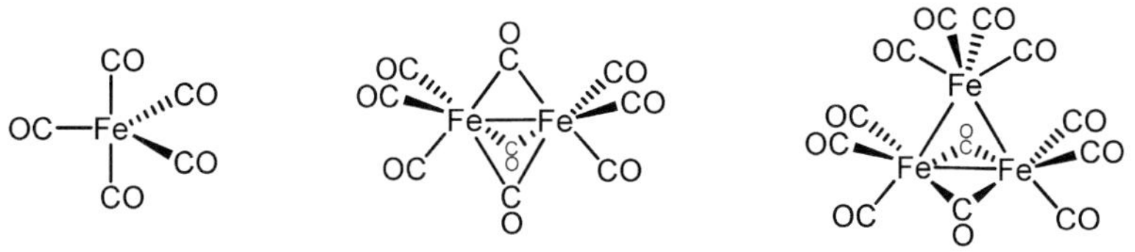

Figure 1.1 Homoleptic ironcarbonyl complexes.

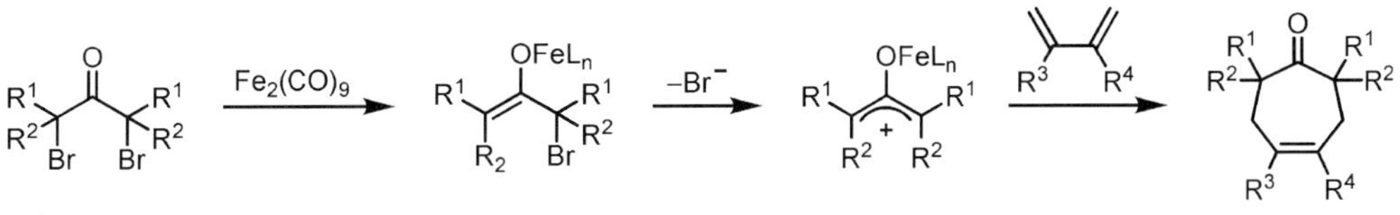

Scheme 1.5

Addition of nucleophiles to a carbon monoxide ligand of pentacarbonyliron provides anionic acyliron intermediates which can be trapped by electrophiles (H^+ or R—X) to furnish aldehydes or ketones [18]. However, carbonyl insertion into alkyl halides using iron carbonyl complexes is more efficiently achieved with disodium tetracarbonylferrate (Collman's reagent) and provides unsymmetrical ketones (Scheme 1.2) [19, 20]. Collman's reagent is extremely sensitive towards air and moisture, but offers a great synthetic potential as carbonyl transfer reagent. It can be prepared by an *in situ* procedure starting from $Fe(CO)_5$ and Na–naphthalene [20].

The reaction of two alkynes in the presence of pentacarbonyliron affords via a [2 + 2 + 1]-cycloaddition tricarbonyl(η^4-cyclopentadienone)iron complexes (Scheme 1.6) [5, 21–23]. An initial ligand exchange of two carbon monoxide ligands by two alkynes generating a tricarbonyl[bis(η^2-alkyne)]iron complex followed by an oxidative cyclization generates an intermediate ferracyclopentadiene. Insertion of carbon monoxide and subsequent reductive elimination lead to the tricarbonyl(η^4-cyclopentadienone)iron complex. These cyclopentadienone-iron complexes are fairly stable but can be demetallated to their corresponding free ligands (see Section 1.2.2). The [2 + 2 + 1]-cycloaddition requires stoichiometric amounts of iron as the final 18-electron cyclopentadienone complex is stable under the reaction conditions.

The iron-mediated [2 + 2 + 1]-cycloaddition to cyclopentadienones has been successfully applied to the synthesis of corannulene [24] and the yohimbane alkaloid (±)-demethoxycarbonyldihydrogambirtannine [25]. A [2 + 2 + 1]-cycloaddition of an alkene, an alkyne and carbon monoxide mediated by pentacarbonyliron, related to the well-known Pauson–Khand reaction [26], has also been described to afford cyclopentenones [27].

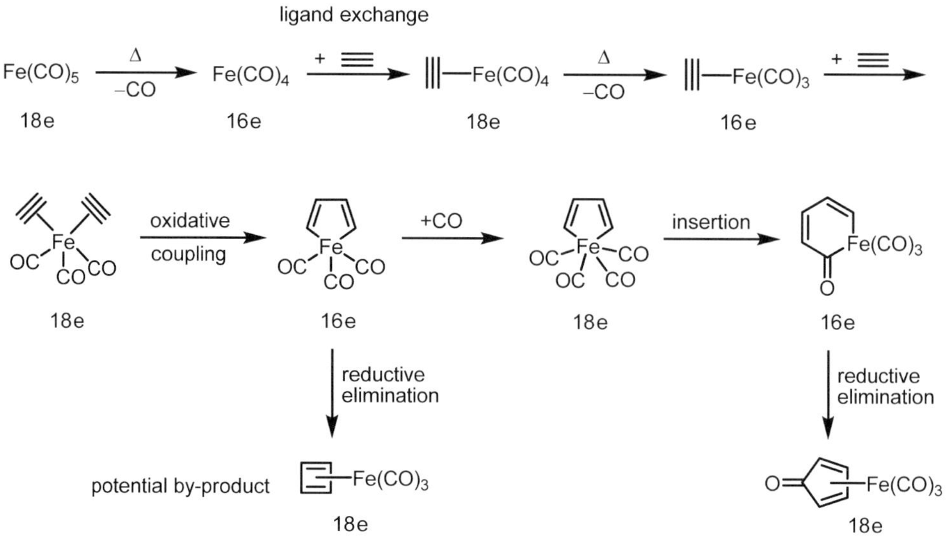

Scheme 1.6

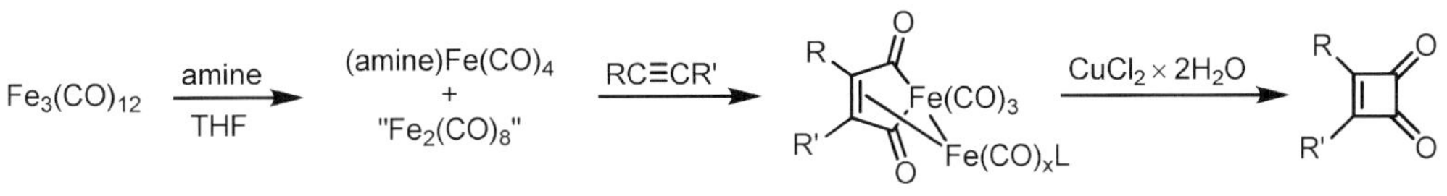

Scheme 1.7

A double insertion of carbon monoxide has been observed for the amine-induced reaction of alkynes with dodecacarbonyltriiron leading to cyclobutenediones (Scheme 1.7) [28].

In the presence of an excess of a primary amine, this reaction has been applied to the synthesis of cyclic imides [29].

1.3.2 Alkene–Iron Complexes

Neutral η^2-alkene–tetracarbonyliron complexes can be prepared from the corresponding alkene and nonacarbonyldiiron via a dissociative mechanism. The organic ligand in the alkene–iron complex is more easily attacked by nucleophiles than the corresponding free alkene due to the acceptor character of the tetracarbonyliron fragment. The reaction principle is demonstrated in Scheme 1.8 [30].

Malonate anions react with the η^2-ethylene–$Fe(CO)_4$ complex to afford after demetallation ethyl malonate derivatives. Reaction of nucleophiles with tetracarbonyliron-activated α,β-unsaturated carbonyl compounds leads after protonation of the intermediate alkyl–$Fe(CO)_4$ anions to the products of Michael addition.

Cationic alkene complexes of the type $[\eta^2\text{-alkene–Fp}]^+$ $[Fp = CpFe(CO)_2]$ are available by reaction of the alkenes with $CpFe(CO)_2Br$. Alternatively, several indirect routes to these complexes are provided by using $CpFe(CO)_2Na$. Both reagents can be prepared from the dimer $[Cp(CO)_2Fe]_2$. The Fp fragment serves as protecting group for alkenes and tolerates bromination and hydrogenation of other double bonds present in the molecule. Due to their positive charge, $[\eta^2\text{-alkene–Fp}]^+$ complexes react with a wide range of nucleophiles such as enamines, enolates, silyl enol ethers, phosphines, thiols and amines. The addition proceeds stereoselectively with the nucleophile approaching *anti* to the Fp group, but often shows poor regioselectivity. This drawback is overcome by using vinyl ether complexes, which are attacked by nucleophiles exclusively at the alkoxy-substituted carbon (Scheme 1.9). The intermediate alkyl–Fp complexes undergo elimination of alcohol and demetallation.

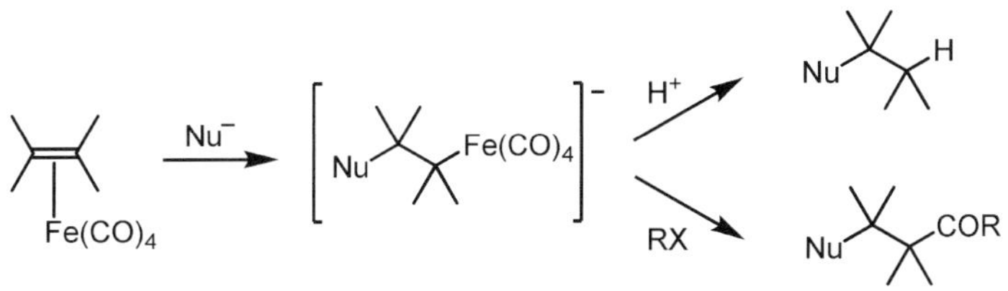

Scheme 1.8

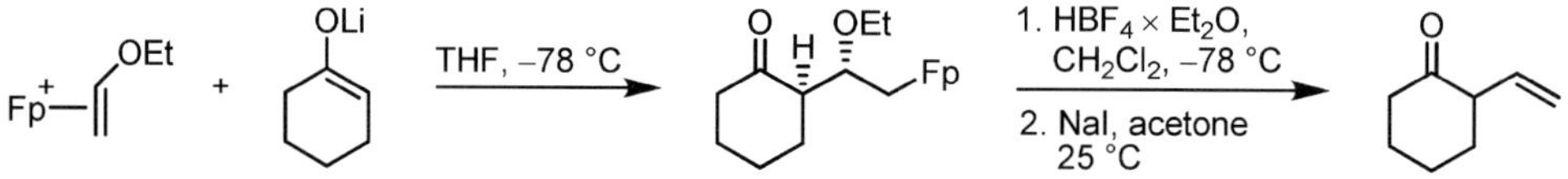

Scheme 1.9

Thus, $[\eta^2\text{-alkene–Fp}]^+$ complexes represent useful cationic synthons for the vinylation of enolates [31].

Related $[\text{alkyne–Fp}]^+$ complexes can be obtained by ligand exchange of $[\text{isobutylene–Fp}]^+$ complexes with alkynes [32].

1.3.3
Allyl– and Trimethylenemethane–Iron Complexes

Allyl complexes of the type η^1-allyl–Fp are prepared by reaction of $[Cp(CO)_2Fe]^-Na^+$ with allyl halides or, alternatively, by deprotonation of $[\eta^2\text{-alkene–Fp}]^+$ complexes. The most important reaction of η^1-allyl–Fp complexes is the $[3+2]$-cycloaddition with electron-deficient alkenes [33]. The reaction proceeds via a non-concerted mechanism, to afford Fp-substituted cyclopentanes (Scheme 1.10).

Removal of the σ-substituted Fp group can be achieved by conversion into the cationic alkene–Fp complex using Ph_3CPF_6 and subsequent treatment with iodide, bromide or acetonitrile. Oxidative cleavage with ceric ammonium nitrate in methanol provides the methyl esters via carbon monoxide insertion followed by demetallation. The $[3+2]$-cycloaddition has been successfully applied to the synthesis of hydroazulenes (Scheme 1.11) [34]. This remarkable reaction takes advantage of the specific nucleophilic and electrophilic properties of η^1-allyl–, cationic η^5-dienyl–, cationic η^2-alkene– and η^4-diene–iron complexes, respectively.

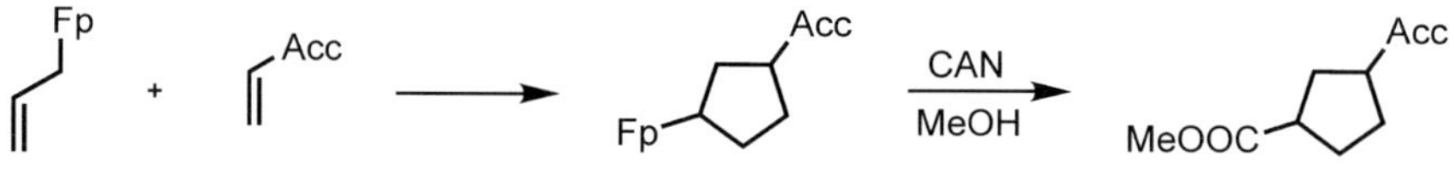

Acc = acceptor

Scheme 1.10

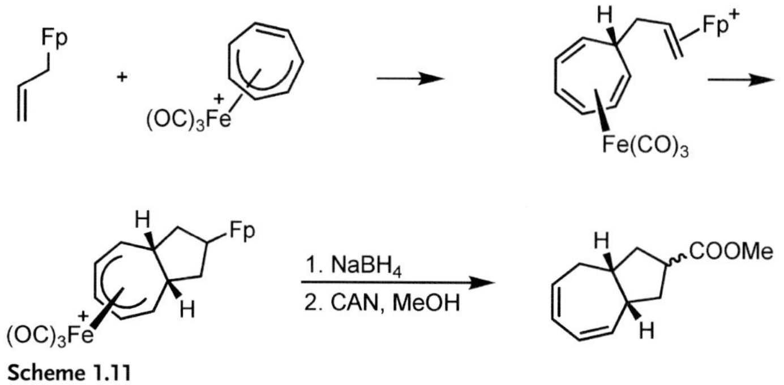

Scheme 1.11

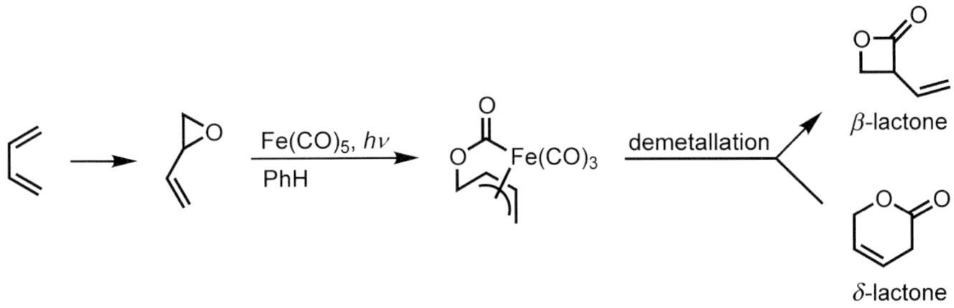

Fe(CO)5, hν
PhH
Fe(CO)3
demetallation
β-lactone
δ-lactone

Scheme 1.12

Cationic η^3-allyltetracarbonyliron complexes are generated by oxidative addition of allyl iodide to pentacarbonyliron followed by removal of the iodide ligand with $AgBF_4$ under a carbon monoxide atmosphere [35]. Similarly, photolysis of vinyl epoxides or cyclic vinyl sulfites with pentacarbonyliron or nonacarbonyldiiron provides π-allyltricarbonyliron lactone complexes. Oxidation with CAN provides by demetallation with concomitant coupling of the iron acyl carbon to one of the termini of the coordinated allyl moiety either β- or δ-lactones (Scheme 1.12) [36, 37]. In a related procedure, the corresponding π-allyltricarbonyliron lactam complexes lead to β- and δ-lactams [37].

In trimethylenemethane complexes, the metal stabilizes an unusual and highly reactive ligand which cannot be obtained in free form. Trimethylenemethanetricarbonyliron (R=H) was the first complex of this kind described in 1966 by Emerson and coworkers (Figure 1.2) [38]. It can be obtained by reaction of bromomethallyl alcohol with $Fe(CO)_5$. Trimethylenemethaneiron complexes have been applied for [3 + 2]-cycloaddition reactions with alkenes [39].

1.3.4
Acyl– and Carbene–Iron Complexes

Acyliron complexes have found many applications in organic synthesis [40]. Usually they are prepared by acylation of $[CpFe(CO)_2]^-$ with acyl chlorides or mixed anhydrides (Scheme 1.13). This procedure affords alkyl, aryl and α,β-unsaturated acyliron complexes. Alternatively, acyliron complexes can be obtained by treatment of $[Fe(C_5Me_5)(CO)_4]^+$ with organolithium reagents. α,β-Unsaturated acyliron complexes can be obtained by reaction of the same reagent with 2-alkyn-1-ols. Deprotonation of acyliron complexes with butyllithium generates the corresponding enolates, which can be functionalized by reaction with various electrophiles [40].

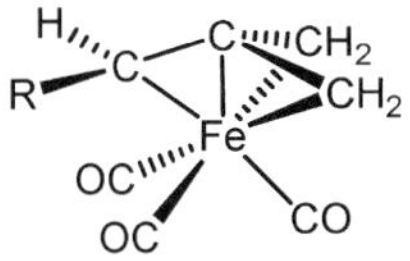

H, C, C, CH2, R, CH2, Fe, OC, OC, CO

Figure 1.2 Trimethylenemethanetricarbonyliron complexes.

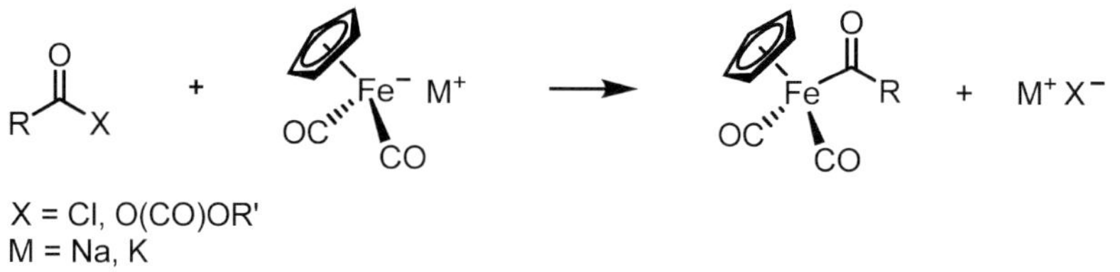

X = Cl, O(CO)OR'
M = Na, K

Scheme 1.13

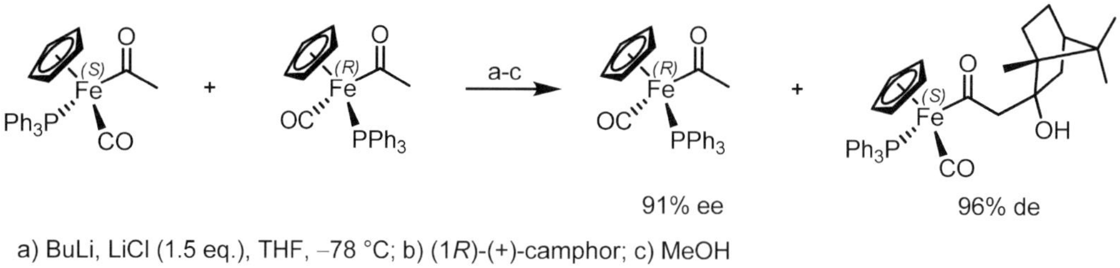

91% ee
96% de

a) BuLi, LiCl (1.5 eq.), THF, –78 °C; b) (1*R*)-(+)-camphor; c) MeOH

Scheme 1.14

Acyliron complexes with central chirality at the metal are obtained by substitution of a carbon monoxide with a phosphine ligand. Kinetic resolution of the racemic acyliron complex can be achieved by aldol reaction with (1*R*)-(+)-camphor (Scheme 1.14) [41]. Along with the enantiopure (R_{Fe})-acyliron complex, the (S_{Fe})-acyliron–camphor adduct is formed, which on treatment with base (NaH or NaOMe) is converted to the initial (S_{Fe})-acyliron complex. Enantiopure acyliron complexes represent excellent chiral auxiliaries, which by reaction of the acyliron enolates with electrophiles provide high asymmetric inductions due to the proximity of the chiral metal center. Finally, demetallation releases the enantiopure organic products.

α,β-Unsaturated acyliron complexes are versatile reagents and show high stereoselectivity in many reactions, e.g. as dienophiles in Diels–Alder reactions [42], as Michael acceptors for heteronucleophiles [43] and in [3 + 2]-cycloadditions with allyltributylstannane to cyclopentanes [44].

Fp-substituted enones and enals undergo cyclocarbonylations on treatment with metal hydrides or metal alkyls to provide γ-lactones (Scheme 1.15) [45]. Similarly, electron-rich primary amines afford dihydropyrrolones with iron-substituted (*Z*)-enals in the presence of titanium tetrachloride and triethylamine [46].

Dicarbonyl(η^5-cyclopentadienyl)iron–alkyl complexes represent useful precursors for iron–carbene complexes [47]. For example, iron–carbene complexes are intermediates in the acid-promoted reaction of Fp–alkyl ether derivatives with alkenes to provide cyclopropanes via a [2 + 1]-cycloaddition (Scheme 1.16).

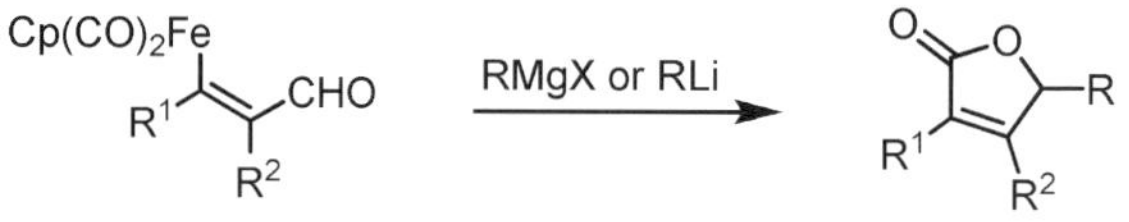

RMgX or RLi

Scheme 1.15

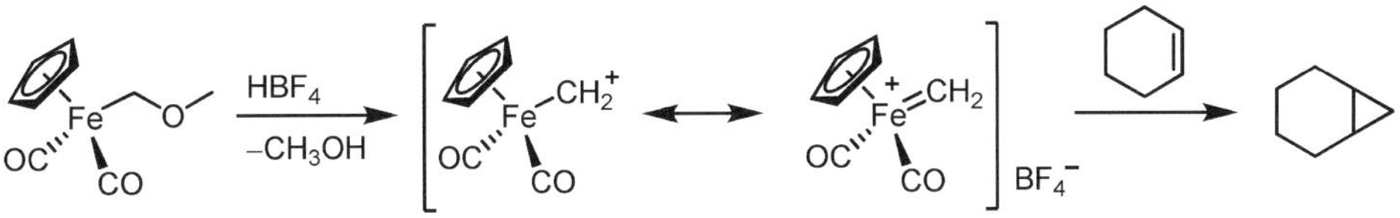

Scheme 1.16

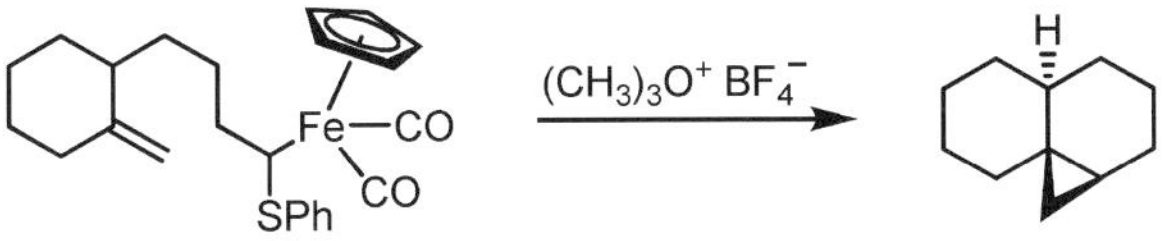

Scheme 1.17

In a related strategy, Helquist and coworkers used thioether-substituted Fp complexes and applied the intermediate iron–carbene complexes to cyclopropanation (Scheme 1.17), C−H insertion and Si−H insertion reactions [48].

A cyclopentane annelation by intramolecular C−H insertion of intermediate cationic iron–carbene complexes has been applied to the synthesis of the fungal metabolite (±)-sterpurene [49].

1.3.5
Diene–Iron Complexes

Acyclic and cyclic diene–iron complexes are stable compounds and have found a wide range of applications in organic synthesis [36, 50, 51]. The reactivity of the 1,3-diene system is altered drastically by coordination to the tricarbonyliron fragment. For example, the coordinated diene moiety does not undergo hydrogenation, hydroboration, dihydroxylation, Sharpless epoxidation, cyclopropanation and Diels–Alder cycloaddition reactions. Hence, the tricarbonyliron fragment has been used as a protecting group for diene systems. The reactivity of the diene unit towards electrophiles is decreased in the complex. However, the reactivity towards nucleophiles is increased due to donation of π-electrons to the metal. Therefore, the coordinated tricarbonyliron fragment may be regarded as an acceptor group. Moreover, the iron carbonyl fragment by its steric demand blocks one face of the diene moiety and serves as a stereo-directing group [51].

The classical protocol for synthesis of iron–diene complexes starts from the homoleptic pentacarbonyliron complex. In a stepwise fashion, via a dissociative mechanism, two carbonyl ligands are displaced by the diene system. However, thermal dissociation of the first CO ligand requires rather harsh conditions (ca. 140 °C). For acyclic 1,3-dienes, the diene ligand adopts an *s-cis* conformation to form stable η^4-complexes (Scheme 1.18).

Non-conjugated dienes isomerize during complexation to afford tricarbonyliron-coordinated conjugated dienes. This isomerization has been applied to a wide range of substituted cyclohexa-1,4-dienes available by Birch reduction from aromatic

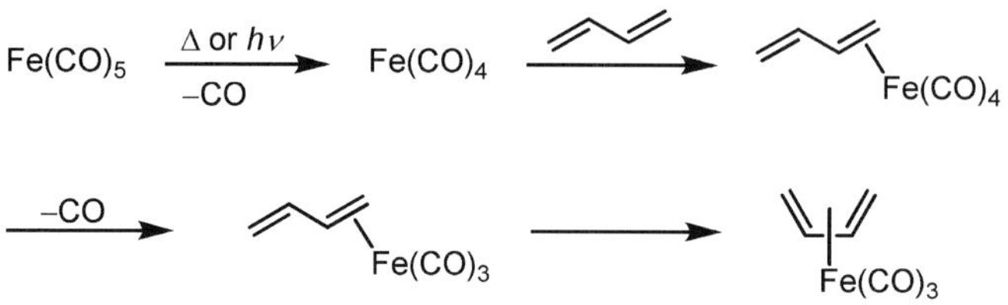

$Fe(CO)_5$ —(Δ or hν, −CO)→ $Fe(CO)_4$ —(butadiene)→ η^2-diene–$Fe(CO)_4$ —(−CO)→ η^2-diene–$Fe(CO)_3$ → η^4-diene–$Fe(CO)_3$

Scheme 1.18

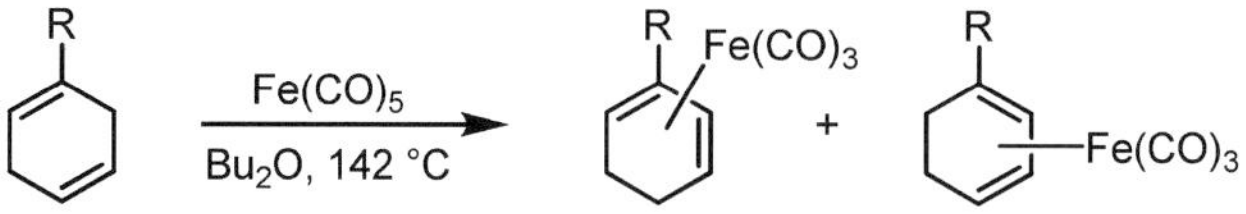

Scheme 1.19

compounds. However, often isomeric mixtures are formed by complexation of substituted cyclohexadienes (Scheme 1.19).

For introduction of the tricarbonyliron fragment under mild reaction conditions, tricarbonyliron transfer reagents have been developed [52]. Among them are tricarbonylbis(η^2-*cis*-cyclooctene)iron (Grevels' reagent) [53] and (η^4-benzylideneacetone) tricarbonyliron [54]. Grevels' reagent is prepared by photolytic reaction of pentacarbonyliron with *cis*-cyclooctene and transfers the tricarbonyliron fragment at temperatures below 0 °C (Scheme 1.20). Although the solid compound can be handled at room temperature, in solution the complex is very labile and stable only at temperatures below −35 °C.

(η^4-Benzylideneacetone)tricarbonyliron has been used for the synthesis of the tricarbonyliron complex of 8,8-diphenylheptafulvalene, which could not be prepared by reaction with pentacarbonyliron and dodecacarbonyltriiron owing to the sensitivity of the substrate to both heat and UV light. The mechanism which has been proposed for the transfer of the tricarbonyliron fragment using this reagent involves an initial η^4 to η^2 haptotropic migration (Scheme 1.21) [54b,c].

Because of the high lability of the reagents described above, (η^4-1-azabuta-1,3-diene)tricarbonyliron complexes have been developed as alternative tricarbonyliron transfer reagents. They are best prepared by an ultrasound-promoted reaction of 1-azabuta-1,3-dienes with nonacarbonyldiiron in tetrahydrofuran at room temperature. Using (η^4-1-azabuta-1,3-diene)tricarbonyliron complexes the transfer of the tricarbonyliron unit proceeds in refluxing tetrahydrofuran in high yields [55a,b].

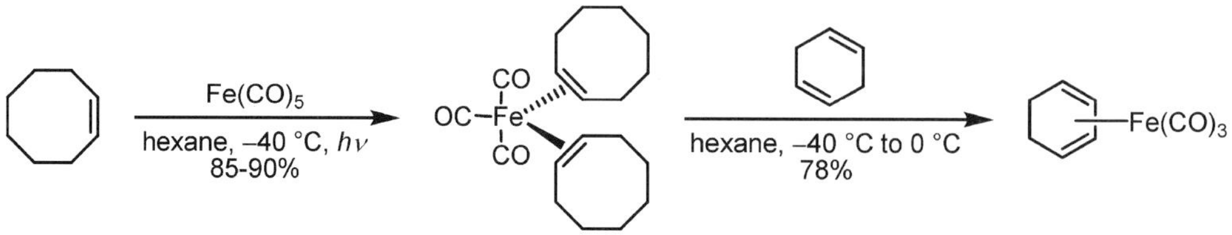

Scheme 1.20

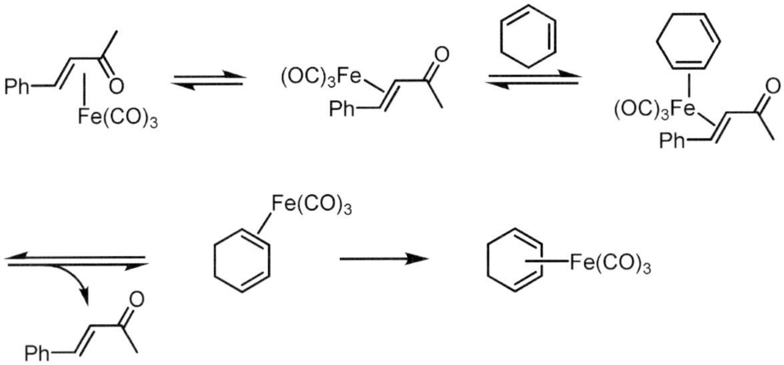

Scheme 1.21

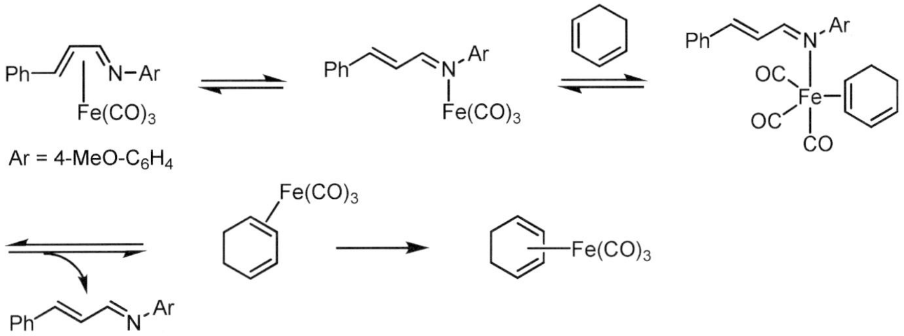

Ar = 4-MeO-C_6H_4

Scheme 1.22

Moreover, after transfer the free ligand can be recovered by crystallization. The mechanistic proposal for the transfer reaction is based on an initial η^4 to η^1 haptotropic migration (Scheme 1.22).

A catalytic process for the complexation of cyclohexadiene with pentacarbonyliron using 0.125 equivalents of the 1-azabuta-1,3-diene in refluxing dioxane affords quantitatively the corresponding tricarbonyliron complex [55c]. Supported by additional experimental evidence, the mechanism shown in Scheme 1.23 has been proposed for the 1-azadiene-catalyzed complexation [52].

1-Azabutadiene reacts with pentacarbonyliron by nucleophilic attack at a carbon monoxide ligand to form a σ-carbamoyliron complex. Subsequent intramolecular displacement of a carbon monoxide ligand affords an (η^3-allyl)carbamoyliron complex. Two consecutive haptotropic migrations (η^3 to η^2 and η^2 to η^1) provide a tetracarbonyl(η^1-imine)iron complex. Release of a second carbon monoxide generates tricarbonyl(η^1-imine)iron, a reactive 16-electron species. Via haptotropic migration (η^1 to η^4), this intermediate converts to the 18-electron (η^4-1-azabuta-1,3-diene) tricarbonyliron complex, the stoichiometric transfer reagent. At the stage of the reactive 16-electron intermediate, a double bond of the diene system can be coordinated at the metal. Regeneration of the 1-azadiene catalyst followed by haptotropic migration (η^2 to η^4) leads to the stable 18-electron complex,

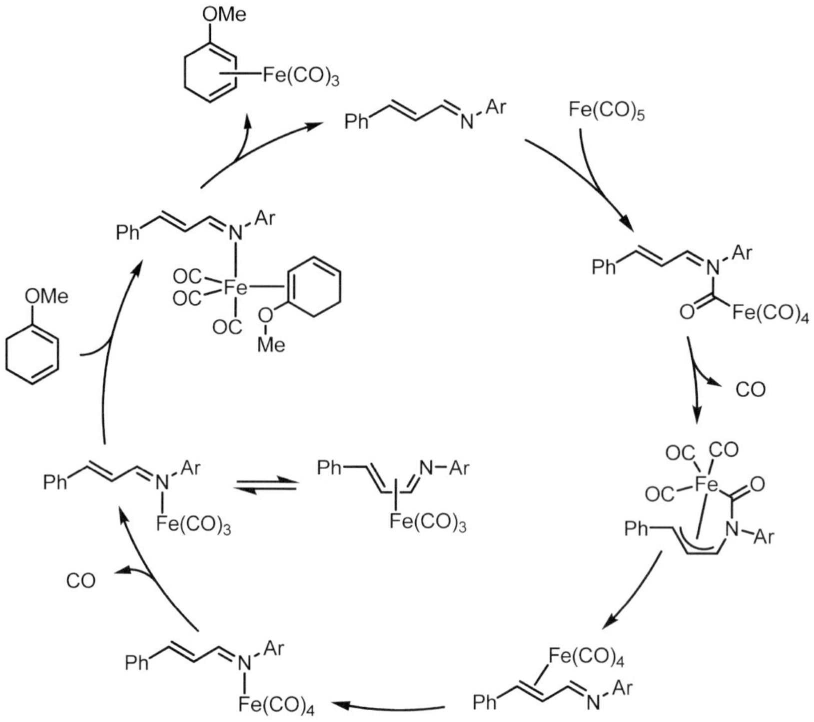

Ar = 4-MeO-C_6H_4

Scheme 1.23

tricarbonyliron(η^4-cyclohexa-1,3-diene)iron. This catalytic cycle contains metal complexes as substrate, product and catalytically active species.

The catalytic system described above has been further developed to an asymmetric catalytic complexation of prochiral 1,3-dienes (99% yield, up to 86% *ee*) using an optically active camphor-derived 1-azabutadiene ligand [56]. This method provides for the first time planar-chiral transition metal π-complexes by asymmetric catalysis.

Chemoselective oxidation of 4-methoxyanilines to quinonimines can be achieved in the presence of tricarbonyl(η^4-cyclohexadiene)iron complexes. This transformation has been used for the synthesis of carbazoles via intermediate tricarbonyliron-coordinated 4b,8a-dihydrocarbazol-3-one complexes (Scheme 1.24) [57].

The *p*-anisidine moiety is oxidized by commercial (water-containing) manganese dioxide to the non-cyclized quinonimine. Iron-mediated oxidative cyclization by treatment with *very active* manganese dioxide affords the tricarbonyliron-coordinated 4b,8a-dihydrocarbazol-3-one. The cyclohexadiene–iron complex is stable even in the presence of the adjacent quinonimine without any aromatization of both systems in an intramolecular redox reaction. The function of the tricarbonyliron fragment as protecting group becomes evident by demetallation with trimethylamine *N*-oxide leading to instantaneous aromatization of both rings. A number of 3-oxygenated carbazole alkaloids have been obtained by this route [58].

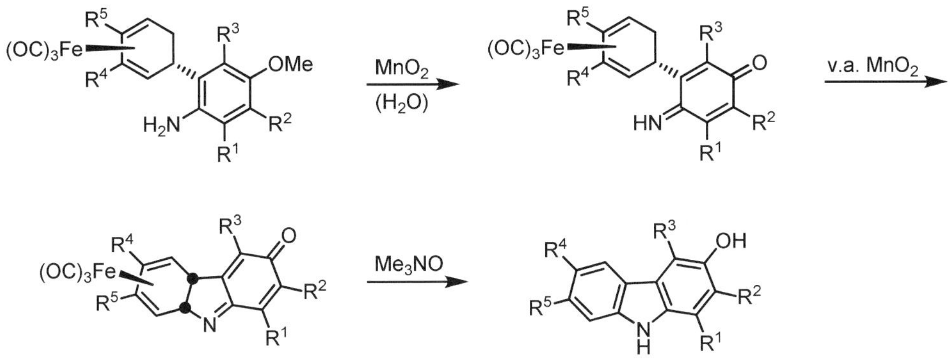

Scheme 1.24

Kinetic resolution can be accomplished by addition of allyl boronates to aldehyde groups adjacent to the tricarbonyliron fragment [59]. For the synthesis of ikarugamycin, Roush and Wada developed an impressive asymmetric crotylboration of a prochiral *meso* complex using a chiral diisopropyl tartrate-derived crotylborane (Scheme 1.25) [60]. In the course of this synthesis, the stereo-directing effect of the tricarbonyliron fragment has been exploited twice to introduce stereospecifically a crotyl and a vinyl fragment.

Cyclohexadienylium–tricarbonyliron complexes are readily available by hydride abstraction from cyclohexadiene–tricarbonyliron complexes using triphenylcarbenium tetrafluoroborate [61]. They are stable and can be handled in air. The hydride ion is removed at one of the non-coordinated carbon atoms from the face opposite to the metal fragment. The resulting cyclohexadienylium system is stabilized as an η^5-ligand by the tricarbonyliron fragment (Scheme 1.26).

Cyclohexadienylium–tricarbonyliron complexes represent the most versatile iron complexes applied as building blocks in synthetic organic chemistry. Because of their positive charge, a large variety of nucleophiles undergo nucleophilic attack at the

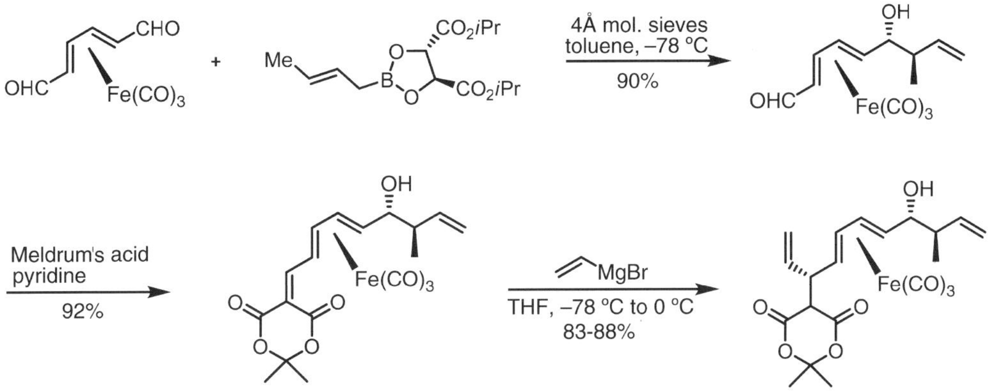

Scheme 1.25

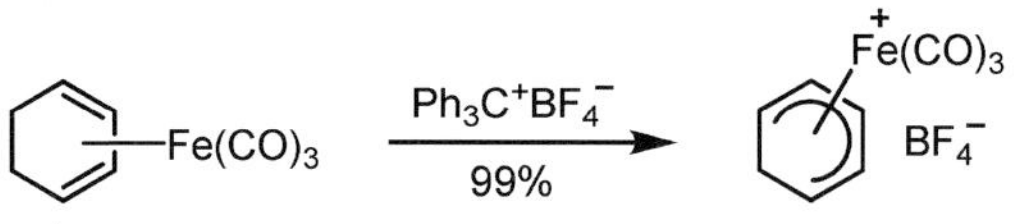

Scheme 1.26

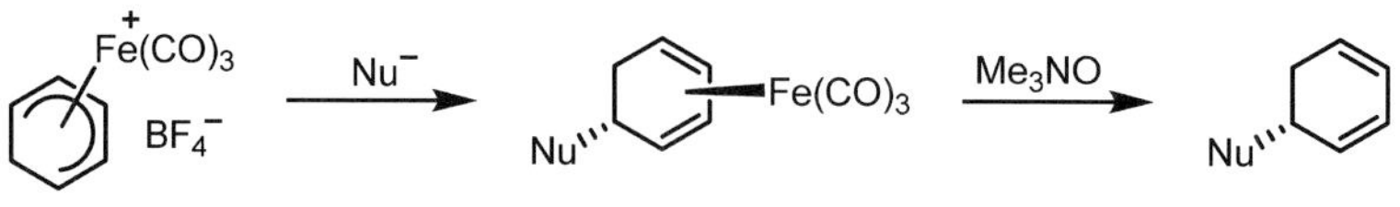

Scheme 1.27

coordinated ligand (Scheme 1.27). The attack of nucleophiles generally proceeds in high yields and takes place regioselectively at the terminus of the coordinated dienyl system (Davies–Green–Mingos rules) [62] and also stereoselectively *anti* to the tricarbonyliron fragment. Demetallation of the resulting functionalized cyclohexadienes to the corresponding free ligands can be achieved with different oxidizing agents (e.g. trimethylamine *N*-oxide).

Additions to substituted dienyl systems, depending on the position of the substituents, their electronic properties and the steric demand of the nucleophile, may lead to a variety of regioisomeric products. However, in most cases the regiochemical outcome of the reaction can be predicted [63]. Addition of nucleophiles to tricarbonyliron-coordinated 1-alkyl-4-methoxy-substituted cyclohexadienyl cations permits the stereoselective construction of quaternary carbon centers (Scheme 1.28). The selectivity of this addition is governed by the regio-directing effect of the methoxy group, which directs the incoming nucleophile in the *para*-position, and the stereo-directing effect of the tricarbonyliron fragment (*anti* selectivity). These building blocks have been used for synthetic approaches to several natural products, e.g. (±)-limaspermine [64], the spirocyclic discorhabdin and prianosin alkaloids [65] and *O*-methyljoubertiamine [66].

Using cationic tricarbonyl(η^5-cyclohexadienyl)iron complexes as starting materials, different synthetic routes to a large number of carbazole alkaloids have been developed [51, 58, 67]. The first step is an electrophilic substitution of a substituted arylamine using the cyclohexadienyliron complex and provides the corresponding 5-aryl-substituted cyclohexadiene–iron complexes (Scheme 1.29).

The construction of the carbazole framework is completed by an iron-mediated oxidative cyclization which proceeds via an initial single electron transfer to generate a 17-electron radical cation intermediate. Iron-mediated oxidative

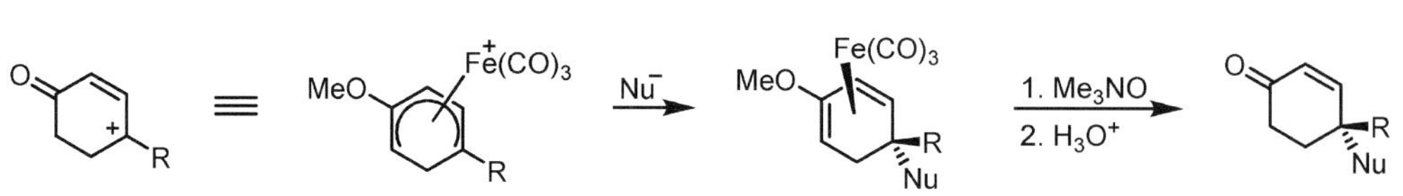

Scheme 1.28

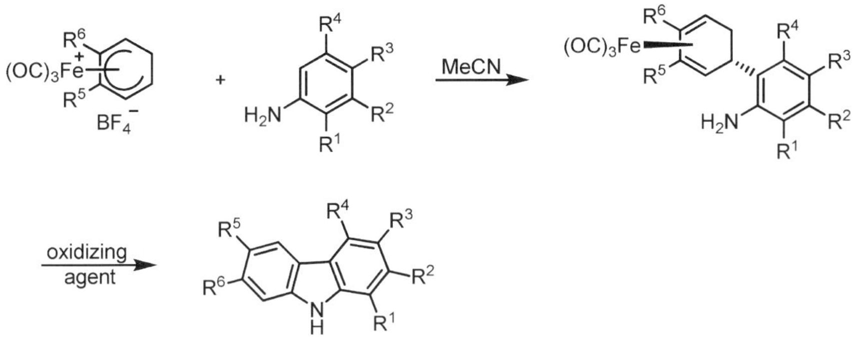

Scheme 1.29

cyclization and subsequent aromatization can be accomplished in a one-pot procedure by using several oxidizing agents (e.g., *very active* manganese dioxide, iodine in pyridine, ferricenium hexafluorophosphate–sodium carbonate). This method has been applied to the total synthesis of a wide variety of carbazole alkaloids [67]. Alternatively, an iron-mediated oxidative cyclization in air leading to stable 4a,9a-dihydrocarbazole–iron complexes has been developed. Final demetallation and dehydrogenation of these complexes afford the carbazoles (Scheme 1.30) [67, 68].

A further route, via initial oxidation of the arylamine to a quinonimine followed by oxidative cyclization to an iron-coordinated 4b,8a-dihydrocarbazol-3-one and demetallation to a 3-hydroxycarbazole, has been described above (Scheme 1.24).

Acyclic pentadienyliron complexes show a similar reactivity towards nucleophiles but have found less application so far. Donaldson and coworkers reported an interesting cyclopropanation starting from a pentadienyliron complex (Scheme 1.31) [69]. This procedure has been used for the stereoselective synthesis of cyclopropylglycines [70], the preparation of the C_9-C_{16} alkenylcyclopropane segment of ambruticin [71] and the synthesis of hydrazulenes via divinylcyclopropanes [72].

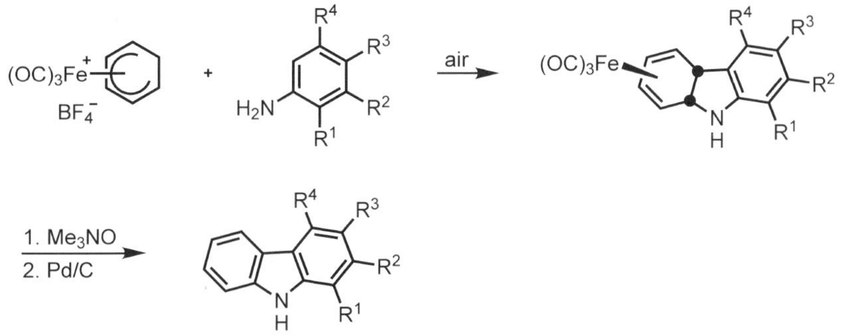

Scheme 1.30

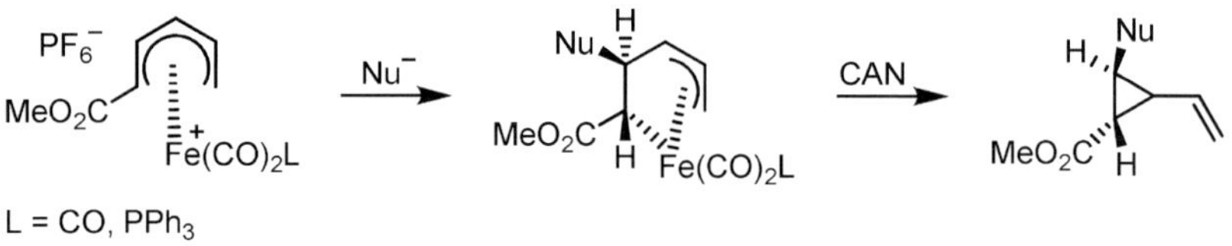

L = CO, PPh_3

Scheme 1.31

1.3.6
Ferrocenes

Ferrocene is an iron(II) sandwich complex with two cyclopentadienyl ligands (Figure 1.3). Since its discovery in 1951 [4], ferrocene has been the subject of extensive investigations due to its reactivity, structural features and potential for applications [73].

Ferrocene is air-stable, can be sublimed without decomposition and reacts with electrophiles by substitution at the cyclopentadienyl ring. The mechanism differs from classical electrophilic aromatic substitution in the way that the electrophile first attacks at the metal center and is subsequently transferred to the ligand followed by deprotonation. Oxidation of ferrocene gives the blue ferricenium ion $[Fe(C_5H_5)_2]^+$, which is used as an oxidizing agent. Ferrocenes which are homoannular disubstituted with two different substituents are planar chiral. This feature has been widely exploited for applications in various asymmetric catalytic reactions. Especially ferrocenylphosphanes represent useful chiral ligands [74]. Also planar chiral ferrocenes with additional chiral substituents have been applied in asymmetric catalysis, and even a combination of planar, central and axial chirality was employed (Figure 1.4). An example of asymmetric synthesis using chiral ferrocene ligands is described below (Section 1.4.2).

1.3.7
Arene–Iron Complexes

Two types of arene–iron complexes are known in the literature, monocationic arene–FeCp and dicationic bis(arene)Fe complexes [75]. The former type is more stable and shows a more useful chemistry. Arene–FeCp complexes can be prepared

Figure 1.3 Ferrocene.

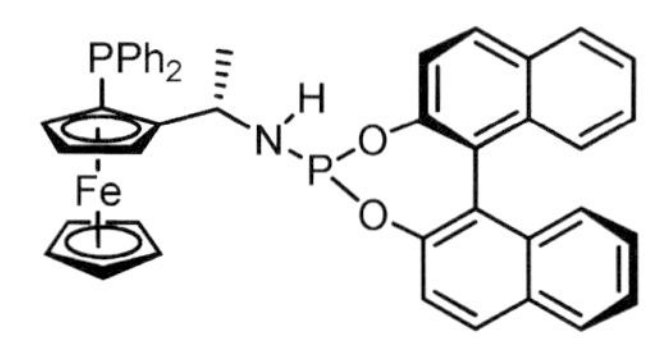

Figure 1.4 A ferrocene with planar, central and axial chirality.

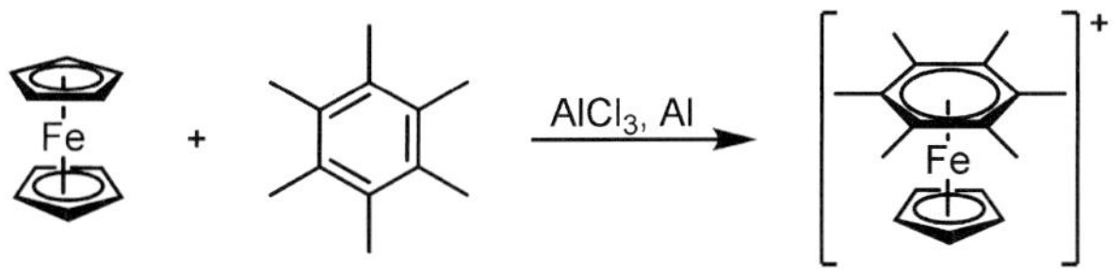

Scheme 1.32

by ligand exchange from ferrocene with the corresponding aromatic compound in the presence of aluminum trichloride and aluminum powder (Scheme 1.32).

The metal reduces the electron density at the arene ligand, thus making it more susceptible to nucleophilic attack (Scheme 1.33). Arene–FeCp complexes react with a great variety of nucleophiles following the Davies–Green–Mingos rules [62] preferentially at the arene ring and stereoselectively *anti* to the metal. Alkyllithium compounds add readily at the arene ligand. Only in the case of steric hindrance addition at the Cp ligand is observed.

Electron-donating substituents direct the incoming nucleophile predominantly to the *meta*-position and electron-withdrawing substituents to the *ortho*-position. Oxidative demetallation (DDQ, iodine) is applied to reoxidize the cyclohexadienyl ligand, releasing a substituted arene. Addition of nucleophiles to halobenzene–FeCp complexes leads to nucleophilic substitution of the halo substituent (Scheme 1.34). Demetallation of the product complexes is achieved by irradiation with sunlight or UV light in acetone or acetonitrile.

This reaction is a powerful tool and represents an alternative for the synthesis of substituted arenes difficult to prepare via classical electrophilic or nucleophilic aromatic substitution. Using bi- or polyfunctional arenes as starting materials, this reaction affords novel organoiron polymers [76] (Scheme 1.35).

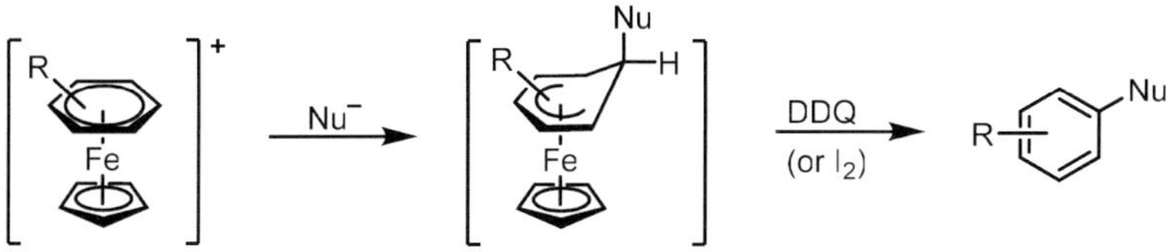

Scheme 1.33

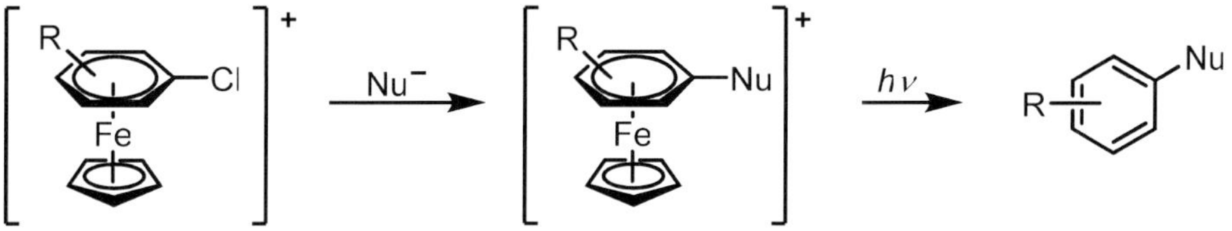

Scheme 1.34

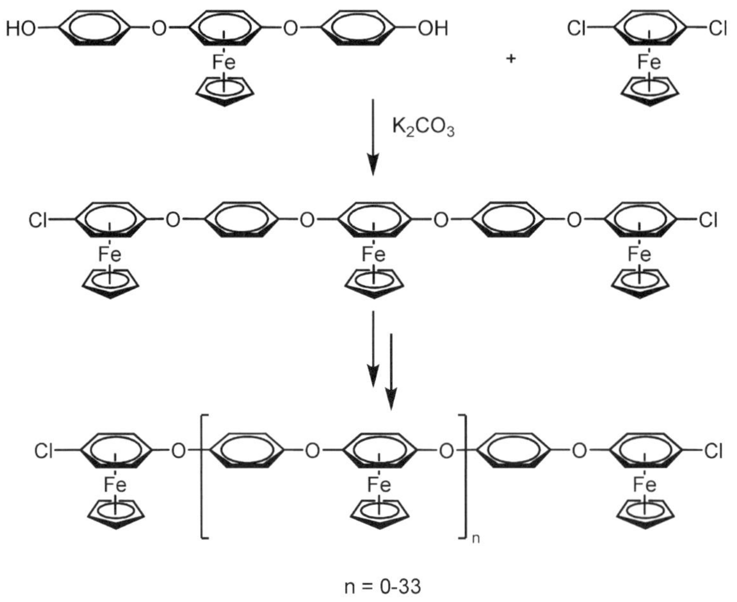

HO–O–O–OH + Cl–Cl; K_2CO_3; Fe; n = 0-33

Scheme 1.35

Dicationic bis(arene)iron complexes are prepared from Fe(II) salts using the corresponding arene as solvent in the presence of aluminum trichloride at elevated temperatures. Because of the double positive charge, they easily add two nucleophiles at one arene ring.

1.4
Catalysis Using Iron Complexes

The difficulty in removing metal residues from the product induced the search for iron-catalyzed reactions [77]. However, iron complexes can play different roles in catalytic processes:

1. as substrate and/or product;
2. as ligands for other transition metal catalysts to achieve activation and stereocontrol;
3. as catalytically active species.

1.4.1
Iron Complexes as Substrates and/or Products in Catalytic Reactions

The first aspect is illustrated by the synthesis of tricarbonyl(η^4-1,3-diene)iron complexes from pentacarbonyliron in the presence of catalytic amounts of a 1-azabutadiene which forms *in situ* during the catalytic cycle a tricarbonyliron transfer reagent (Scheme 1.23) [52]. The 1-azabutadiene represents the catalytically active species in this process. Electrocatalysis with ferrocenes also belongs to this category [78, 79].

1.4.2
Iron Complexes as Ligands for Other Transition Metal Catalysts

Organoiron complexes have been applied as ligands for processes catalyzed by other transition metals in order to activate them or to achieve stereocontrol. This principle has been utilized by readily available ferrocenes bearing additional coordinating groups at their cyclopentadienyl rings [74]. Due to their planar chirality, substituted ferrocenes, often in combination with additional central chirality of pendant groups, have been extensively investigated as ligands for asymmetric catalysis. A few examples of commercially available ferrocene ligands are shown in Figure 1.5.

A broad variety of asymmetric reactions has been studied using chiral ferrocene ligands, e.g. asymmetric hydrogenation, asymmetric metal-catalyzed coupling reactions and enantioselective nucleophilic additions to aldehydes and imines. An example of such a catalytic process is the synthesis of the peroxime proliferator activated receptor (PPAR) agonist, applying a rhodium-catalyzed hydrogenation of a cinnamic acid derivative in 78% yield and 92% *ee* (Scheme 1.36) [80]. In this particular case, the ferrocene ligand Walphos proved to be most efficient.

1.4.3
Iron Complexes as Catalytically Active Species

This section provides only a brief insight into iron-catalyzed reactions. Iron complexes as catalytically active species undergo typical steps of transition metal catalysis

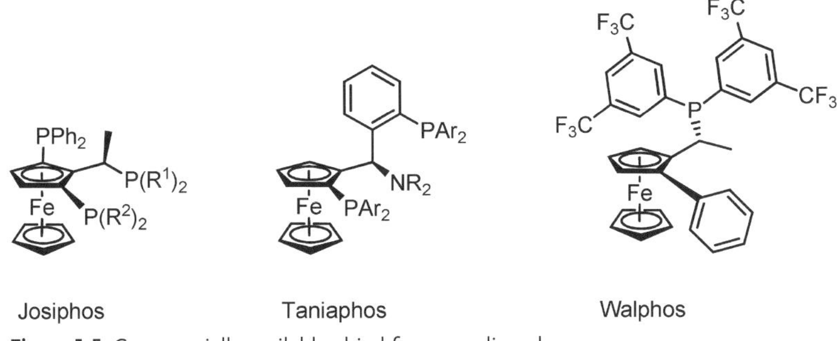

Figure 1.5 Commercially available chiral ferrocene ligands.

BnO, CO_2H, OEt

2 mol% $[Rh(nbd)_2]BF_4$
2 mol% Walphos
H_2, NaOMe, MeOH

BnO, CO_2H, OEt

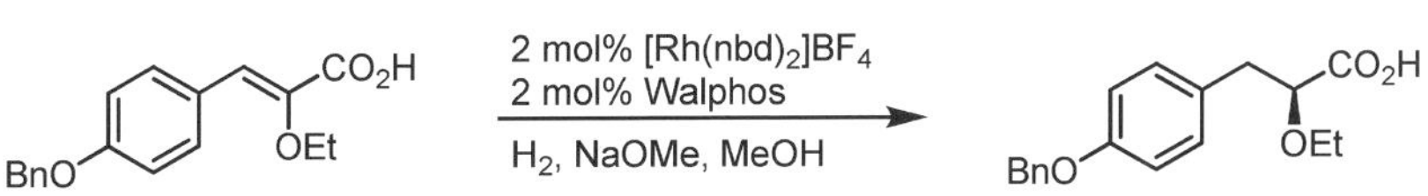

Scheme 1.36

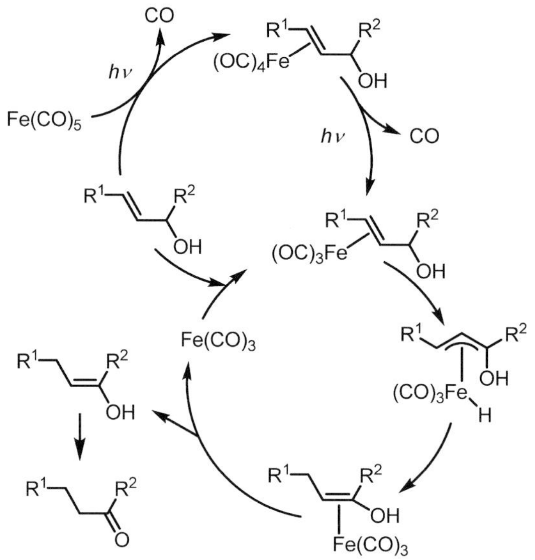

Scheme 1.37

such as oxidative addition and reductive elimination, thus leading to a reversible change of the formal oxidation state of the metal.

Pentacarbonyliron can catalyze the isomerization of double bonds under photochemical conditions. Using catalyst loadings as low as 1–5 mol%, this process proceeds smoothly for allyl alcohols, which isomerize to the corresponding saturated carbonyl compounds.

The mechanism of the catalytic cycle is outlined in Scheme 1.37 [11]. It involves the formation of a reactive 16-electron tricarbonyliron species by coordination of allyl alcohol to pentacarbonyliron and sequential loss of two carbon monoxide ligands. Oxidative addition to a π-allyl hydride complex with iron in the oxidation state +2, followed by reductive elimination, affords an alkene–tricarbonyliron complex. As a result of the [1, 3]-hydride shift the allyl alcohol has been converted to an enol, which is released and the catalytically active tricarbonyliron species is regenerated. This example demonstrates that oxidation and reduction steps can be merged to a one-pot procedure by transferring them into oxidative addition and reductive elimination using the transition metal as a reversible switch. Recently, this reaction has been integrated into a tandem isomerization-aldolization reaction which was applied to the synthesis of indanones and indenones [81] and for the transformation of vinylic furanoses into cyclopentenones [82].

In a similar reaction, allylamines can be isomerized to afford enamines. Photochemical isomerization of the silylated allylamine in the presence of catalytic amounts of pentacarbonyliron provided exclusively the *E*-isomer of the enamine, whereas a thermally induced double bond shift provided a 4:1 mixture of the *E*- and *Z*-enamines (Scheme 1.38) [13].

Pentacarbonyliron has also been applied as catalyst for the reduction of nitroarenes by carbon monoxide and water to afford anilines [17, 83].

	E		*Z*
$Fe(CO)_5$, $h\nu$	100	:	0
$Fe(CO)_5$, Δ	80	:	20

Scheme 1.38

A comparison of the electronic configuration of iron with that of nickel suggests that iron systems which are isoelectronic to the redox couple Ni(0)/Ni(II) could have potential as catalysts. This would apply to the Fe(–II)/Fe(0) system. The increased nucleophilicity of the Fe(–II) species should facilitate oxidative addition reactions, which often represent the limiting step. In fact, several iron-catalyzed cross-coupling reactions of Grignard or organomanganese reagents with alkenyl halides or aryl chlorides, tosylates and triflates have been reported recently [84, 85]. In these examples, the Fe(–II)/Fe(0) redox system appears to drive the catalytic cycle. A mechanism involving a catalytically active Fe(–II) species has been postulated by Fürstner *et al.* for this cross-coupling reaction (Scheme 1.39) [77, 86].

The "inorganic Grignard" species $[Fe(MgX)_2]$, which has not yet been structurally confirmed, is regarded as the propagating agent. Oxidative addition of an aryl halide generates an Fe(0) complex, which is alkylated by another Grignard reagent. Reductive elimination provides the organic product and regenerates the catalytically active species. The Fe(–II)/Fe(0) redox-couple appears to have great potential for further applications in organic synthesis.

$$FeCl_2 + 4\ RCH_2CH_2MgX \longrightarrow [Fe(MgX)_2] + 2\ MgX_2$$

$$2\ RCH_2CH_3 + RCH{=}CH_2 + R(CH_2)_4R$$

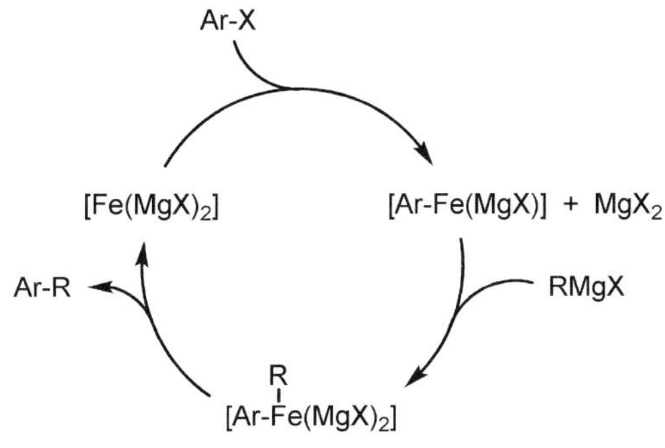

Scheme 1.39

References

1 (a) A. J. Pearson, *Metallo-organic Chemistry*, Wiley, Chichester, 1985; (b) J. P. Collman, L. S. Hegedus, J. R. Norton, R. G. Finke, *Principles and Application of Organotransition Metal Chemisty*, University Science Books, Mill Valley, CA, 1987; (c) S. G. Davies, *Organotransition Metal Chemistry: Application to Organic Synthesis*, Pergamon Press, Oxford, 1989; (d) L. S. Hegedus, *Transition Metals in the Synthesis of Complex Organic Molecules*, 2nd edn., University Science Books, Sausalito, CA, 1999; (e) J. Tsuji, Transition Metal Reagents and Catalysts, *Innovations in Organic Synthesis*, Wiley, Chichester, 2000.

2 L. Mond, F. Quinke, *J. Chem. Soc.* 1891, **59**, 604.

3 M. Berthelot, *C. R. Hebd. Seances Acad. Sci.* 1891, **112**, 1343.

4 T. J. Kealy, P. L. Pauson, *Nature* 1951, **168**, 1039.

5 W. Reppe, H. Vetter, *Justus Liebigs Ann. Chem.* 1953, **582**, 133.

6 (a) M. Tamura, J. Kochi, *Synthesis* 1971, 303; (b) M. Tamura, J. Kochi, *J. Am. Chem. Soc.* 1971, **93**, 1487.

7 A. J. Pearson, *Iron Compounds in Organic Synthesis*, Academic Press, London, 1994.

8 H.-J. Knölker, H. Goesmann, R. Klauss, *Angew. Chem.* 1999, **111**, 727; *Angew. Chem. Int. Ed.* 1999, **38**, 702.

9 H.-J. Knölker, E. Baum, H. Goesmann, R. Klauss, *Angew. Chem.* 1999, **111**, 2196. *Angew. Chem. Int. Ed.* 1999, **38**, 2064.

10 E. J. Corey, G. Moinet, *J. Am. Chem. Soc.* 1973, **95**, 7185.

11 (a) H. Cherkaoui, M. Soufiaoui, R. Grée, *Tetrahedron* 2001, **57**, 2379; (b) R. Uma, C. Crevisy, R. Grée, *Chem. Rev.* 2003, **103**, 27.

12 R. J. P. Corriu, V. Huynh, J. J. E. Moreau, M. Pataud-Sat, *J. Organomet. Chem.* 1983, **255**, 359.

13 K. Paulini, H.-U. Reissig, *Liebigs Ann. Chem.* 1991, 455.

14 R. Noyori, *Acc. Chem. Res.* 1979, **12**, 61.

15 A. A. O. Sarhan, *Curr. Org. Chem.* 2001, **5**, 827.

16 (a) R. Noyori, S. Makino, T. Okita, Y. Hayakawa, *J. Org. Chem.* 1975, **40**, 806; (b) H. Takaya, Y. Hayakawa, S. Makino, R. Noyori, *J. Am. Chem. Soc.* 1978, **100**, 1778.

17 J. M. Landesberg, L. Katz, C. Olsen, *J. Org. Chem.* 1972, **37**, 930.

18 M. Ryang, I. Rhee, S. Tsutsumi, *Bull. Chem. Soc. Jpn.* 1964, **37**, 341.

19 (a) J. P. Collman, *Acc. Chem. Res.* 1975, **8**, 342; (b) M. P. Cooke, Jr., R. M. Parlman, *J. Am. Chem. Soc.* 1975, **97**, 6863.

20 M. Periasamy, C. Rameshkumar, U. Radhakrishnan, A. Devasagayaraj, *Curr. Sci.* 2000, **78**, 1307.

21 (a) E. Weiss, W. Hübel, *J. Inorg. Nucl. Chem.* 1959, **11**, 42; (b) E. Weiss, R. Merenyi, W. Hübel, *Chem. Ind. (London)* 1960, 407; (c) E. Weiss, R. Merenyi, W. Hübel, *Chem. Ber.* 1962, **95**, 1170.

22 (a) H.-J. Knölker, J. Heber, C. H. Mahler, *Synlett* 1992, 1002; (b) H.-J. Knölker, J. Heber, *Synlett* 1993, 924; (c) H.-J. Knölker, E. Baum, J. Heber, *Tetrahedron Lett.* 1995, **36**, 7647.

23 (a) A. J. Pearson, R. J. Shively, Jr., R. A. Dubbert, *Organometallics* 1992, **11**, 4096; (b) A. J. Pearson, R. J. Shively, Jr., *Organometallics* 1994, **13**, 578; (c) A. J. Pearson, X. Yao, *Synlett* 1997, 1281.

24 H.-J. Knölker, A. Braier, D. J. Bröcher, P. G. Jones, H. Piotrowski, *Tetrahedron Lett.* 1999, **40**, 8075.

25 H.-J. Knölker, S. Cämmerer, *Tetrahedron Lett.* 2000, **41**, 5035.

26 (a) P. L. Pauson, *Tetrahedron* 1985, **41**, 5855; (b) J. Blanco-Urgoiti, L. Añorbe, L. Pérez-Serrano, G. Domínguez, J. Pérez-Castells, *Chem. Soc. Rev.* 2004, **33**, 32.

27 A. J. Pearson, R. A. Dubbert, *J. Chem. Soc., Chem. Commun.* 1991, 202.

28 C. Rameshkumar, M. Periasamy, *Tetrahedron Lett.* 2000, **41**, 2719.

29 M. Periasamy, C. Rameshkumar, A. Mukkanti, *J. Organomet. Chem.* 2002, **649**, 209.

30 (a) B. W. Roberts, J. Wong, *J. Chem. Soc., Chem. Commun.* 1977, 20; (b) B. W. Roberts, M. Ross, J. Wong, *J. Chem. Soc., Chem. Commun.* 1980, 428.

31 M. Rosenblum, A. Bucheister, T. C. T. Chang, M. Cohen, M. Marsi, S. B. Samuels, D. Scheck, N. Sofen, J. C. Watkins, *Pure Appl. Chem.* 1984, **56**, 129.

32 (a) J. Benaim, A. L'Honore, *J. Organomet. Chem.* 1980, **202**, C53; (b) D. J. Bates, M. Rosenblum, S. B. Samuels, *J. Organomet. Chem.* 1981, **209**, C55.

33 M. Rosenblum, *Acc. Chem. Res.* 1974, **7**, 122.

34 N. Grenco, D. Marten, S. Raghu, M. Rosenblum, *J. Am. Chem. Soc.* 1976, **98**, 848.

35 H. D. Murdoch, E. Weiss, *Helv. Chim. Acta* 1962, **45**, 1927.

36 L. R. Cox, S. V. Ley, *Chem. Soc. Rev.* 1998, **27**, 301.

37 S. V. Ley, L. R. Cox, G. Meek, *Chem. Rev.* 1996, **96**, 423.

38 (a) G. F. Emerson, K. Ehrlich, W. P. Giering, P. C. Lauterbur, *J. Am. Chem. Soc.* 1966, **88**, 3172; (b) K. Ehrlich, G. Emerson, *J. Am. Chem. Soc.* 1972, **94**, 2464.

39 M. Franck-Neumann, A. Kastler, *Synlett* 1995, 61.

40 For reviews, see: (a) K. Rück-Braun, M. Mikulas, P. Amrhein, *Synthesis* 1999, 727; (b) K. Rück-Braun, in *Transition Metals for Organic Synthesis – Building Blocks and Fine Chemicals*, Vol. 1, ed. M. Beller, C. Bolm, Wiley-VCH, Weinheim, 1998, Chap. 3.12, p. 523;(c) K. Rück-Braun, in *Transition Metals for Organic Synthesis – Building Blocks and Fine Chemicals*, Vol.1, 2nd edn., ed. M. Beller, C. Bolm, Wiley-VCH, Weinheim, 2004, Chap. 3.10, p. 575.

41 S. C. Case-Green, J. F. Costello, S. G. Davies, N. Heaton, C. J. R. Hedgecock, V. M. Humphreys, M. R. Metzler, J. C. Prime, *J. Chem. Soc., Perkin Trans. 1* 1994, 933.

42 (a) J. W. Herndon, *J. Org. Chem.* 1986, **51**, 2853; (b) K. Rück-Braun, J. Kühn, D. Scholl-Meyer, *Chem. Ber.* 1996, **129**, 1057.

43 S. G. Davies, N. M. Garrido, P. A. McGee, J. P. Shilvock, *J. Chem. Soc., Perkin Trans. 1* 1999, 3105.

44 (a) J. W. Herndon, *J. Am. Chem. Soc.* 1987, **109**, 3165; (b) J. W. Herndon, C. Wu, *Synlett* 1990, 411; (c) J. W. Herndon, C. Wu, J. J. Harp, *Organometallics* 1990, **9**, 3157.

45 C. Möller, M. Mikulas, F. Wierchem, K. Rück-Braun, *Synlett* 2000, 182.

46 K. Rück-Braun, T. Martin, M. Mikulas, *Chem. Eur. J.* 1999, **5**, 1028.

47 (a) M. Brookhart, W. B. Studabaker, *Chem. Rev.* 1987, **87**, 411; (b) P. Helquist, in *Advances in Metal–Organic Chemistry*, Vol. 2, ed. L. S. Liebeskind, JAI Press, London, 1991, p. 143; (c) W. Petz, *Iron–Carbene Complexes*, Springer, Berlin, 1993.

48 E. J. O'Connor, S. Brandt, P. Helquist, *J. Am. Chem. Soc.* 1987, **109**, 3739.

49 (a) S.-K. Zhao, P. Helquist, *J. Org. Chem.* 1990, **55**, 5820; (b) S. Ishii, S. Zhao, G. Mehta, C. J. Knors, P. Helquist, *J. Org. Chem.* 2001, **66**, 3449.

50 (a) A. J. Pearson, *Acc. Chem. Res.* 1980, **13**, 463; (b) R. Grée, *Synthesis* 1989, 341; (c) A. J. Pearson, in *Comprehensive Organic Synthesis*, Vol. 4, ed. B. M. Trost, Pergamon Press, Oxford, 1991, Chap. 3.4, p. 663; (d) W. A. Donaldson, *Aldrichim. Acta* 1997, **30**, 17. (e) W. A. Donaldson, *Curr. Org. Chem.* 2000, **4**, 837; (f) W. A. Donaldson, in *The Chemistry of Dienes and Polyenes*, Vol. 2, ed. Z. Rappoport, Wiley, Chichester, 2000, Chap. 11, p. 885.

51 (a) H.-J. Knölker, *Synlett* 1992, 371; (b) H.-J. Knölker, in *Transition Metals for Organic Synthesis – Building Blocks and Fine Chemicals*, Vol. 1, ed. M. Beller, C. Bolm, Wiley-VCH, Weinheim, 1998, Chap. 3.13, p. 534; (c) H.-J. Knölker, *Chem. Soc. Rev.* 1999, **28**, 151; (d) H.-J. Knölker, A. Braier, D. J. Bröcher, S. Cämmerer, W. Fröhner, P. Gonser, H. Hermann, D. Herzberg, K. R. Reddy, G. Rohde, *Pure Appl. Chem.* 2001, **73**, 1075; (e) H.-J. Knölker, in *Transition Metals for Organic Synthesis – Building Blocks and Fine Chemicals*, Vol. 1, 2nd edn., ed. M. Beller, C. Bolm, Wiley-VCH, Weinheim, 2004, Chap. 3.11, p. 585.

52 H.-J. Knölker, *Chem. Rev.* 2000, **100**, 2941.

53 H. Fleckner, F.-W. Grevels, D. Hess, *J. Am. Chem. Soc.* 1984, **106**, 2027.

54 (a) J. A. S Howell, B. F. G Johnson, P. L. Josty, J. Lewis, *J. Organomet. Chem.* 1972, **39**, 329; (b) C. R. Graham, G. Scholes, M. Brookhart, *J. Am. Chem. Soc.* 1977, **99**, 1180; (c) M. Brookhart, G. O. Nelson, *J. Organomet. Chem.* 1979, **164**, 193.

55 (a) H.-J. Knölker, G. Baum, N. Foitzik, H. Goesmann, P. Gonser, P. G. Jones, H. Röttele, *Eur. J. Inorg. Chem.* 1998, 993; (b) H.-J. Knölker, B. Ahrens, P. Gonser, M. Heininger, P. G. Jones, *Tetrahedron* 2000, **56**, 2259; (c) H.-J. Knölker, E. Baum, P. Gonser, G. Rohde, H. Röttele, *Organometallics* 1998, **17**, 3916.

56 (a) H.-J. Knölker, H. Herrmann, *Angew. Chem.* 1996, **108**, 363; *Angew. Chem., Int. Ed. Engl.* 1996, **35**, 341; (b) H.-J. Knölker, H. Goesmann, H. Hermann, D. Herzberg, G. Rohde, *Synlett* 1999, 421; (c) H.-J. Knölker, H. Hermann, D. Herzberg, *Chem. Commun.* 1999, 831.

57 H.-J. Knölker, M. Bauermeister, J.-B. Pannek, M. Wolpert, *Synthesis* 1995, 397.

58 (a) H.-J. Knölker, G. Schlechtingen, *J. Chem. Soc., Perkin Trans. 1* 1997, 349; (b) H.-J. Knölker, E. Baum, T. Hopfmann, *Tetrahedron* 1999, **55**, 10391; (c) H.-J. Knölker, T. Hopfmann, *Tetrahedron* 2002, **58**, 8937.

59 W. R. Roush, J. C. Park, *Tetrahedron Lett.* 1990, **31**, 4707.

60 W. R. Roush, C. K. Wada, *J. Am. Chem. Soc.* 1994, **116**, 2151.

61 E. O. Fischer, R. D. Fischer, *Angew. Chem.* 1960, **72**, 919.

62 S. G. Davies, M. L. H. Green, D. M. P. Mingos, *Tetrahedron* 1978, **34**, 3047.

63 W. A. Donaldson, L. Shang, C. Tao, Y. K. Yun, M. Ramaswamy, V. G. Young, Jr., *J. Organomet. Chem.* 1997, **539**, 87.

64 (a) A. J. Pearson, D. C. Rees, *J. Am. Chem. Soc.* 1982, **104**, 1118; (b) A. J. Pearson, D. C. Rees, *J. Chem. Soc., Perkin Trans 1* 1982, 2467.

65 (a) H.-J. Knölker, R. Boese, K. Hartmann, *Angew. Chem.* 1989, **101**, 1745; *Angew. Chem. Int. Ed. Engl.* 1989, **28**, 1678. (b) H.-J. Knölker, K. Hartmann, *Synlett* 1991, 428.

66 (a) A. J. Pearson, I. C. Richards, D. V. Gardner, *J. Org. Chem.* 1984, **49**, 3887; (b) G. R. Stephenson, D. A. Owen, H. Finch, S. Swanson, *Tetrahedron Lett.* 1991, **32**, 1291.

67 (a) H.-J. Knölker, K. R. Reddy, *Chem. Rev.* 2002, **102**, 4303; (b) H.-J. Knölker, *Top. Curr. Chem.* 2005, **244**, 115; (c) H.-J. Knölker, K. R. Kethiri, in *The Alkaloids*, Vol. 65, ed. G. A. Cordell, Academic Press, Amsterdam, 2008, p. 1.

68 (a) H.-J. Knölker, W. Fröhner, *Tetrahedron Lett.* 1999, **40**, 6915; (b) H.-J. Knölker, E. Baum, K. R. Reddy, *Chirality* 2000, **12**, 526; (c) H.-J. Knölker, M. Wolpert, *Tetrahedron* 2003, **59**, 5317; (d) R. Czerwonka, K. R. Reddy, E. Baum, H.-J. Knölker, *Chem. Commun.* 2006, 711.

69 (a) Y. K. Yun, W. A. Donaldson, *J. Am. Chem. Soc.* 1997, **119**, 4084. (b) Y. K. Yun, K. Godula, Y. Cao, W. A. Donaldson, *J. Org. Chem.* 2003, **68**, 901.

70 (a) K. Godula, W. A. Donaldson, *Tetrahedron Lett.* 2001, **42**, 153; (b) N. J. Wallock, W. A. Donaldson, *Tetrahedron Lett.* 2002, **43**, 4541.

71 J. M. Lukesh, W. A. Donaldson, *Chem. Commun.* 2005, 110.

72 N. J. Wallock, D. W. Bennett, T. Siddiquee, D. T. Haworth, W. A. Donaldson, *Synthesis* 2006, 3639.

73 A. Togni, T. Hayashi, *Ferrocenes*, VCH, Weinheim, 1995.

74 (a) L.-X. Dai, T. Tu, S.-L. You, W.-P. Deng, X.-L. Hou, *Acc. Chem. Res.* 2003, **36**, 659; (b) R. Gomez Arrayas, J. Adrio, J. C. Carretero, *Angew. Chem.* 2006, **118**, 7836. *Angew. Chem. Int. Ed.* 2006, **45**, 7674.

75 (a) D. Astruc, *Tetrahedron* 1983, **39**, 4027; (b) A. S. Abd-El-Aziz, S. Bernardin, *Coord. Chem. Rev.* 2000, **203**, 219.

76 A. S. Abd-El-Aziz, E. K. Todd, *Coord. Chem. Rev.* 2003, **246**, 3.

77 C. Bolm, J. Legros, J. Le Paih, L. Zani, *Chem. Rev.* 2004, **104**, 6217.

78 D. Astruc, *Angew. Chem.* 1988, **100**, 662; *Angew. Chem. Int. Ed. Engl.* 1988, **27**, 643.

79 E. Steckhan, *Top. Curr. Chem.* 1997, **185**.

80 I. N. Houpis, L. E. Patterson, C. A. Alt, J. R. Rizzo, T. Y. Zhang, M. Haurez, *Org. Lett.* 2005, **7**, 1947.

81 J. Petrignet, T. Roisnel, R. Grée, *Chem. Eur. J.* 2007, **13**, 7374.

82 J. Petrignet, I. Prathap, S. Chandrasekhar, J. S. Yadav, R. Grée, *Angew. Chem.* 2007, **119**, 6413; *Angew. Chem. Int. Ed.* 2007, **46**, 6297.

83 K. Cann, T. Cole, W. Slegeir, R. Pettit, *J. Am. Chem. Soc.* 1978, **100**, 3969.

84 A. Fürstner, R. Martin, *Chem. Lett.* 2005, **34**, 624.

85 G. Cahiez, V. Habiak, C. Duplais, A. Moyeux, *Angew. Chem.* 2007, **119**, 4442; *Angew. Chem. Int. Ed.* 2007, **46**, 4364.

86 A. Fürstner, A. Leitner, M. Mendez, H. Krause, *J. Am. Chem. Soc.* 2002, **124**, 13856.

2
Iron Catalysis in Biological and Biomimetic Reactions

2.1
Non-heme Iron Catalysts in Biological and Biomimetic Transformations

Jens Müller

2.1.1
Introduction: Iron in Biological Processes

Being the second most abundant metal after aluminum and hence the most abundant transition metal in the Earth's crust, iron is exploited for biological processes by numerous organisms [1]. Under the reducing atmosphere that existed prior to the advent of photosynthesis more than 2 billion years ago, iron(II) was most likely abundant also in seawater and could easily be utilized during the first billion years of biological evolution. The fact that iron can exist in several oxidation states made (and still makes) it an ideal candidate for the incorporation into the active center of metalloenzymes that catalyze redox reactions or that are involved in electron transfer. However, the presence of photosynthetically generated oxygen led to dramatic changes for the early organisms. First, iron(II) was now oxidized to iron(III), which rapidly precipitated in the form of its hydroxide salts and was no longer easily available for biological processes. Since then, evolutionary pressure has led to the appearance of sophisticated mechanisms to salvage the poorly soluble iron(III) as an iron source. Second, oxidative damage due to dioxygen influenced and disturbed numerous cellular processes, leading to the development of enzymes such as superoxide dismutase and superoxide reductase that are capable of "detoxifying" the harmful oxygen species. Nonetheless, iron is still being used to perform or catalyze vital biological transformations, e.g. oxygen transport, mono- and dioxygenations, hydroxylations, electron transfer and many more.

Because of the broad variety of iron-containing proteins, they are usually divided into three distinct classes, based on their structural composition. Iron–sulfur proteins, which contain clusters of iron and inorganic sulfur and in which the metal

Iron Catalysis in Organic Chemistry. Edited by Bernd Plietker

ISBN: 978-3-527-31927-5

ions are typically coordinated by cysteine or in some instances by histidine amino acid residues, are involved in electron transfer, in the reduction of dinitrogen and in sensing oxygen and iron [2]. This class of proteins will not be discussed in this book. In heme proteins, the iron is coordinated by four nitrogen ligands that are located in a macrocyclic porphyrin ring system. These proteins are engaged mainly in oxygen transfer and in oxidation reactions. Section 2.2 gives an overview of the reactions that are catalyzed by heme-containing iron proteins and their model systems [3]. Finally, the remaining non-heme iron proteins are still so diverse that they need to be subdivided further. In this chapter, a general division into proteins with mononuclear active sites and proteins with dinuclear active sites will be adhered to. As the various classes of non-heme iron proteins display entirely different properties and hence reactivities, a discussion of all aspects of all possible iron-catalyzed biological transformations is beyond the scope of this chapter. It will instead give a detailed insight into the mechanisms of the catalytic oxidation reactions of a few selected enzymes and their model complexes. Readers who are interested in a more thorough description are referred to some excellent review articles on the metalloproteins [4–9] and their model complexes [10–12].

2.1.2
Non-heme Iron Proteins

As already mentioned above, the class of non-heme iron proteins is inhomogeneous and therefore often divided into subclasses. A distinction can be made based on the ligands that coordinate to the iron center and concomitantly on the types of potential iron cofactors. In most catalytically active non-heme iron proteins, the metal ions are coordinated via nitrogen and oxygen ligands. These types of proteins will be discussed in this chapter.

Due to the choice of nitrogen- and oxygen-containing ligands, the metal ions are found mostly in an electronic high-spin state in all of the proteins. In the following, a few classes of non-heme iron proteins will be discussed on the basis of selected examples of representative enzymes. Where appropriate, mechanistic details obtained from model structures will also be reviewed.

2.1.2.1 Mononuclear Iron Sites

The archetypal structural motif in non-heme iron enzymes with mononuclear active sites is that of the so-called 2-His-1-carboxylate facial triad. Proteins containing this motif represent a broad class covering five families, i.e. extradiol dioxygenases, Rieske dioxygenases, α-ketoglutarate-dependent hydroxylases, pterin-dependent enzymes and other oxidases [13]. As the name implies, two histidine residues and one carboxylate-containing amino acid (glutamic or aspartic acid) coordinate facially to an iron(II), leaving three adjacent positions available for the binding of cofactor, substrate or solvent. In the resting state of the enzymes, these three positions are typically occupied by aqua ligands. In the following, the catalytic mechanisms of two iron-mediated oxidations will be discussed using two enzymes from different families as examples.

Scheme 2.1 Representation of the reaction catalyzed by αKG-dependent hydroxylases.

Taurine Dioxygenase The largest family of proteins with a 2-His-1-carboxylate triad is that of the α-ketoglutarate-dependent (αKG) enzymes [8, 14], which are widespread in microorganisms (including viruses!), animals and plants. In these enzymes, the sacrificial cofactor αKG serves as an electron donor in the reduction of dioxygen and concomitantly becomes oxidized to succinate and carbon dioxide (Scheme 2.1). The reactions catalyzed by these enzymes include, among others, key steps in the biosynthesis of antibiotics and the repair of alkylated nucleic acids (Scheme 2.2). From the chemical point of view and depending on the substrate, the various αKG-dependent enzymes are capable of catalyzing, e.g., hydroxylations, desaturations, ring expansions, ring closures and even halogenations.

Taurine dioxygenase from *Escherichia coli* is the paradigm Fe^{II}/αKG hydroxylase. It is expressed during times of sulfur starvation and catalyzes the hydroxylation of taurine (2-aminoethanesulfonate) to yield aminoacetaldehyde and sulfite, which can then be exploited by the bacterium as a sulfur source (Scheme 2.2a) [8]. Numerous studies involving X-ray crystallography and spectroscopic methods have led to a generally accepted view of the reaction mechanism of the αKG-dependent enzymes. Scheme 2.3 gives an overview of the proposed catalytic cycle of the prototypal protein TauD as discussed in the following [15].

In the resting state, three water molecules occupy the coordination sites of the iron(II) that are located opposite to the three amino acid residues. Introduction of the α-ketoglutarate leads to displacement of two of the solvent molecules and

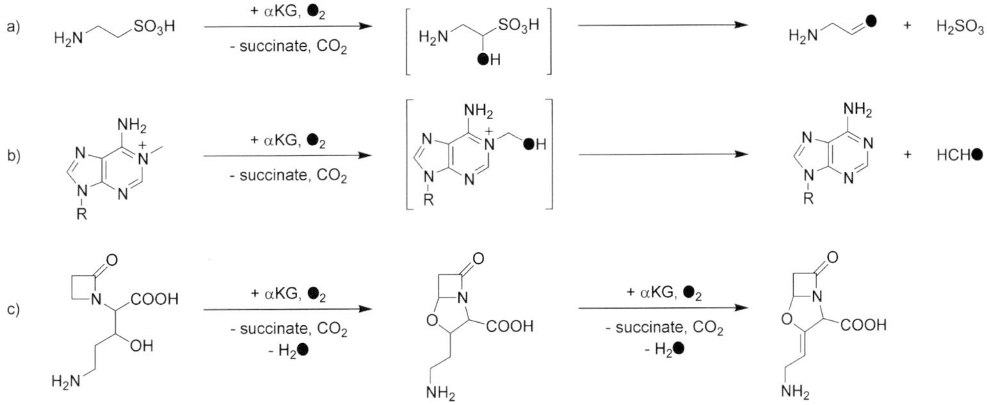

Scheme 2.2 Examples of reactions catalyzed by αKG-dependent enzymes showing the versatility of this type of proteins: (a) hydroxylation of taurine by taurine dioxygenase (TauD) [53]; (b) repair of 1-methyladeninium lesions in DNA and RNA by the protein AlkB [54] (R = sugar phosphate backbone); (c) cyclization and desaturation reaction during the biosynthesis of the β-lactamase inhibitor clavulanic acid by clavaminate synthase (CAS) [55].

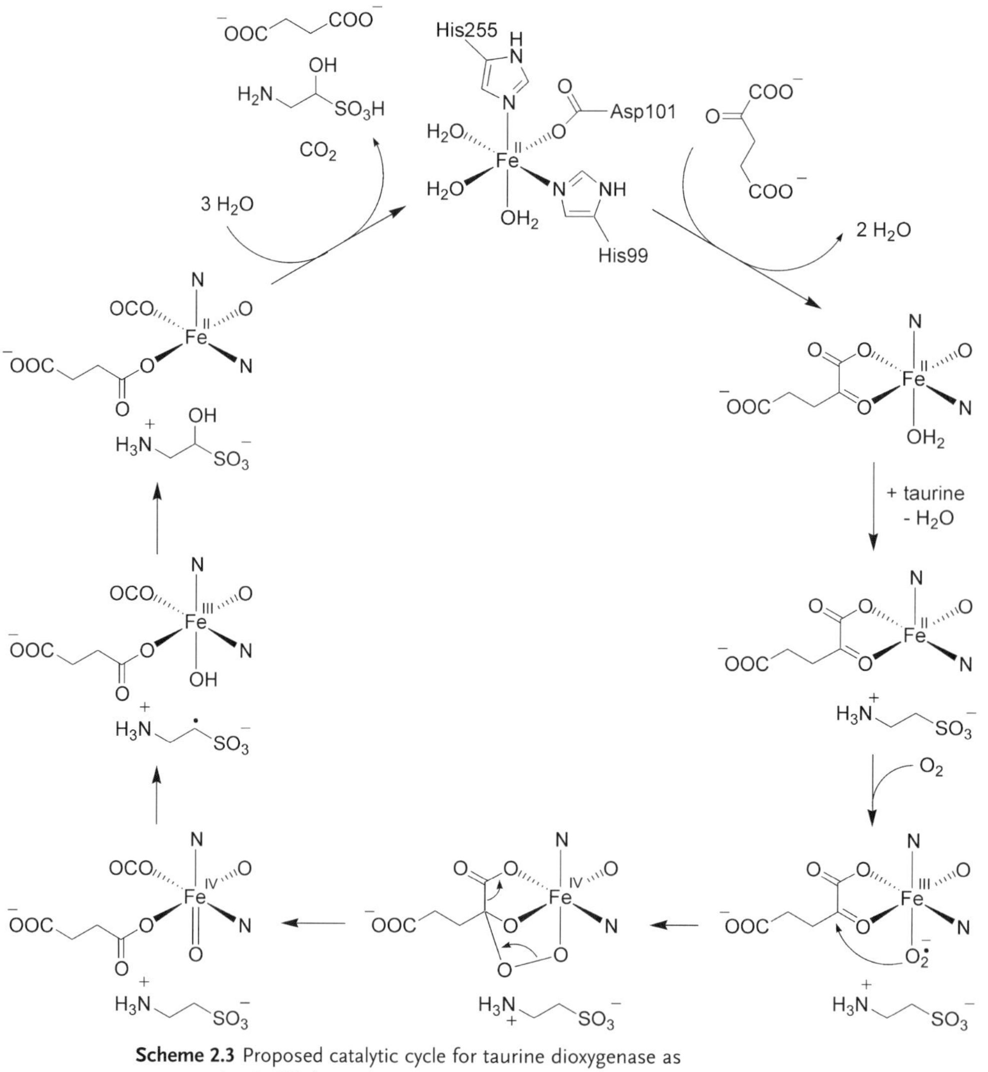

Scheme 2.3 Proposed catalytic cycle for taurine dioxygenase as an example of αKG-dependent enzymes [8, 11, 15]. The amino acids are numbered according to the sequence of taurine dioxygenase from *E. coli* [53].

concomitant binding of the cofactor in a bidentate fashion, with its keto group positioned *trans* with respect to the aspartate residue. The formation of this lilac-colored intermediate can be followed by stopped-flow UV–visible spectroscopy due to its broad absorption band around 530 nm ($\varepsilon_{530} = 140\,M^{-1}\,cm^{-1}$) that is attributed to metal-to-ligand charge transfer [16]. The next step in the catalytic cycle comprises outer-sphere binding of the substrate (i.e. taurine in the case of TauD), which leads to dissociation of the remaining aqua ligand. It has been speculated that the departure of the solvent is facilitated by the loss of a stabilizing hydrogen bond between this

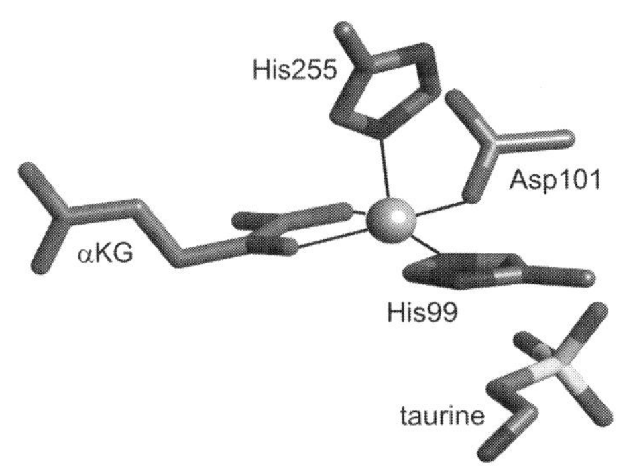

Figure 2.1 X-ray crystal structure of a proposed intermediate of the TauD catalytic cycle (coordinates taken from PDB file 1GQW) [53]. Only the active site is shown and hydrogen atoms have been omitted for clarity. Coordinative bonds are represented by solid lines. The iron(II) is coordinated to three amino acids (only side-chains displayed) and to the cofactor α-ketoglutarate. The substrate taurine is already located in its binding pocket near the metal center, leaving the sixth coordination site of the square pyramidal iron center free for the incoming dioxygen molecule. This figure was prepared using the computer program MOLMOL [58].

aqua ligand and the neighboring aspartate due to a carboxylate shift of this amino acid that is necessary to accommodate the incoming substrate [11]. Spectral changes around 520 nm associated with this binding step were used to verify experimentally the expected stoichiometry of one taurine per TauD subunit and to confirm the high affinity of the substrate (approximately 30 μM) [16]. The concomitantly formed, coordinatively unsaturated metal center is now activated for binding of the dioxygen molecule (Figure 2.1). Interestingly, the enzyme reacts much faster with dioxygen when substrate is present in the active site (as deduced from the decay of the chromophore with a decay rate of $(42 \pm 9)\ s^{-1}$) than in the absence of bound substrate (as evidenced by the lack of spectral changes at 520 nm within the first 100 ms) [16]. This discrimination results in less undesired, non-productive reactions in which the cofactor would be oxidized without the hydroxylation of any substrate, typically leading to the inactivation of the enzyme (see below). The addition of dioxygen is accompanied by the transfer of one electron from the metal ion, resulting in a species with significant Fe^{III}-superoxo character. The assignment as a high-spin iron(III) antiferromagnetically coupled to a superoxide ion is suggested on the basis of computational studies [17] and agrees well with the findings with other iron enzymes. Furthermore, the superoxide radical anion is known to be a strong nucleophile capable of cleaving α-keto acids in an oxidative fashion [18]. Hence the α-ketoglutarate, located directly adjacent to the superoxide and activated by binding to the metal center, is attacked by the radical oxygen species. This leads to cleavage of the C–C bond, release of carbon dioxide (which is not necessarily bonded to the metal center as shown in Scheme 2.3 but also not yet released from the active site at this stage of the catalytic cycle) and formation of a highly reactive Fe^{IV} intermediate. It is this intermediate that is responsible for oxidizing the substrate. Rapid freeze–quench

Mössbauer and EPR spectroscopic studies provided proof of the presence of a high-spin Fe^{IV} transient in the TauD catalytic cycle, representing the first example of such an intermediate in any mononuclear non-heme iron enzyme [19]. It is assumed that product formation occurs via abstraction of a hydrogen atom from the substrate by the Fe^{IV}-oxo intermediate to yield an Fe^{III}-hydroxo species and a substrate radical. This intermediate has never been detected, however [15]. Transfer of the OH group similar to the "oxygen rebound mechanism" proposed for the cytochrome P450 monooxygenases [3] would then lead to product formation and the active site containing an Fe^{II} center. Finally, both the product and the side products succinate and carbon dioxide are released from the active site and the three vacant coordination sites are occupied again by solvent molecules.

Several αKG-dependent enzymes are known to undergo uncoupled reactions in the absence of substrate. In these instances, the cofactor is consumed without concomitant oxidation of the organic substrate (or substrate analog). This undesired reaction typically leads to inactivation of the enzyme, most likely due to the oxidation from Fe^{II} to Fe^{III} in the resting state. A possibility for circumventing uncoupled reactions *in vitro* is the addition of stoichiometric amounts of ascorbate that probably serves as an alternative source of reducing equivalents, preventing the metal center from being oxidized [20]. In many cases, the presence of this antioxidant in substoichiometric amounts is even necessary for maximum activity of the enzymes. It can therefore be assumed that it reactivates inactivated enzyme that was formed inadvertently during an uncoupled reaction [21].

Homoprotocatechuate 2,3-Dioxygenase The enzyme homoprotocatechuate 2,3-dioxygenase (2,3-HPCD) from *Brevibacterium fuscum* belongs to the class of extradiol dioxygenases. These dioxygenases catalyze the ring-opening step in the degradation of various aromatic compounds by cleaving the C–C bond adjacent to the two hydroxyl groups of the catechol substrate. [In contrast to the extradiol dioxygenases, intradiol dioxygenases cleave the C–C bond in between the two hydroxyl groups of their catechol substrate. Whereas extradiol dioxygenases contain an iron(II) active site with a 2-His-1-carboxylate facial triad, intradiol dioxygenases require iron(III). The high regioselectivity in the oxidation reactions suggests two different mechanisms for the respective catechol cleavage.] Due to their relatively relaxed substrate specificity, they are promising targets for the engineering of improved enzymes for industrial bioremediation. They are also prominent members of the 2-His-1-carboxylate facial triad class of iron-containing enzymes. However, as opposed to the αKG-dependent enzymes, there is no need for a cofactor in the catalytic cycle of the extradiol dioxygenases. In the following, the mechanism of substrate oxidation by 2,3-HPCD will be discussed as an example of this class of catalytically active proteins.

By replacing the natural substrate homoprotocatechuate (3,4-dihydroxyphenyl acetate) with the slower reacting substrate analog 4-nitrocatechol, six intermediates of the catalytic reaction cycle of 2,3-HPCD could be observed experimentally, four of which could even be characterized by X-ray crystal structure analysis [22]. Scheme 2.4 gives an overview of the proposed reaction mechanism.

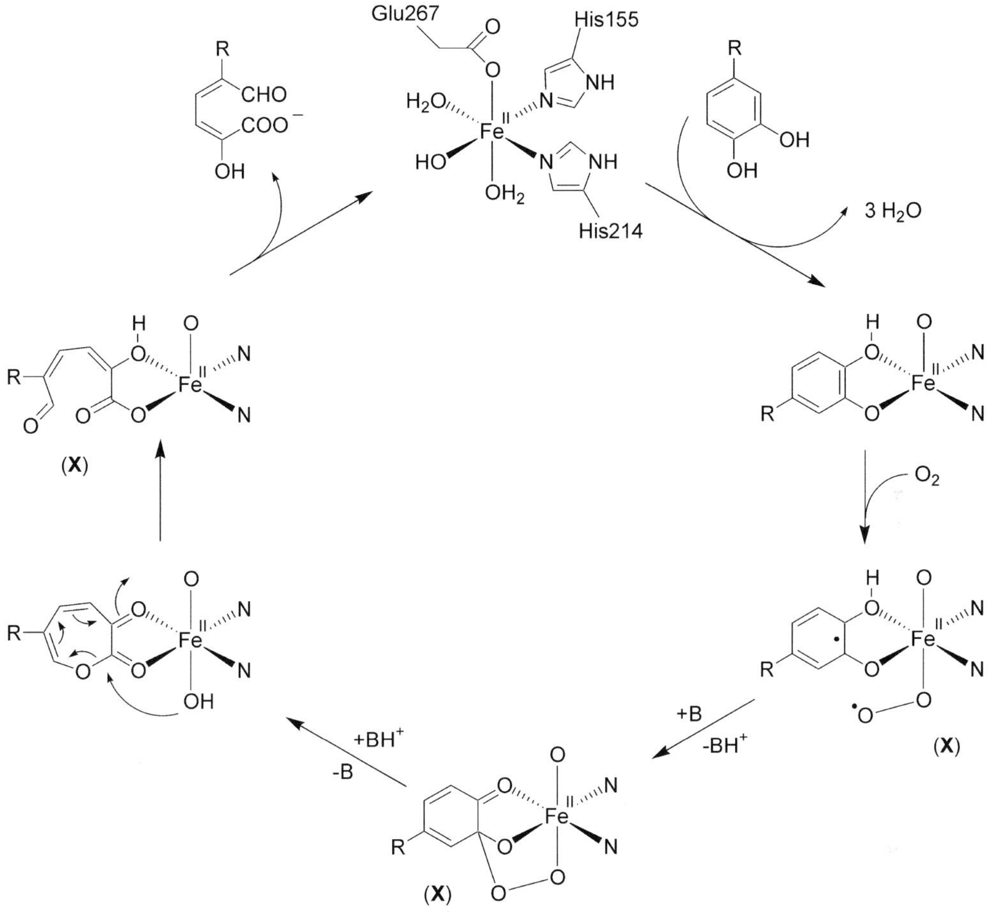

Scheme 2.4 Catalytic cycle for extradiol ring-cleaving dioxygenases (R = CH_2COOH or NO_2) as proposed by Lipscomb and coworkers [22, 56]. The amino acids are numbered according to the sequence of homoprotocatechuate 2,3-dioxygenase from *B. fuscum*. All intermediates marked (**X**) and also the resting state of the protein have been structurally characterized by X-ray crystallography [22]. During the reaction cycle, various protonated and unprotonated amino acid residues near the active site serve as proton donors and acceptors or as hydrogen bonding partners for substrate and dioxygen [56]. The identity of these bases is not known in all cases, so they are merely labeled "B" in this scheme.

In the first step of the reaction, the substrate binds by replacing two of the three facially oriented solvent ligands, while the third dissociates to be rapidly replaced by the incoming dioxygen molecule. The oxygen species that is formed has long been thought to be an Fe^{III}-superoxo species in accord with the findings with numerous other iron enzymes. Recent X-ray crystallographic studies, however, showed that the metal ion is most likely still divalent, and hence can be considered as having Fe^{II}-superoxo character [22]. Here, the electron that is necessary for reducing dioxygen to superoxide comes from the substrate that in turn becomes a radical, as can be deduced from its puckered ring system. In the next step, the two radicals combine to yield an alkylperoxo intermediate. After heterolytic O–O bond cleavage, an intermediate lactone can be formed via a concerted Criegee rearrangement. This lactone is then attacked by the second oxygen of the former dioxygen molecule that still resides on the metal to give the final ring-opened product, which is released in the last step of the catalytic cycle. By using this method, Nature elegantly overcomes the problem of the low reactivity of triplet dioxygen by forming two radicals that can react with each other more easily. This dioxygenation can be considered a rare example of an iron(II)-catalyzed reaction in which the active site metal ion does not necessarily change its oxidation state during the catalytic cycle [23]. Although the presence of higher-valent iron species cannot be fully excluded based on current experimental data, it appears reasonable to assume that the electron transfer takes place directly between dioxygen and the substrate and that the metal ion supports this reaction by facilitating the electron transfer and by arranging the two reactants in the proper geometry [22].

Model Complexes Several of the above-mentioned intermediates in the catalytic cycles have been assigned based on a comparison of various spectral properties of the transients trapped during the reaction with those of structurally well-characterized model complexes [6]. Even high-valent Fe^{IV}-oxo compounds with a non-heme environment could be prepared and characterized by X-ray crystal structure analysis [24]. It is therefore appropriate to discuss also some model compounds. To generate a functional model of a protein with a facially capped iron center, tridentate ligands such as substituted 1,4,7-triazacyclononanes or hydridotris-(pyrazolyl)borates have been used very successfully (Scheme 2.5a, b) [25–28]. However, these ligands are all *N,N,N*-tripods, hence they do not mimic the 2-His-1-carboxylate triad in all its aspects. Therefore, *N,N,O*-tripodal ligands containing an anionic carboxylate donor have been introduced recently (Scheme 2.5c, d) [29, 30]. Provided that the ligands are bulky enough to suppress formation of a complex with 2 : 1 stoichiometry of ligand and metal center, they all bind to the metal ion in a manner that leaves three sites available for the coordination of additional ligands such as substrate or cofactor.

a)

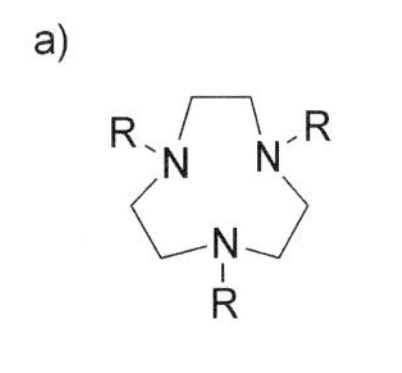

b)

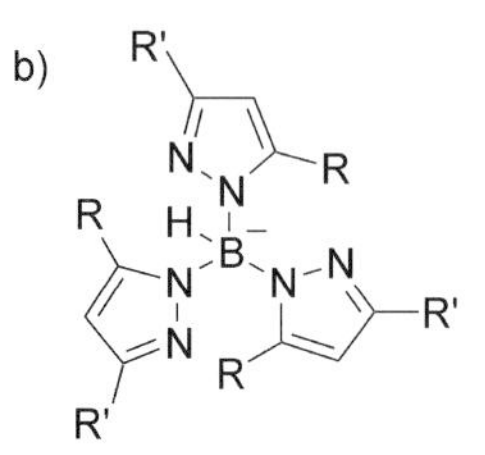

c)

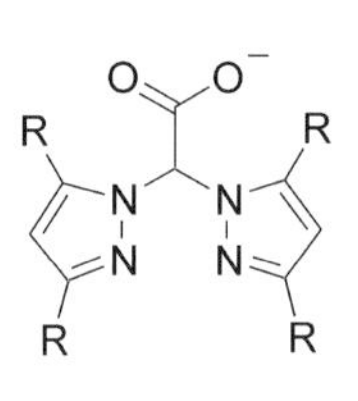

d)

Scheme 2.5 Ligands used to synthesize structural and/or functional models for iron enzymes containing a 2-His-1-carboxylate facial triad: (a) R_3TACN (TACN = 1,4,7-triazacyclononane) [25]; (b) $Tp^{R,R'}$ [differently substituted hydridotris(pyrazol-1-yl)borates] [27]; (c) differently substituted bis(pyrazol-1-yl)acetates [29]; (d) differently substituted 3,3-bis(1-alkylimidazol-2-yl)propionates [30].

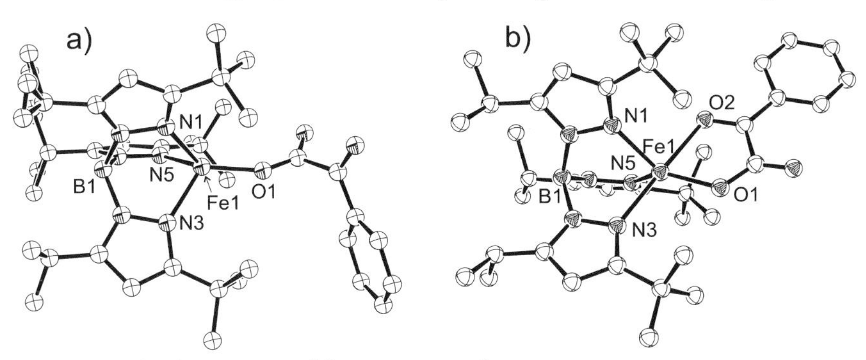

Figure 2.2 Molecular structures of the two isomers of $[Fe(Tp^{iPr,tBu})(bf)]$: (a) colorless isomer with monodentate bf ligand; (b) bluish purple isomer with bidentate bf ligand [31]. Hydrogen atoms have been omitted for clarity.

One interesting example is the iron(II) complex $[Fe(Tp^{iPr,tBu})(bf)]$ (bf = benzoyl formate). This compound was synthesized as a model for an αKG-dependent enzyme. The 2-His-1-carboxylate triad is mimicked by $Tp^{iPr,tBu}$ and benzoyl formate serves as the α-keto carboxylate [31]. Interestingly, two isomers were found for $[Fe(Tp^{iPr,tBu})(bf)]$, i.e. a colorless one and a bluish purple one. X-ray crystal structure analyses showed different binding modes of the benzoyl formate to be responsible for the formation of the two isomers [31]. The bluish purple isomer, with an absorption maximum at 565 nm and reminiscent in its color of the lilac-colored intermediate that is formed upon addition of α-ketoglutarate to taurine dioxygenase (see the section Taurine Dioxygenase above), contains a planar, bidentate benzoyl formate, whereas monodentate binding via a carboxylic oxygen occurs in the colorless isomer (Figure 2.2). The similarities of the spectral features of the colored model compound and the catalytic intermediate suggest that the latter contains the same five-coordinate iron center as the model complex (Scheme 2.3). Unfortunately, no dioxygen adducts of $[Fe(Tp^{iPr,tBu})(bf)]$ could be observed. The dramatically reduced reactivity of the metal center towards dioxygen as compared with the enzyme has been explained in terms of a sterically too demanding tripodal ligand [31]. By applying $Tp^{Ph,Ph}$ with its less bulky phenyl groups, reactivity towards dioxygen was restored: Within 30 min, all of the cofactor benzoyl formate was decarboxylated to benzoate [32]. The $Tp^{Ph,Ph}$ ligand itself served as the substrate in the oxidation reaction, i.e. the *ortho*-carbon of one 3-phenyl group of each complex was hydroxylated. The use of $^{18}O_2$ helped to establish that this model complex indeed reproduces the dioxygenase activity of αKG-dependent enzymes: One ^{18}O atom was incorporated into the benzoate moiety, whereas the other ^{18}O atom was found in the hydroxylated phenyl ring of the $Tp^{Ph,Ph}$ ligand. In a related system, this hydroxylated product could also be characterized by X-ray crystal structure analysis [33]. Although by no means catalytically active, these structural and functional mimics of αKG-dependent enzymes are highly valuable in the investigation of mechanistic details of the catalytically active proteins.

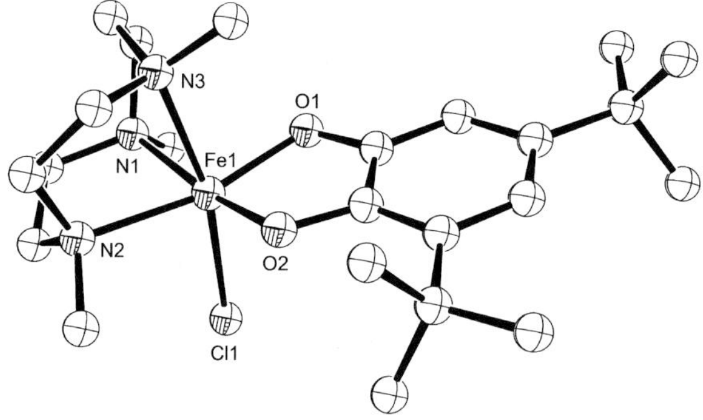

Figure 2.3 Molecular structure of [Fe(Me_3TACN)(DBC)Cl], a model complex for a catechol dioxygenase coordinated to its substrate molecule [28]. Hydrogen atoms have been omitted for clarity. The 1,4,7-trimethyl-1,4,7-triazacyclononane (Me_3TACN) ligand coordinates facially to the iron center. The remaining three coordination sites are occupied by 3,5-di-*tert*-butylcatecholate (DBC) and a chlorido ligand.

As an example for a complex modeling the active site of a catechol dioxygenase, Figure 2.3 provides the molecular structure of [Fe(Me_3TACN)(DBC)Cl]. The iron(III) in this model complex is coordinated to the tridentate 1,4,7-trimethyl-1,4,7-triazacyclononane (Me_3TACN) ligand that resembles the endogenous amino acid ligands, to the doubly deprotonated substrate molecule 3,5-di-*tert*-butylcatecholate (DBC) and to a chlorido ligand that stands in as a surrogate for dioxygen [28]. Treatment with silver(I) was shown to remove the chlorido ligand, leading to the generation of a coordinatively unsaturated metal center and thereby enhancing its reactivity towards dioxygen. The following oxidation reaction proceeds with almost quantitative yield (97%) and gives only extradiol ring cleavage products. However, this relatively simple complex lacks the excellent regioselectivity displayed by its natural counterparts, as it leads to the formation of both 3,5- and 4,6-di-*tert*-butyl-2-pyrone (83 and 14%, respectively) [28]. Also, the complex contains iron(III), whereas natural extradiol dioxygenases contain iron(II). Obviously, modeling only the first coordination sphere of the active site metal ion in this case is not sufficient for obtaining biomimetic reactivity. The steric constraints imposed by the binding pocket of an enzyme and interactions with second-sphere residues also have a large influence on the precise outcome of the respective reaction. This assumption is corroborated by experimental results obtained by using iron(III) complexes of the newly developed 3,3-bis(1-alkylimidazol-2-yl) propionate ligands (Scheme 2.5d) with an *N,N,O*-donor set as the 2-His-1-carboxylate surrogate [30]. Again using DBC as the catechol substrate, different product ratios of extradiol cleavage, intradiol cleavage and auto-oxidation to the quinone could be observed, depending on the substituents of the tripodal ligand, the solvent and the presence or absence of a proton donor. The quinone was formed almost quantitatively in coordinating solvents, showing that the availability of a vacant site for dioxygen coordination is a prerequisite for biomimetic extradiol-type catechol cleavage [28, 30]. As expected, the use of the non-coordinating solvent dichloromethane led to a completely changed product distribution with a dramatic decrease in the formation of quinone. Interestingly, extradiol and intradiol cleavage products were formed in a more or less equimolar ratio. Addition of a proton donor made it possible to increase the selectivity towards extradiol cleavage [30]. Taken together, these data suggest that second-sphere residues in the active site, which are not included in these model complexes, must exert a decisive influence on the regiospecificity of the cleavage. For example, the site-selective mutation of a conserved histidine residue that might act as an acid-base catalyst and that is located close to the active site of 2,3-HPCD (see the section Homoprotocatechuate 2,3-Dioxygenase above) to a phenylalanine that cannot accept or donate protons is able to change the product selectivity of the enzyme from extradiol to intradiol cleavage [34]. The development of even more sophisticated functional model systems that take into account also second-sphere interactions is therefore highly desirable if biomimetic transformations are to be performed.

2.1.2.2 Dinuclear Iron Sites

Numerous examples exist of proteins carrying dinuclear iron active sites. Interestingly, despite containing similar carboxylate-bridged dinuclear iron centers, a wide variety of functions is exerted, providing evidence for an impressive flexibility of the dinuclear core. Reactions catalyzed by these enzymes include, among others, the hydroxylation of aliphatic or aromatic substrates [7, 9, 35], the reduction [the dinuclear iron site is not directly involved in the reduction process; it rather starts a reaction cascade by generating a stable tyrosyl (phenolate) radical, finally leading to the reduction of a ribonucleotide] of ribonucleotides [36] and the desaturation of fatty acids (Scheme 2.6) [37]. Similarly to the mononuclear iron sites discussed above,

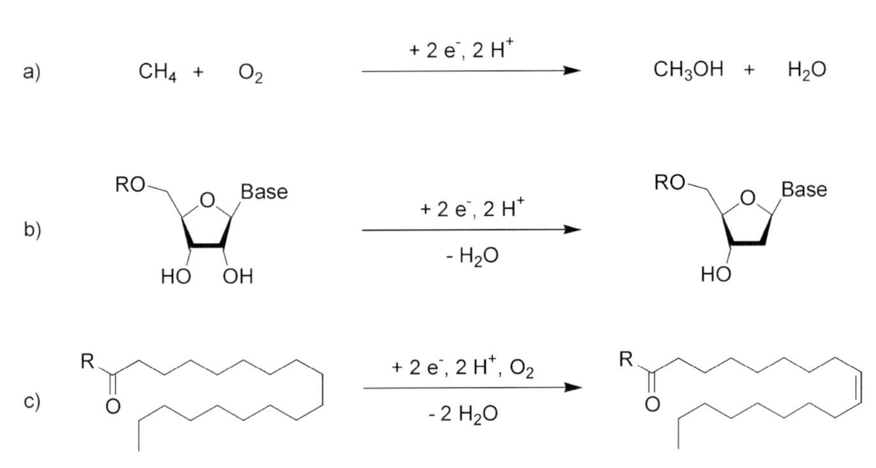

Scheme 2.6 Examples of reactions catalyzed by enzymes that carry a dinuclear iron active site: (a) hydroxylation of methane by soluble methane monooxygenase (sMMO) [7]; (b) reduction of ribonucleotides by class I ribonucleotide reductase (RNR) ($R = (PO_3)_n^{(n+1)-}$) [36]; (c) α,β-dehydrogenation of fatty acids by soluble stearoyl-ACP Δ^9-desaturase (Δ9D) (R = holo-acyl carrier protein) [37].

the metal ions in dinuclear iron sites are coordinated by histidine and by carboxylate-containing amino acids. In the following, the extensively studied soluble methane monooxygenase system will be discussed in more detail.

Methane Monooxygenase Soluble methane monooxygenase (sMMO) from *Methylococcus capsulatus* (Bath) is a complex enzyme system comprising a hydroxylase protein (MMOH) that houses the dinuclear iron active site, a reductase protein that provides the electrons necessary for dioxygen activation, a regulatory protein that is required for efficient catalysis and a fourth component of as yet unknown function that is a potent inhibitor of activity and that has been proposed to play a role in the assembly of the dinuclear iron center [7, 38]. This enzyme system is capable of converting methane to methanol and thereby catalyzes the first step in the metabolic pathway of methane, which represents the sole source for carbon and energy for *M. capsulatus* (Bath). Because the catalytic conversion of methane to methanol is "a holy grail of the petrochemical industry" [39], the methane monooxygenase system has generated substantial interest.

Several X-ray crystal structure analyses exist of the hydroxylase component MMOH with its dinuclear active site, varying in the iron oxidation states and in the identity of additional small molecules bound to the active site (i.e. water, hydroxide, methoxide, ethanol, acetate, etc.) [7]. Figure 2.4 gives two representative examples of these structures, showing the geometry of the active site in (a) the $Fe^{II}{}_2$ form and (b) the $Fe^{III}{}_2$ form with bound methoxide. Both structures have in common the distorted octahedral environment of the metal ions with at least two bridging ligands, including the unsymmetrically bridging carboxylate (Glu144). In addition, Fe1 is invariantly

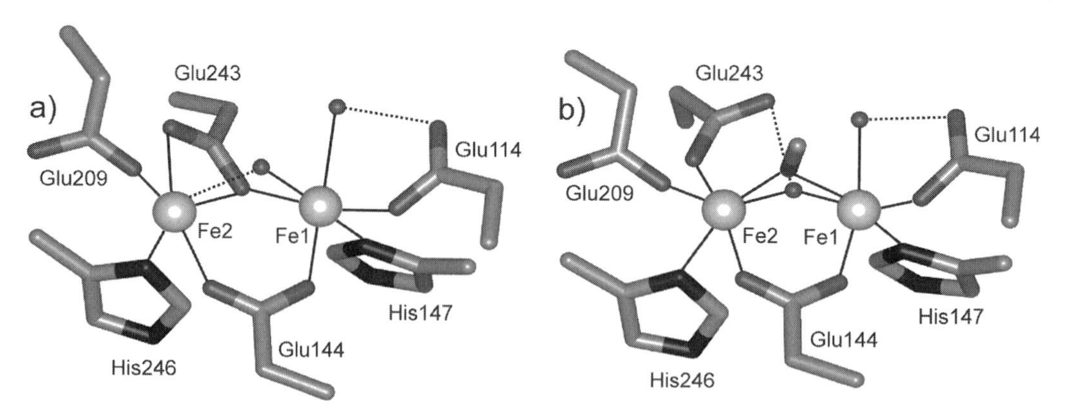

Figure 2.4 X-ray crystal structures of different forms of MMOH from *M. capsulatus* (Bath) {coordinates taken from PDB files 1FYZ (a) [59] and 1FZ6 (b) [60]}. Coordinative bonds are represented by solid lines and hydrogen bonds or weak bonds by dotted lines. Only the amino acid side-chains of the active site and coordinated aqua/hydroxido/methoxido ligands are shown. Hydrogen atoms have been omitted for clarity. Clearly visible is the different coordination mode of Glu243 in the $Fe^{II}{}_2$ form (a) and the $Fe^{III}{}_2$ form (b). The latter structure also contains one methoxide coordinated to the active site and is believed to resemble the point of product release in the catalytic cycle. The third bridging ligand is (a) an aqua ligand and (b) a hydroxido ligand, respectively. The figures were prepared using the computer program MolMol [58].

coordinated to one histidine (His147), one monodentate glutamate (Glu114) and one aqua ligand in all structures. Fe2 has a more flexible coordination environment than Fe1. It is always coordinated to one histidine (His246) and one monodentate glutamate (Glu209). The flexibility occurs in part due to the positioning of Glu243, which can bind to Fe2 in a monodentate or bidentate fashion and which can also serve as a bridging ligand.

Numerous theoretical studies have been performed in addition to the synthesis of model complexes to identify intermediates during the catalytic cycle of sMMO, including both the dioxygen activation step and the actual substrate oxidation [40]. The currently accepted view of the dioxygen activation is summarized in Scheme 2.7 [41].

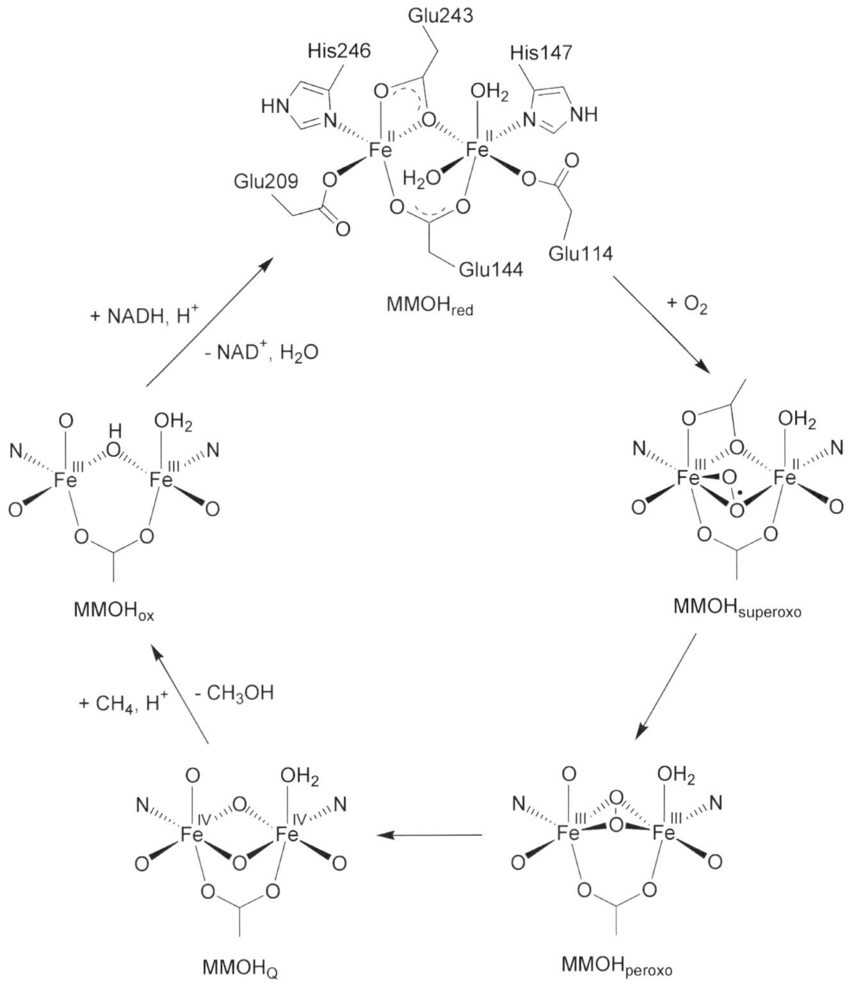

Scheme 2.7 The catalytic cycle of MMOH based on experimental and computational results [41]. The amino acids are numbered according to the sequence of *M. capsulatus* (Bath).

In the resting state, the active site is in its Fe^{III}_2 form ($MMOH_{ox}$). Once reduction by the reductase protein of sMMO has taken place, the system is primed for dioxygen activation. In its fully reduced form (Figure 2.4a), the high-spin iron(II) ions are ferromagnetically coupled. Reaction with dioxygen leads to a superoxo intermediate, in which Fe2, the iron atom with the more flexible coordination environment, undergoes a one-electron oxidation. This intermediate has not yet been observed spectroscopically, but its formation is in agreement with kinetic studies and is has been predicted based on computational results [41]. Next, Fe1 is also oxidized from iron(II) to iron(III), giving rise to the formation of a peroxo intermediate. In Scheme 2.7, this species is depicted with a μ-η^2,η^2 coordination mode ("butterfly structure"), as proposed on the basis of DFT calculations [42]. Based on a comparison of the spectroscopic properties of $MMOH_{peroxo}$ with those of structurally characterized model compounds, a *cis*-μ-1,2-peroxo species was suggested previously [7]. This apparent discrepancy could be explained by assuming that both peroxo species are formed along the dioxygen activation pathway, and hence that the species observed spectroscopically is an intermediate between $MMOH_{superoxo}$ and $MMOH_{peroxo}$. In the next step of the dioxygen activation cycle, both iron sites undergo further oxidation. Transfer of two electrons to the superoxide leads to a cleavage of the O–O bond. The concurrently formed so-called intermediate $MMOH_Q$ has a diamond core structure and comprises two antiferromagnetically coupled high-spin iron(IV) ions in addition to two bridging oxido ligands in an almost symmetric environment. In the final step of the catalytic cycle, substrate oxidation takes place. $MMOH_Q$ reacts with methane, concomitantly being reduced again to the resting Fe^{III}_2 state. [A few non-natural substrates such as propylene, ethyl vinyl ether and diethyl ether were shown to react directly with $MMOH_{peroxo}$ instead of $MMOH_Q$ [43]. This shows that, depending on the substrate, reaction pathways other than the one shown in Scheme 2.7 are possible.] The kinetics of this step of the reaction can easily be monitored due to the bright yellow color of $MMOH_Q$ ($\varepsilon_{350} = 3600\ M^{-1}\ cm^{-1}$ and $\varepsilon_{420} = 7200\ M^{-1}\ cm^{-1}$). It has been postulated that the generation of the high-valent Fe^{IV}_2 intermediate $MMOH_Q$ is mandatory for oxidizing the very stable C–H bond in methane. For example, in toluene 4-monooxygenase and phenol hydroxylase, which both contain active sites closely related to that of sMMO, no dinuclear iron(IV) intermediates could be detected [9]. In contrast to MMOH, these enzymes are only capable of oxidizing the less stable aromatic C–H bonds. Along these lines, the active intermediate of class I ribonucleotide reductase has a structure similar to that of $MMOH_Q$, but only reaches the mixed-valent $Fe^{III}Fe^{IV}$ oxidation state [44].

Whereas several transient species have been observed for dioxygen activation by MMOH, no intermediates were found by rapid-mixing spectroscopic methods for the actual methane hydroxylation step. Mechanistic probes, i.e. certain non-natural substrates that are transformed into rearranged products only if the reaction proceeds via a specific intermediate such as a radical or a cation, give ambivalent results: Some studies show that products according to a pathway via cationic intermediates are obtained in sMMO hydroxylations and at least one study suggests the presence of a radical intermediate [40]. Computational analyses of the reaction of $MMOH_Q$ with methane suggest a so-called radical recoil/rebound mechanism in which $MMOH_Q$

abstracts a hydrogen atom from methane with a linearly oriented C–H bond. The resulting methyl radical recoils away and rebounds to form the new C–O bond [42]. An additional pathway of similar energy was found to proceed via a concerted oxygen atom insertion [45]. In summary, it appears as if different substrates can be differently activated and the reaction pathway hence proceeds via different intermediates. Clearly, more studies are necessary before the mechanism of methane hydroxylation by MMOH can be fully understood.

Model Complexes The synthesis of structural and functional model systems for dinuclear iron enzymes proves to be a challenging field in many ways. For example, a non-negligible influence of the protein matrix on the catalytic pathway has been suggested for MMOH: According to computational results, the protein matrix destabilizes $MMOH_{red}$ by compressing the Fe1–Fe2 distance. This strain is released along the reaction pathway until $MMOH_{peroxo}$ is reached and might be a relevant driving force for the reaction [41]. Furthermore, and not less important, ligands have to be found that support the assembly of a compound modeling the carboxylate-, histidine- and water-containing core with asymmetrically substituted iron centers in various oxidation states. Numerous ligands have been used to achieve this goal [10], and Scheme 2.8 gives two examples, including a complex organic scaffold that geometrically pre-organizes the ligands and a more simple ligand with an increased steric bulk. This bulk is necessary because sterically less hindered carboxylate-based complexes prefer to adopt extended structures. Typically, different ligands are used for the stabilization of the different possible iron oxidation states. Although no structural model for $MMOH_Q$ with its dioxo-bridged $Fe^{IV}{}_2$ center could be obtained so far (the first structurally characterized non-heme iron complex containing an $Fe^{IV}{}_2$ center with a *single* oxo bridge has been reported recently [46]), numerous model complexes exist for $Fe^{II}{}_2$, $Fe^{III}{}_2$ and even $Fe^{III}Fe^{IV}$ centers [10]. In the following, a few examples of complexes with an $Fe^{II}{}_2$ core will be presented.

An elegant and efficient approach for the synthesis of structural models of carboxylate-rich non-heme dinuclear iron proteins involves the use of bulky ligands of the terphenyl carboxylate family [47, 48]. The resulting dinuclear iron(II) complexes can adopt two conformations, called the windmill and the paddlewheel motif (Scheme 2.9).

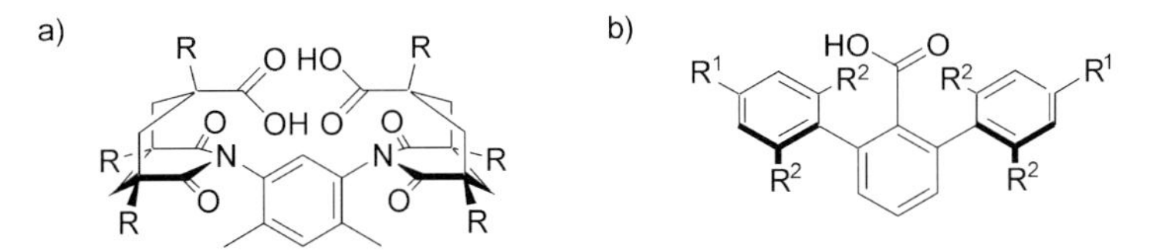

Scheme 2.8 Examples of ligands used to model the carboxylate core of carboxylate-bridged dinuclear iron enzymes: (a) R_3XDK [for $R = CH_3$: H_2XDK = *m*-xylenediamine bis(Kemp's triacid) imide] [57]; (b) differently substituted terphenyl-based ligands (e.g. for $R^1 = CH_3$, $R^2 = H$: $^{Tol}ArCOOH$) [48].

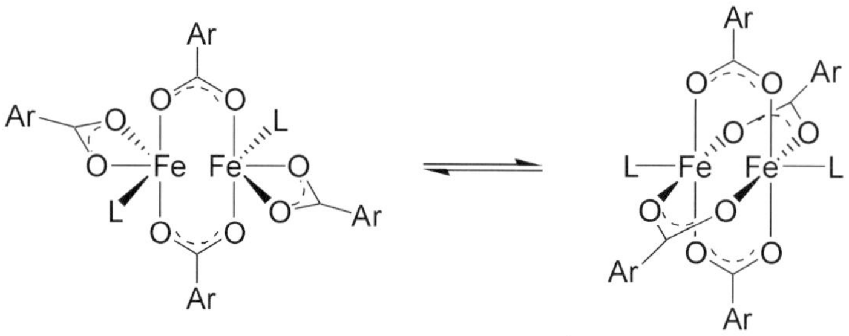

Scheme 2.9 Equilibrium between windmill (left) and paddlewheel (right) structures of dinuclear iron(II) complexes comprising ligands from the terphenyl carboxylate family (Scheme 2.8b).

The desired windmill motif comprises two terminal and two bridging carboxylates in addition to two monodentate ligands and hence resembles the coordination pattern found in the active site of the fully reduced form of the protein. Interestingly, these and related complexes are indeed able to activate dioxygen. When incorporating the designated substrates as ancillary ligands L (according to Scheme 2.9) into the complexes, some substrates become oxidized in the presence of dioxygen. Reported reactions include, among others, oxidative *N*-dealkylations (e.g. *N,N*-dibenzylethylenediamine to benzaldehyde), oxygenation of 2-ethylpyridine to 2-acetylpyridine, oxidation of 2-pyridylphenyl sulfide to 2-pyridylphenyl sulfoxide and oxidation of 2-diphenylphosphinopyridine (2-Ph_2Ppy) to 2-$Ph_2P(O)py$ [49]. Especially the last three examples demonstrate that close proximity of the position to be oxidized and the dinuclear iron center is of great importance to achieve high reactivity. In addition, the substituents on the carboxylate ligands (Scheme 2.8b) exert a non-negligible influence on the reactivity of these systems: Carboxylate ligands with a higher electron-releasing capability lead to an increased C–H activation [49]. These results show that various not only structural but also functional models of dinuclear iron sites exist that necessitate further investigation of their intriguing chemical properties.

A related dinuclear iron complex, $[Fe_2(\mu\text{-}^{Tol}ArCOO)_2(Me_3TACN)_2(MeCN)_2]^{2+}$, assembled from terphenyl carboxylate-related and triazacyclononane-derived ligands, was recently reported to catalyze the reaction of triarylphosphines and dioxygen to triarylphosphine oxides with several thousand turnovers [50]. This conversion, however, is coupled to the oxidation of the solvent THF to 3-hydroxypropyl formate. It is most likely initiated by hydrogen atom abstraction from THF by an oxo-bridged dinuclear iron(III) complex that is formed from $[Fe_2(\mu\text{-}^{Tol}ArCOO)_2(Me_3TACN)_2(MeCN)_2]^{2+}$ in the presence of dioxygen and proceeds by a radical pathway without any further participation of the metal complexes. Despite the involvement of a metal complex that models the active site of a dinuclear iron enzyme, this catalytic process cannot be considered a biomimetic one.

In a different set of experiments, the oxidation of the 2-diphenylphosphinopyridine ligand in $[Fe_2(\mu\text{-}^{Tol}ArCOO)_3(^{Tol}ArCOO)(2\text{-}Ph_2Ppy)]$ as mentioned above was also shown to proceed catalytically (Figure 2.5) [49, 51]. These experiments showed that 2-$Ph_2P(O)py$ is formed when an excess of 2-Ph_2Ppy is present in solution. However, the "turnover numbers" are rather low, with 17 equivalents of 2-Ph_2Ppy being oxidized

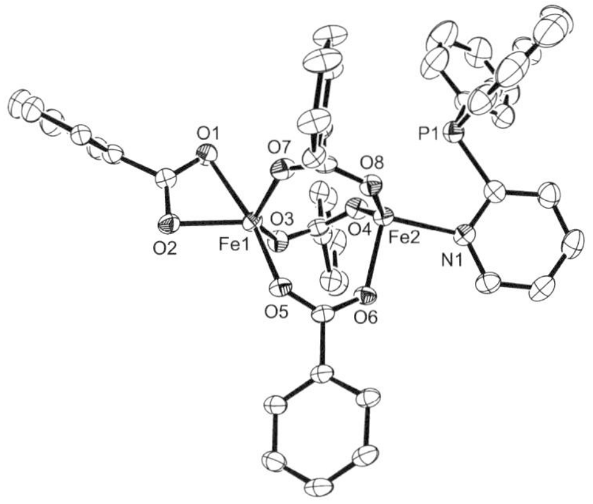

Figure 2.5 Molecular structure of $[Fe_2(\mu\text{-}^{Tol}ArCOO)_3(^{Tol}ArCOO)(2\text{-}Ph_2Ppy)]$, a catalytically active model complex for an enzyme with a dinuclear iron active site [51]. Hydrogen atoms and all atoms of the $^{Tol}ArCOO^-$ ligand except for the central $ArCOO^-$ moiety have been omitted for clarity. Clearly visible is the close proximity of metal center and the phosphorus atom appended to the ancillary pyridine ligand.

in 17 h in a 0.3 mM solution of the dinuclear iron complex. Although auto-oxidation could not be fully ruled out, several control experiments hint towards the metal center having a key role in this catalysis. First, a substitution of $2\text{-}Ph_2Ppy$ by PPh_3 which lacks the pendant coordinating pyridine ligand leads to a dramatic decrease in reactivity. This suggests that the oxidation occurs in an intramolecular fashion after coordination of the substrate to the metal center. Second, the reaction proceeds in a variety of solvents (dichloromethane, acetonitrile, benzene), diminishing the possibility of the solvent actually being involved in the reaction. Third, in the absence of the dinuclear iron complex, no oxidation can be observed within the same time frame. Hence the characterization of the complex $[Fe_2(\mu\text{-}^{Tol}ArCOO)_3(^{Tol}ArCOO)(2\text{-}Ph_2Ppy)]$ represents a promising step on the way towards a truly biomimetic catalyst.

2.1.3 Summary

As shown in this chapter by providing a few selected examples, reduced transition metal ions such as iron(II) are ideal catalysts for oxidation reactions that involve dioxygen. Although oxidations by dioxygen are thermodynamically favored, they usually require the presence of a suitable catalyst to proceed at an acceptable rate. In a typical catalytic cycle, iron(II) activates dioxygen by initially forming an iron(III) superoxo intermediate, followed by further reduction of the superoxide to a peroxide. The electron necessary for this second step may be supplied by an external cofactor, by the substrate itself or in the case of the dinuclear enzymes by the second metal ion. Subsequent O–O bond cleavage leads to a high-valent iron-oxo transient that carries out the substrate oxidation. Obviously, the extent to which an enzyme adheres to this generalized mechanism of dioxygen activation varies from enzyme to enzyme [52].

The nature of the enzyme also determines the fate of the two oxygen atoms: In monooxygenases (e.g. sMMO), one oxygen atom from dioxygen is transferred to the substrate while the other is reduced and forms water. In dioxygenases (e.g. TauD), both oxygen atoms are transferred to the substrate, whereas in oxidases (e.g. Δ9D), both oxygen atoms are reduced to water.

Compared with the heme proteins discussed in Section 2.2, the non-heme iron proteins presented here have a much more flexible coordination geometry. Taken together with the differences in electronic properties – heme enzymes contain mostly low-spin iron whereas non-heme enzymes contain mostly a high-spin iron – this is responsible for the more diverse chemistry found for the non-heme iron proteins. The great versatility of these enzymes makes them a treasure trove for the development of iron-based catalysts. Inspired by their biological archetypes, numerous catalytic reactions await to be reproduced by iron catalysts in organic synthesis.

References

1 J. J. R. Fraústo da Silva, R. J. P. Williams, *The Biological Chemistry of the Elements: the Inorganic Chemistry of Life*, 2nd edn., Oxford University Press, Oxford, 2001.

2 H. Beinert, P. J. Kiley, *Curr. Opin. Chem. Biol.* 1999, **3**, 152–157.

3 M. Bröring, in *Iron Catalysis in Organic Chemistry: Reactions and Applications*, ed. B. Plietker, Wiley-VCH, Weinheim, 2008, pp. 48–72.

4 P. Nordlund, in *Handbook on Metalloproteins*, ed. I. Bertini, A. Sigel, H. Sigel, Marcel Dekker, New York, 2001, pp. 461–570.

5 B. G. Fox, in *Comprehensive Biological Catalysis: a Mechanistic Reference*, Vol. 3, ed. M. Sinnott, Academic Press, San Diego, 1998, pp. 261–348.

6 E. I. Solomon, T. C. Brunold, M. I. Davis, J. N. Kemsley, S.-K. Lee, N. Lehnert, F. Neese, A. J. Skulan, Y.-S. Yang, J. Zhou, *Chem. Rev.* 2000, **100**, 235–349.

7 M. Merkx, D. A. Kopp, M. H. Sazinsky, J. L. Blazyk, J. Müller, S. J. Lippard, *Angew. Chem.* 2001, **113**, 2860–2888; *Angew. Chem. Int. Ed.* 2001, **40**, 2782–2807.

8 R. P. Hausinger, *Crit. Rev. Biochem. Mol. Biol.* 2004, **39**, 21–68.

9 L. J. Murray, S. J. Lippard, *Acc. Chem. Res.* 2007, **40**, 466–474.

10 E. Y. Tshuva, S. J. Lippard, *Chem. Rev.* 2004, **104**, 987–1012.

11 M. Costas, M. P. Mehn, M. P. Jensen, L. Que, Jr., *Chem. Rev.* 2004, **104**, 939–986.

12 T. A. Jackson, L. Que, Jr., in *Concepts and Models in Bioinorganic Chemistry*, ed. H.-B. Kraatz, N. Metzler-Nolte, Wiley-VCH, Weinheim, 2006, pp. 259–286.

13 K. D. Koehntop, J. P. Emerson, L. Que, Jr., *J. Biol. Inorg. Chem.* 2005, **10**, 87–93.

14 V. Purpero, G. R. Moran, *J. Biol. Inorg. Chem.* 2007, **12**, 587–601.

15 J. M. Bollinger, Jr., J. C. Price, L. M. Hoffart, E. W. Barr, C. Krebs, *Eur. J. Inorg. Chem.* 2005, 4245–4254.

16 M. J. Ryle, R. Padmakumar, R. P. Hausinger, *Biochemistry* 1999, **38**, 15278–15286.

17 C. A. Brown, M. A. Pavlosky, T. E. Westre, Y. Zhang, B. Hedman, K. O. Hodgson, E. I. Solomon, *J. Am. Chem. Soc.* 1995, **117**, 715–732.

18 J. San Filippo, Jr., C.-I. Chern, J. S. Valentine, *J. Org. Chem.* 1976, **41**, 1077–1078.

19 J. C. Price, E. W. Barr, B. Tirupati, J. M. Bollinger, Jr., C. Krebs, *Biochemistry* 2003, **42**, 7497–7508.

20 R. Myllylä, K. Majamaa, V. Günzler, H. Hanauske-Abel, K. I. Kivirikko, *J. Biol. Chem.* 1984, **259**, 5403–5405.

21 L. Tuderman, R. Myllylä, K. I. Kivirikko, *Eur. J. Biochem.* 1977, **80**, 341–348.

22 E. G. Kovaleva, J. D. Lipscomb, *Science* 2007, **316**, 453–457.
23 P. E. M. Siegbahn, F. Haeffner, *J. Am. Chem. Soc.* 2004, **126**, 8919–8932.
24 L. Que, Jr., *Acc. Chem. Res.* 2007, **40**, 493–500.
25 G. Lin, G. Reid, T. D. H. Bugg, *J. Am. Chem. Soc.* 2001, **123**, 5030–5039.
26 N. Kitajima, N. Tamura, H. Amagai, H. Fukui, Y. Moro-oka, Y. Mizutani, T. Kitagawa, R. Mathur, K. Heerwegh, C. A. Reed, C. R. Randall, L. Que, Jr., K. Tatsumi, *J. Am. Chem. Soc.* 1994, **116**, 9071–9085.
27 N. Kitajima, W. B. Tolman, *Prog. Inorg. Chem.* 1995, **43**, 419–531.
28 D.-H. Jo, L. Que, Jr., *Angew. Chem.* 2000, **112**, 4454–4457; *Angew. Chem. Int. Ed.* 2000, **39**, 4284–4287.
29 A. Beck, A. Barth, E. Hübner, N. I. Burzlaff, *Inorg. Chem.* 2003, **42**, 7182–7188.
30 P. C. A. Bruijnincx, M. Lutz, A. L. Spek, W. R. Hagen, B. M. Weckhuysen, G. van Koten, R. J. M. Klein Gebbink, *J. Am. Chem. Soc.* 2007, **129**, 2275–2286.
31 S. Hikichi, T. Ogihara, K. Fujisawa, N. Kitajima, M. Akita, Y. Moro-oka, *Inorg. Chem.* 1997, **36**, 4539–4547.
32 E. L. Hegg, R. Y. N. Ho, L. Que, Jr., *J. Am. Chem. Soc.* 1999, **121**, 1972–1973.
33 M. P. Mehn, K. Fujisawa, E. L. Hegg, L. Que, Jr., *J. Am. Chem. Soc.* 2003, **125**, 7828–7842.
34 S. L. Groce, J. D. Lipscomb, *J. Am. Chem. Soc.* 2003, **125**, 11780–11781.
35 B. J. Wallar, J. D. Lipscomb, *Chem. Rev.* 1996, **96**, 2625–2657.
36 B.-M. Sjöberg, *Struct. Bonding* 1997, **88**, 139–173.
37 B. G. Fox, K. S. Lyle, C. E. Rogge, *Acc. Chem. Res.* 2004, **37**, 421–429.
38 M. Merkx, S. J. Lippard, *J. Biol. Chem.* 2002, **277**, 5858–5865.
39 J. B. van Beilen, E. G. Funhoff, *Curr. Opin. Biotechnol.* 2005, **16**, 308–314.
40 M.-H. Baik, M. Newcomb, R. A. Friesner, S. J. Lippard, *Chem. Rev.* 2003, **103**, 2385–2419.
41 D. Rinaldo, D. M. Philipp, S. J. Lippard, R. A. Friesner, *J. Am. Chem. Soc.* 2007, **129**, 3135–3147.
42 P. E. M. Siegbahn, *J. Biol. Inorg. Chem.* 2001, **6**, 27–45.
43 L. G. Beauvais, S. J. Lippard, *J. Am. Chem. Soc.* 2005, **127**, 7370–7378.
44 N. Mitić, M. D. Clay, L. Saleh, J. M. Bollinger, Jr., E. I. Solomon, *J. Am. Chem. Soc.* 2007, **129**, 9049–9065.
45 B. F. Gherman, B. D. Dunietz, D. A. Whittington, S. J. Lippard, R. A. Friesner, *J. Am. Chem. Soc.* 2001, **123**, 3836–3837.
46 A. Ghosh, F. Tiago de Oliveira, T. Yano, T. Nishioka, E. S. Beach, I. Kinoshita, E. Münck, A. D. Ryabov, C. P. Horwitz, T. J. Collins, *J. Am. Chem. Soc.* 2005, **127**, 2505–2513.
47 W. B. Tolman, L. Que, Jr., *J. Chem. Soc, Dalton Trans.* 2002, 653–660.
48 D. Lee, S. J. Lippard, *J. Am. Chem. Soc.* 1998, **120**, 12153–12154.
49 E. C. Carson, S. J. Lippard, *Inorg. Chem.* 2006, **45**, 837–848, and references cited therein.
50 R. F. Moreira, E. Y. Tshuva, S. J. Lippard, *Inorg. Chem.* 2004, **43**, 4427–4434.
51 E. C. Carson, S. J. Lippard, *J. Am. Chem. Soc.* 2004, **126**, 3412–3413.
52 L. Que, Jr., Y. Watanabe, *Science* 2001, **292**, 651–653.
53 J. M. Elkins, M. J. Ryle, I. J. Clifton, J. C. Dunning Hotopp, J. S. Lloyd, N. I. Burzlaff, J. E. Baldwin, R. P. Hausinger, P. L. Roach, *Biochemistry* 2002, **41**, 5185–5192.
54 B. Yu, W. C. Edstrom, J. Benach, Y. Hamuro, P. C. Weber, B. R. Gibney, J. F. Hunt, *Nature* 2006, **439**, 879–884.
55 E. G. Pavel, J. Zhou, R. W. Busby, M. Gunsior, C. A. Townsend, E. I. Solomon, *J. Am. Chem. Soc.* 1998, **120**, 743–753.
56 S. L. Groce, J. D. Lipscomb, *Biochemistry* 2005, **44**, 7175–7188.
57 S. Herold, S. J. Lippard, *J. Am. Chem. Soc.* 1997, **119**, 145–156.
58 R. Koradi, M. Billeter, K. Wüthrich, *J. Mol. Graphics* 1996, **14**, 51–55.
59 D. A. Whittington, S. J. Lippard, *J. Am. Chem. Soc.* 2001, **123**, 827–838.
60 D. A. Whittington, M. H. Sazinsky, S. J. Lippard, *J. Am. Chem. Soc.* 2001, **123**, 1794–1795.

2.2
Organic Reactions Catalyzed by Heme Proteins

Martin Bröring

2.2.1
Classification and General Reactivity Schemes of Heme Proteins Used in Organic Synthesis

Heme proteins introduced to organic synthesis can be roughly divided into two classes, the monoxygenases [61, 62] and the peroxidases [63, 64]. Both classes have enjoyed multifaceted interest from a large number of research groups in many fields for over half a century and many aspects of their chemical, biological, physical and medicinal properties have been extensively and repeatedly reviewed in the past. Although a huge body of work has been devoted to this field and the overwhelming number of different heme-containing oxidases and oxygenases known to date would render a comprehensive review of their reactivity simply impossible, the development of heme enzymes as catalysts for synthetic chemists is still rather in its infancy. Most of the work dedicated to the development of the use of heme enzymes in organic transformations has been performed using prototypical examples of these two classes, the soluble cytochromes P450CAM and P450BM-3 on the one hand and horseradish peroxidase (HRP), chloroperoxidase (CPO), microperoxidase (MP11), myeloperoxidase and Coprinus peroxidase (CiP) on the other. The individual reactivity of these and related heme enzymes can easily be explained by the setting of the heme active site, which differs significantly from one class to another. Figure 2.6 illustrates this point.

The monooxygenases [65–70] are characterized by the axial attachment of the heme cofactor to a thiolate functionality from a cysteine residue of the protein matrix. The binding pockets of monooxygenases are usually organized in such a way as to lead the substrate directly to the active FeO subunit of the porphyrin. This

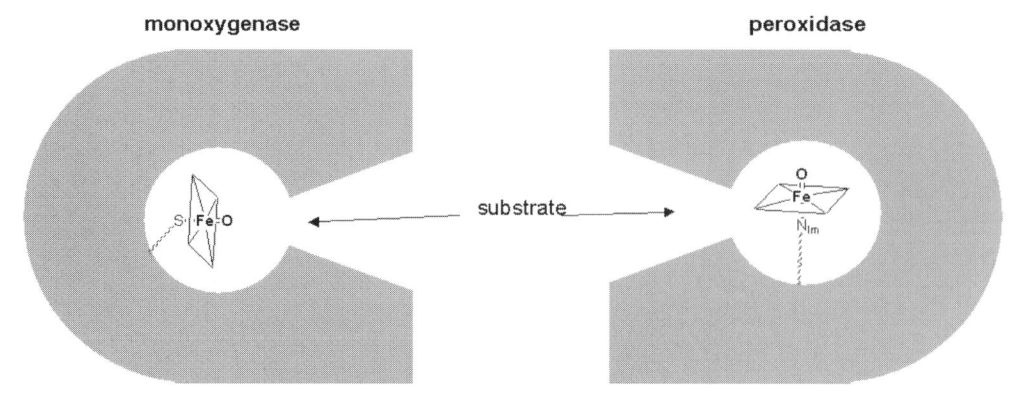

Figure 2.6 Schematic representation of the heme group orientation and different substrate accessibility in the heme active sites in monoxygenases and peroxidases.

is in agreement with the finding that so-called suicide substrates such as alkyl- or phenylhydrazines irreversibly inhibit the reactivity of monooxygenases upon the formation of N_{heme}–R or Fe–R species. Terminal alkenes and alkynes can also react with this class of heme enzymes by a similar mechanism-based inhibition process. The binding pockets of most monooxygenases are usually non-polar and often optimized by shape and depth for a particular substrate. A high degree of chemo-, regio- and stereoselectivity has frequently been observed, whereas the acceptance of alternative substrates is usually rather restricted for those cases. As a general rule, monooxygenases react upon the introduction of a single oxygen atom to a substrate. Mechanistically, these reactions have received much attention for over half a century and key steps of the mechanistic schemes are still a topic of lively and controversial debate. In order to develop the application of heme-containing monooxygenases to organic syntheses, however, the knowledge of a simplified and generally accepted catalytic cycle as depicted in Scheme 2.10 provides sufficient details [71].

Most characteristic for the catalytic cycle of heme monooxygenases is the activation of molecular dioxygen to an active FeO moiety and water. Only two out of four oxidation equivalents are thus used for the synthesis of oxygenated products. This fact is often used as a mechanistic possibility of a short-cut, the so-called peroxide shunt, where a

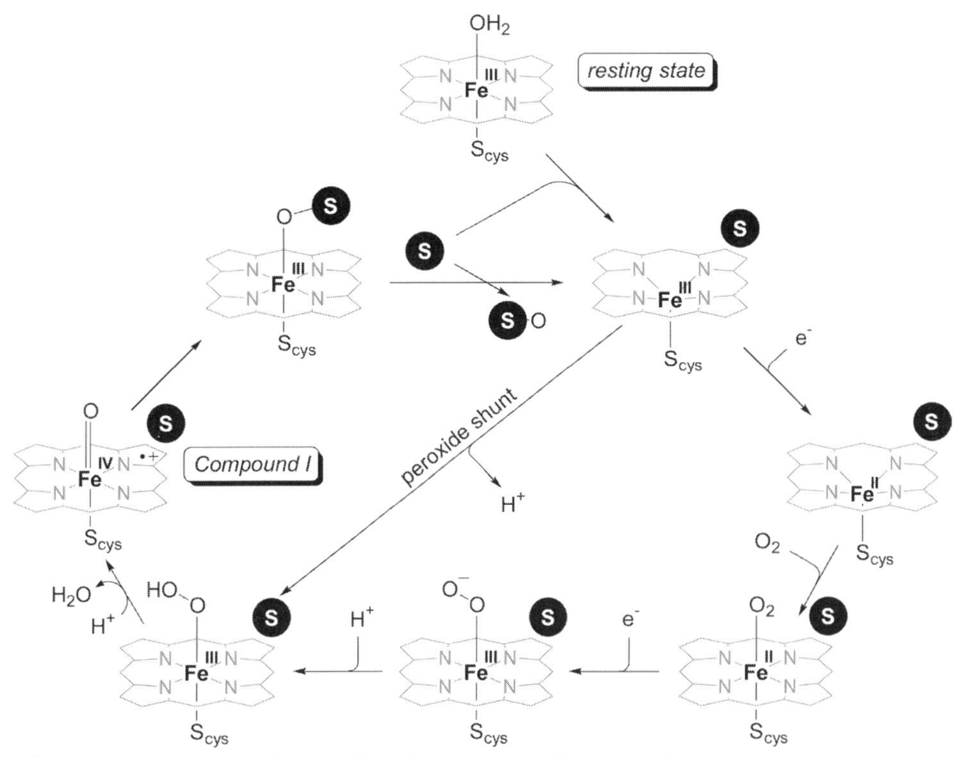

Scheme 2.10 Consensus mechanism of cytochrome P450 catalysis (S = substrate) [61, 62, 72].

two-electron acceptor such as hydrogen peroxide is used instead of dioxygen as the terminal oxidant. The active species of heme monooxygenases is the ferryl(IV) radical cation called Compound I. Compound I forms exclusively in the presence of a substrate, thereby reducing the risk of autocatalytic degradation of the enzyme.

A major drawback for the development of an application of isolated cytochromes P450 in organic syntheses lies in the membrane-bound nature and thus insolubility of most of the ca. 5500 known natural members of this enzyme class. Soluble exceptions are rare, but exist in the form of the microbial P450BM-3 from *Bacillus megaterium*, which catalyzes the hydroxylation of fatty acids, amides or alcohols with little selectivity at the (ω-1), (ω-2) and (ω-3) positions, and P450CAM from *Pseudomonas putida*, which hydroxylates polyisocyclic and *N*-heterocyclic alkanes with remarkable regio- and stereoselectivity. Much work has been devoted to both membrane-bound oxygenases in whole cells and soluble P450 enzymes, and several advances and applications in the fields of biotechnology and substrate activation have recently emerged from these efforts. Another significant problem for the development of P450-based catalysts is the complexity of the reduction system, which usually contains the three components NAD(P)H, ferredoxin and an NAD(P)H-dependent ferredoxin reductase. This rather large number of necessary components renders efficient catalysis outside a cell membrane impossible and is, in particular with respect to biotechnological applications, very cost intensive. P450BM-3 is again an exception to this rule, since the reduction system is integrated as a subdomain to the holoenzyme, which makes this monooxygenase an efficient catalyst even in homogeneous solution [73].

While the intense current endeavors in the development of P450 monooxygenases as catalysts for alkyl hydroxylations point to many future applications, the group of heme peroxidases already provides a spectrum of different enzymes which have proven useful in chemical syntheses. Besides the well-known horseradish peroxidase (HRP) and chloroperoxidase (CPO), for which by far the most investigations have been reported, several less prominent enzymes such as Cyprinus peroxidase (CiP), myeloperoxidase, microperoxidase (MP11) and others have found entry into studies in the past aimed at their catalytic potential. The immediate surrounding of the heme group in peroxidases is characteristically distinct from those found for monooxygenases. Heme in peroxidases is linked to the enzyme by an iron–histidine bond and the seating of the porphyrin in the binding pocket is such that a substrate is directed to an edge of the cofactor (Figure 2.6). In general, the direct contact between the FeO subunit and a substrate is inhibited and the reactivity at the active site is therefore restricted to a simple one-electron charge-transfer step. Suicide substrates deactivate the enzyme after binding to a *meso*-carbon atom of the porphyrin ring. Oxygenations, on the other hand, are rare and result from an initial one-electron activation step of the substrate, followed by an O-atom transfer reaction. A generally accepted mechanistic scheme exists for peroxidase activity and is shown in Scheme 2.11.

Other than for the monooxygenases, a two-electron acceptor such as hydrogen peroxide is required as the terminal oxidant for peroxidases. In the so-called resting state the Fe ion is situated in the oxidation state +3. Reaction with hydrogen peroxide proceeds with loss of water and yields a ferryl(IV) radical cation called Compound I,

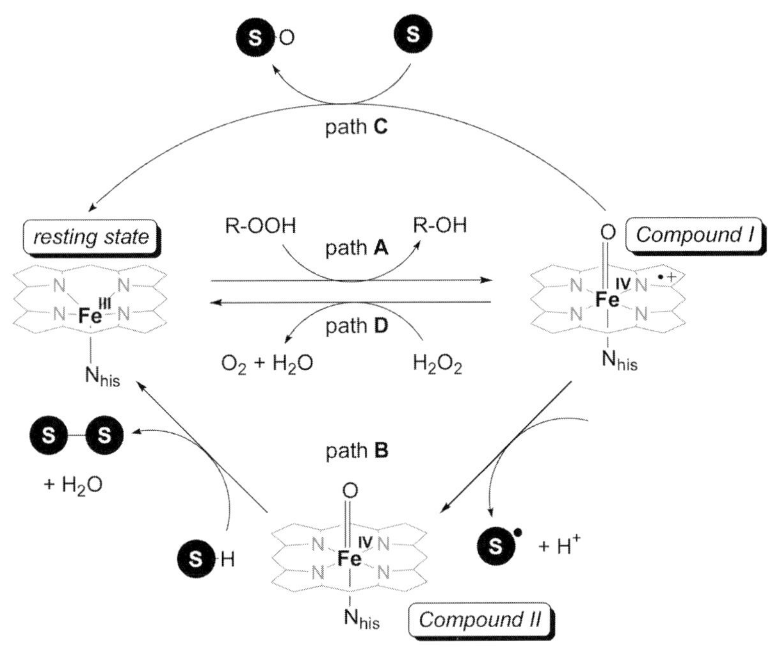

Scheme 2.11 General catalytic cycle of heme peroxidase action (S = substrate) [63, 74, 75].

which is identical with the monooxygenase Compound I except for the proximal donor. The back reaction from Compound I to the resting state can occur via different routes **A–D**. This fact results an a plethora of potential reactions catalyzed by heme peroxidases and thus in great potential for synthetic chemistry. It should be noted that both Compound I and the reduced ferryl(IV) species Compound II may serve as key targets for selective one-electron transfer reactions. In catalytic transformations CPO is a special case and shows monooxygenase rather than peroxidase behavior. This peculiarity is believed to be based on a structural similarity of CPO and P450 enzymes and also on the fact that the heme moiety in CPO is bonded to the protein by an iron–sulfur link.

2.2.2
Organic Reactions Catalyzed by Cytochromes P450

The P450 enzyme family is known to catalyze a wide range of organic transformations, as shown in Scheme 2.12 [61], with arguably the most intriguing process being the regio- and stereoselective hydroxylation of non-activated aliphatic hydrocarbons such as steroids [76–78], terpenes [79–83] and other alicyclic species [84–93]. C=C double bonds are transformed via an initial epoxidation which ultimately leads to (*inter alia*) epoxides from alkenes, carboxylic acids from alkynes and phenols from aryls. Amine and imine nitrogen and also thioether sulfur atoms are easily oxygenated or hydroxylated and O-, N- and S-bound alkyl groups often undergo oxidative dealkylation processes with P450. Many other deamination, dehalogenation,

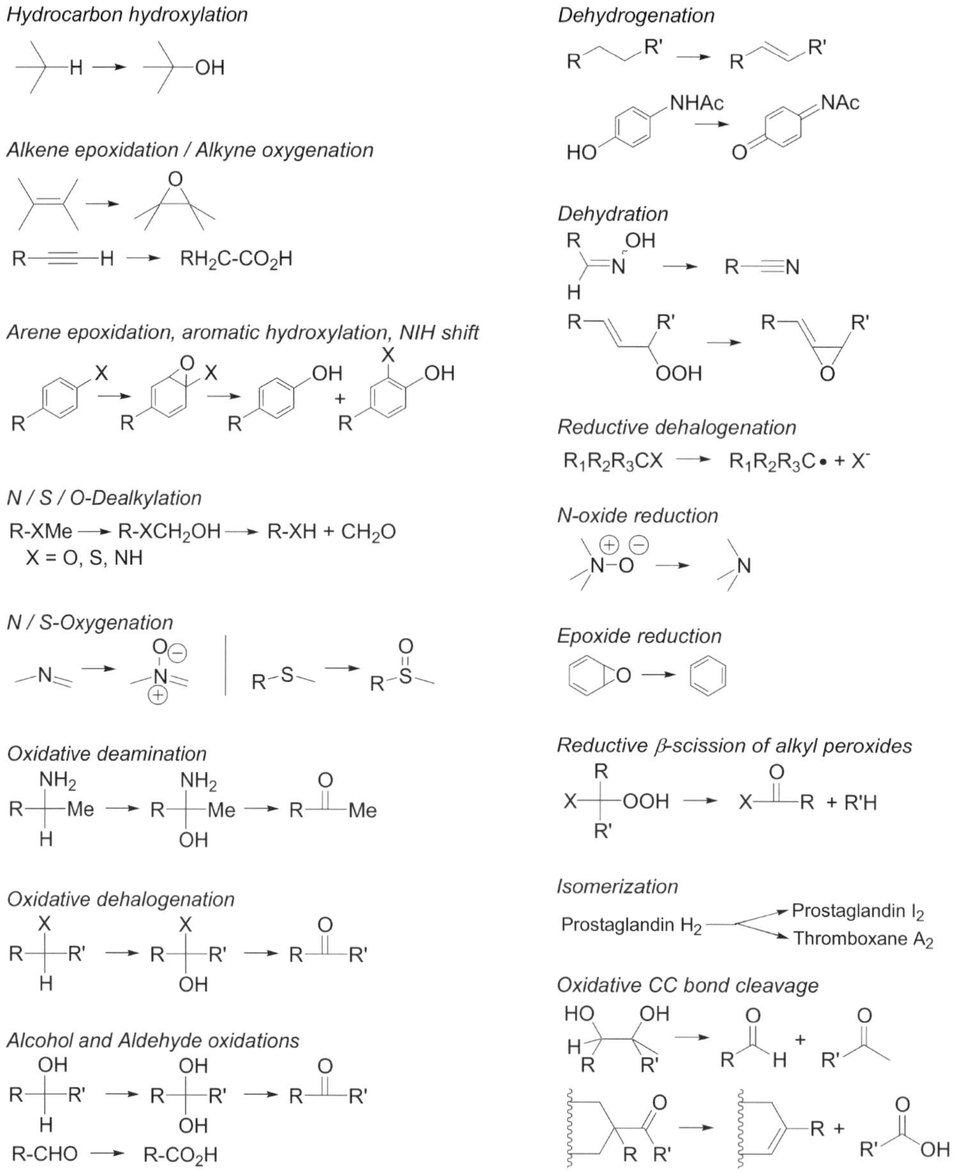

Hydrocarbon hydroxylation

Alkene epoxidation / Alkyne oxygenation

R—≡—H → $RH_2C\text{-}CO_2H$

Arene epoxidation, aromatic hydroxylation, NIH shift

N / S / O-Dealkylation

R-XMe → $R\text{-}XCH_2OH$ → R-XH + CH_2O
X = O, S, NH

N / S-Oxygenation

Oxidative deamination

Oxidative dehalogenation

Alcohol and Aldehyde oxidations

R-CHO → $R\text{-}CO_2H$

Dehydrogenation

Dehydration

Reductive dehalogenation

$R_1R_2R_3CX \rightarrow R_1R_2R_3C\bullet + X^-$

N-oxide reduction

Epoxide reduction

Reductive β-scission of alkyl peroxides

Isomerization

Prostaglandin H_2 → Prostaglandin I_2 / Thromboxane A_2

Oxidative CC bond cleavage

Scheme 2.12 Summary of P450-catalyzed reactions [61].

alcohol/aldehyde oxygenation and dehydrogenation reactions have been described and interestingly a group of reductive processes and of isomerizations are also occasionally catalyzed by this enzyme class.

As mentioned before, the application of P450-catalyzed monoxygenations in organic synthesis is largely hampered by the insolubility/instability of most of the isolated enzymes of the family and by the complexity of the reduction system. In order to circumvent these drawbacks, most of the development has been devoted to biotechnological processes in which whole cells are used as catalysts. Examples are found in the use of "blue roses" [94, 95], the production of dicarboxylic acids by oxidation of alkanes using yeasts (*Candida tropicalis, Yarrowina lipolytica* [96, 97]), the preparation of aromatic precursors in the agrochemical industry using the fungus *Beauveria bassiana* [98] and the oxidative transformation of steroids by different microorganisms [99, 100]. Several reviews have appeared recently on selected topics in this field [101–106] which cover the most important aspects for the future development of P450 enzymes in biotechnology.

Protein engineering has been proven to be successful in the development of valuable hydroxylation catalysts for organic laboratory synthesis from the soluble monoxygenases P450BM-3, P450CAM [73] and others [107–109]. The very active P450BM-3 usually hydroxylates medium- to long-chain alkyl groups at the subterminal (ω-1)-, (ω-2)- and (ω-3) positions and needs improved substrate acceptance and better selectivity (regio-, stereo-). In addition, the exchange of NADPH for a cheaper terminal reductant would be appreciated. A number of improved species were obtained from site-directed mutagenesis of one or two amino acid residues at the active site [110]. The R47E mutant of P450BM-3 was found to be 11–26 times as efficient in the hydroxylation of $C_{12}/C_{14}/C_{16}$-alkyl trimethylammonium [111] and the L181K mutant and also the L75T/L181K double mutant showed an about 15-fold improved efficiency for the hydroxylation of butyrate and hexanoate [112]. Laurate and myristate could be hydroxylated at the terminal positions using an F87A mutant, although the selectivity is rather poor [113].

Site-specific saturation at key residues, followed by a high-throughput activity assay, produced the triple mutant F78V/L188Q/A74G, which shows a greatly increased 700-fold efficiency for the hydroxylation of *n*-octane and a 200-fold increase in the hydroxylation of β-ionone at the 3-position [114]. Stereoselectivity is poor, however, and all diastereomers are present in the products in comparable amounts. A five-fold mutant, the so-called (F87V)LARV, was developed in a related fashion [115] and shown to hydroxylate capric acid, a substrate not attacked by the wild-type enzyme [116].

The most successful route to functional mutants of P450BM-3 [117, 118] and related CYP102 monoxygenases [119] uses directed evolution in combination with site-selective mutagenesis. Selective hydroxylations of *n*- and cycloalkanes have been in the focus of investigations using this approach. In addition, examples of aromatic hydroxylations and epoxidations have occasionally been reported. Typical results for alkane hydroxylations are given in Table 2.1. It should be mentioned that this method can also be employed successfully to prepare enzymes insensitive to organic co-solvents such as THF and DMSO and thus to optimize reactions with substrates showing little or no solubility in water [120].

The efficiency of *n*-octane hydroxylation was optimized and a mutant called variant 139-3 was found to perform this oxygenation 38 times faster than the wild-type enzyme. This variant is also very active for other substrates such as propane, hexane, cyclohexane, heptane, nonane and decane. Even ethane can be hydroxylated selectively with this engineered biocatalyst [121]. First reports on the control of

Table 2.1 Selected P450 mutants with modified catalytic performance [109].

Enzyme	Mutant	Preparation	Substrates/effects
P450BM-3	Wild type		Saturated fatty acids at (ω-1), (ω-2), (ω-3) [126, 127]
	F87A	Site-directed mutagenesis	Laurate, myristate (ω-position) [128]
	F87A(LARV)	Rational evolution	Shorter chain-length substrates [120]
		Site-specific randomization mutagenesis of F87A(LARV)	Shorter chain fatty acid derivatives [129]
	F87A5F5	Random mutations on F87A	Increased resistance to organic co-solvents [130]
	F87A/L188Q/A74G	Rational evolution	Indole to indigo [131], broad range of non-natural substrates [132]
	139-3	Five generations of mutagenesis	*n*-Octane, *n*-hexane, cyclohexane, *n*-pentane (hydroxylation) [133], benzene, styrene, cyclohexene, 1-hexene (epoxidation) [134]
	9-10A	Rational evolution	Propane [135]
	53-5H	Directed evolution of 9-10A	Propane, ethane [136]
	77-9H	Saturation mutagenesis of 9-10A	Terminal alkane hydroxylation and epoxidation [137, 138]
	P15S	Directed evolution	SDS, lauric acid, 1,4-naphthoquinone, ethacrynic acid [139]
	9-10A–F87A	Site-directed mutagenesis	2-Arylacetic acid derivatives, buspirone [140]
P450CAM	Wild type		Camphor (5-*exo*)
	Y96F, Y96F/F193L	Site-directed mutagenesis	Adamantane [141]
	Y96F/V247L	Site-directed mutagenesis	(+)-α-Pinene [142], (*S*)-limonene [143]
	Y96A	Site-directed mutagenesis	Benzylcyclohexane, benzoylcyclohexanol [144]
	Y96A, Y96F	Site-directed mutagenesis	Styrene [145], linear alkanes [146], branched alkanes [147]
	F87W/Y96F/T101L/V247L(EB)	Site-directed mutagenesis	Butane, propane [148, 149]
	EB/L294M/T185M/L1358P/G248A	Directed evolution	Ethane [150]
CYP1A2	E225I, E225N, F226Y	Random mutagenesis	7-Ethoxyresorufin [151], phenacetin [152]
	E163K/V193M/167Q	Directed evolution	7-Methoxyresorufin [153]
CYP2A6	L240C/N297Q	Rational evolution	4- and 5-benzyloxyindole [154], indigo production [155]
CYP2B1		Directed evolution	H_2O_2-supported hydroxylation of progesterone [156], 7-benzyloxyresorufin, benzphetamine, testosterone, cyclophosphamide, ifosfamide [157]

stereoselectivity for hydroxylation in 2-position have also appeared in the literature. For 2-octanol, both the *S*- and the *R*-enantiomers can be produced with *ee*s of about 85% using different mutants of P450BM-3 [122]. The regioselectivity of *n*-octanol hydroxylation was found to differ significantly from the action of the wild-type P450BM-3. This fact was used as the starting point for optimization studies towards terminal hydroxylation processes. Increased selectivities for the formation of 1-octanol have been reported [123, 124]. Very recently a special class of P450 enzymes was found which selectively hydroxylates *n*-alkanes at the 1-position. These bacterial enzymes are expressed in alkane-rich growth media only and most probably long-chain alkanes such as dodecane are their natural substrates [125]. Other reactions, especially epoxidations and aromatic hydroxylations, have occasionally been reported to be catalyzed by isolated P450BM-3 (Table 2.1). As for the aliphatic hydroxylations, promising results have been obtained which point to possible future applications [71]. Similarly, other P450 enzymes have most recently been successfully engineered for selective hydroxylation of special substrates.

Much work in protein engineering has also been devoted to the soluble P450CAM from *P. putida*, which naturally catalyzes the hydroxylation of camphor to 5-*exo*-hydroxycamphor, in order to make use of the excellent regio- and stereochemistry shown by this enzyme [129]. As before, most of the mutant production and screening were performed using site-directed mutagenesis and focused on alkane hydroxylation [134–136, 139], albeit the majority of reports deals with cyclic and polycyclic alkane targets for this enzyme [132, 133, 141–144]. Variants were found which catalyze the hydroxylation of adamantane, (+)-α-pinene and (*S*)-limonene more efficiently than the wild-type enzyme. Selective hydroxylation at C-4 was achieved with yet another variant for several monosubstituted cyclohexane derivatives. The catalytic performance for acyclic isoalkanes as substrates did not improve significantly upon mutagenesis, but the efficiency for the hydroxylation of C_5–C_7 *n*-alkanes could be increased by this approach by about 20-fold. If large residues are present in the binding pocket, even small alkanes such as *n*-butane, propane and ethane are efficiently hydroxylated by engineered P450CAMs.

In addition to solubility, the major drawback of the use of P450 enzymes in organic preparations is still the requirement for a costly source of reduction equivalents. Five general attempts are currently being pursued in order to circumvent the use of expensive NADPH as the sole terminal reductant. Transfer of electrons to the heme moiety by an electrode was accomplished using immobilized enzymes, enzyme/surfactant films or a transfer agent [156–182]. Alternatively, light was used in combination with a photo-oxidant in order to transfer the necessary electrons to the heme cofactor [183–188]. H_2O_2 and alkyl peroxides have been used as the terminal oxidant making use of the peroxide-shunt [189–197] and further terminal oxidants such as periodate, iodosylbenzene and others have also been applied for this purpose [198–202]. These approaches, however, suffer significantly from stability issues. The most successful tool so far follows the idea of *in situ* NADPH regeneration [151–155]. In addition, the exchange of NADPH with the much cheaper NADH allows for a low-cost reduction system with substantial efficiency that may be broadly applied in this field. There still remains, however, much room for improvements.

2.2.3 Organic Reactions Catalyzed by Heme Peroxidases

Peroxidases are ubiquitous in the plant and animal kingdoms. They are usually classified by their origin, i.e. into those found in bacteria (class 1), fungi (class 2) and plants (class 3). The amino acid residues of the active site, which are important for the catalytic performance of the respective peroxidase, change from case to case, with the P450-like CPO being an extreme example. The layout of the binding pocket also has a profound influence on the reaction catalyzed by the individual enzyme. Since peroxidases are generally soluble and their function is not coupled to a specific reduction system, the *ex vivo* use of these enzymes in chemical preparations is straightforward and has been documented in many cases. Biotransformations promoted by heme peroxidases and applications to organic synthesis can be roughly organized into five schemes.

2.2.3.1 Dehydrogenations ("Peroxidase Reactivity")

In the peroxidase reaction, Compound I is reduced to the Fe^{III} resting state via Compound II in two consecutive one-electron transfer steps which results in the formation of two substrate radicals (path **B** in Scheme 2.11). With respect to the whole catalytic cycle, this behavior can be described by the following equation:

$$2AH + ROOH \rightarrow 2A\bullet + ROH + H_2O \tag{2.1}$$

Typical substrates AH for this reaction are electron-rich aromatic compounds such as guaiacol, and the dimerization of these radicals usually leads to the formation of the final products. The reaction is known for all heme peroxidases; however, the biological purpose differs considerably with enzyme and organism. Lignin peroxidase, for example, is responsible for the oxidative degradation of lignin in plants and uses hydrogen peroxide as the terminal oxidant [203]. Ascorbate peroxidase, on the other hand, uses ascorbate as a reductant and serves the cell in the destruction of toxic excess hydrogen peroxide [63]. Horseradish peroxidase (HRP) finally produces radicals as intermediates for the biosynthesis of phytohormones [204]. *N*- and *O*-dealkylations have also occasionally been observed with peroxidases [205]. These reactions can be regarded as a special case of the oxidative dehydrogenation reaction and thus belong in this group.

The peroxidase reaction of heme peroxidases has found technical uses mainly for the polymerization of phenols and anilines (Scheme 2.13) [206–216] and for waste water treatment [217–220], whereas phenol coupling reactions in the biosynthesis of e.g. vancomycin are carried out by P450-type cytochromes [221].

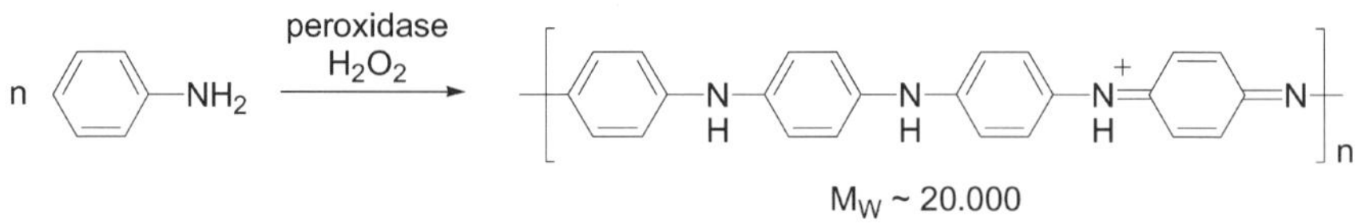

n (C6H5)–NH2 —peroxidase, H_2O_2→ polyaniline, n

$M_W \sim 20.000$

Scheme 2.13 Peroxidase-catalyzed polymerization of aniline.

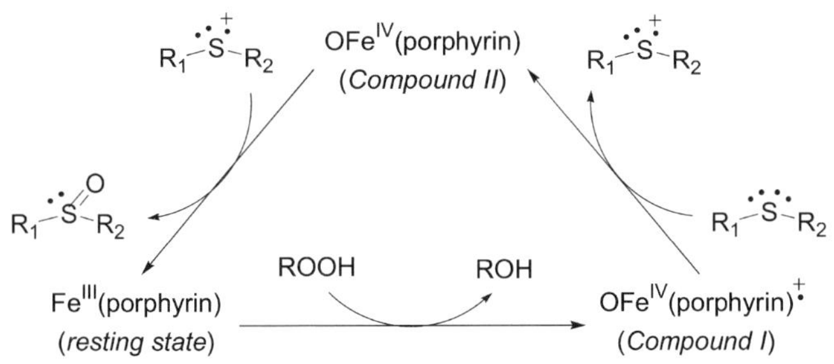

Scheme 2.14 Catalytic cycle of peroxidase-catalyzed sulfoxidations.

2.2.3.2 **Sulfoxidations ("Peroxygenase Reactivity")**

Many heme peroxidases accept an organic sulfide as substrate for the oxygen transfer reaction from Compound I (path **C** in Scheme 2.11) with formation of a sulfoxide. As before, this reaction consists of two individual one-electron transfer steps and is therefore designated a "peroxygenase reaction". In a first step, the sulfide is oxidized by Compound I to a sulfur radical cation and Compound II (Scheme 2.14) [222, 223].

Two possible mechanistic alternatives are discussed for the following oxygen transfer step, the so-called oxygen rebound. The oxygen atom might be transferred to the sulfur radical directly from Compound II [224] or a hydroxyl radical is split off protonated Compound II and reacts with the S-radical cation [225].

After the first discovery of the asymmetric sulfoxidation by Kobayashi *et al.* [226], it could be shown that a large number of aryl alkyl sulfides are oxygenated with enantiomeric excesses higher than 98% [227–229]. Other peroxidases also catalyze this reaction. Interestingly, the plant peroxidase HRP [230] yields the (*S*)-sulfoxide, whereas mammalian myeloperoxidase [223] and lactoperoxidase [231] catalyze the formation of the *R*-enantiomers. The stereospecific sulfoxidation of aryl alkyl sulfides by purified toluene dioxygenase (TDO) from *P. putida* was also studied in this context [232] and showed that sulfoxidation yielded the (*S*)-sulfoxides in 60–70% yield, whereas CPO under the same conditions yielded 98% (*R*)-sulfoxides (Scheme 2.15). CPO is thus again an exception from the rule in that it produces *R*-enantiomeric sulfoxides, besides its bacterial origin [227]. The reason for this behavior lies in the

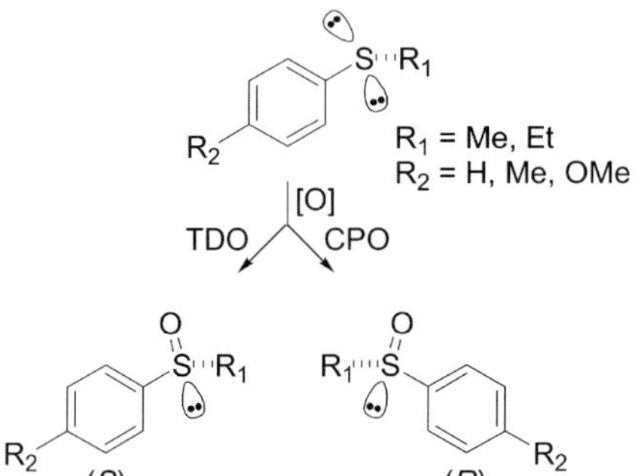

R_1 = Me, Et
R_2 = H, Me, OMe

[O]

TDO CPO

(*S*) (*R*)

Scheme 2.15 Stereospecific sulfoxidation of aryl alkyl sulfides by different peroxidases.

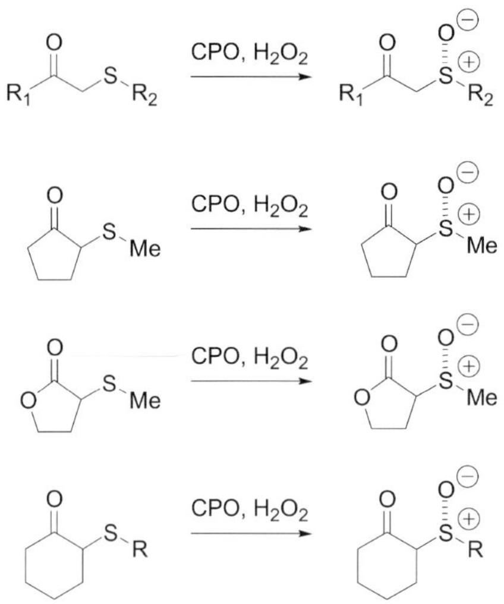

Scheme 2.16 Stereospecific sulfoxidation of β-carbonyl sulfides.

better accessibility of the FeO active site in CPO as supported by site-directed mutagenesis studies on HRP [233, 234]. HRP mutants containing sterically less enforcing amino acid residues in the binding pocket show significantly increased turnover numbers and higher enantiomeric excesses than the wild-type enzyme. Methyl naphthyl sulfoxide, for example, was obtained with 99% *ee* compared with 60% *ee* for wild-type HRP. The real synthetic potential of these mutants is unknown, however, since no efforts have been made to suppress the uncatalyzed sulfoxidation. Excellent enantioselectivities with wild-type heme peroxidases have been reported exclusively for CPO [235, 236] and many efforts have been directed in recent years in order to lift the preparative potential of this enzyme [64, 237]. Hydrogen peroxide has almost universally been used as the oxidant. The slow, uncatalyzed oxidation that takes place in the background can be reduced to a minimum by keeping the hydrogen peroxide concentration as low as possible. The reaction proceeds without any further oxidation to the sulfone and is often performed to demonstrate the oxygen-transfer capabilities of peroxidases.

A series of β-carbonyl sulfides was studied as substrates for CPO-catalyzed oxygenation (Scheme 2.16) [238]. The corresponding dialkyl sulfoxides formed quantitatively if R_2 is methyl or ethyl, but the yields drop dramatically for larger substituents. The steric control was present also in cyclic derivatives where the cyclohexanone residue results in about a 50% reduction in yield with respect to the smaller cyclopentanone. Surprisingly, the γ-butyrolactone produces the sulfoxide in quantitative amounts [239]. A similar result was obtained with benzo[*b*]thiophenes as substrates [240].

2.2.3.3 Peroxide Disproportionation ("Catalase Reactivity")

Another possible back-reaction to the resting state is realized in the function of the catalases (path **D** in Scheme 2.11) and therefore denoted "catalase reaction".

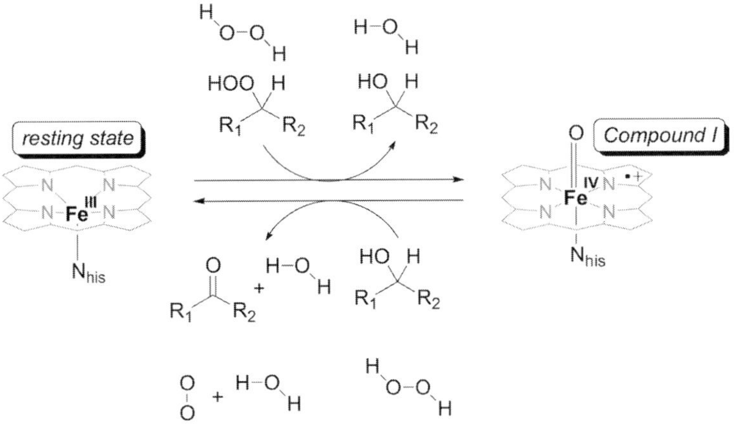

Scheme 2.17 Catalytic cycle of catalases.

Catalases are structural relatives of the peroxidases in that they contain a heme tetramer at the active site [241]. Their biological function is to control the cellular concentration of hydrogen peroxide by the following disproportionation reaction:

$$2H_2O_2 \rightarrow 2H_2O + O_2 \tag{2.2}$$

As for the peroxidases, Compound I and water are formed in the first step from one equivalent of hydrogen peroxide and the resting state of the catalase. The back-reaction, however, does not proceed via Compound II but rather via a two-electron–two-proton transfer cascade, in which both hydrogen atoms of a second molecule of hydrogen peroxide are transferred to the ferryl subunit of the porphyrin cofactor. Due to the similarity of catalases and peroxidases, it is not too surprising that this reaction is also catalyzed by most peroxidases. Alternatively, catalases and some peroxidases react with alkyl hydroperoxides via the respective alkanol to an aldehyde or ketone (Scheme 2.17). A requirement for this reaction is an easily accessible active site for the hydroperoxide, so that only those peroxidases with open access such as CPO or CiP are able to promote this reaction.

CPO catalyzes the oxidation of 2-alkynes to aldehydes in the presence of H_2O_2 or *t*BuOOH via an alcoholic intermediate as depicted in Scheme 2.18 [242]. Propargylic alcohols are rapidly oxidized to the corresponding aldehydes [243] and there is a report about highly enantioselective propargylic hydroxylations catalyzed by CPO [244]. In addition, a number of primary alcohols are selectively oxidized to aldehydes in a biphasic mixture of hexane and a buffer (Scheme 2.18) [245, 246].

Benzylic and allylic positions are hydroxylated by CPO in halide-dependent catalytic transformations. Toluene and *p*-xylene are oxidized to the respective aldehydes and carboxylic acids [247, 248]. Ethylbenzene and other substrates with longer alkyl chains form the respective benzylic/allylic alcohols with high enantioselectivity. Straight-chain aliphatic and cyclic (*Z*)-alkenes are hydroxylated, favoring small unsubstituted substrates in which the double bond is not more than two carbon atoms from the terminus. Steric control is observed for benzylic hydroxylations.

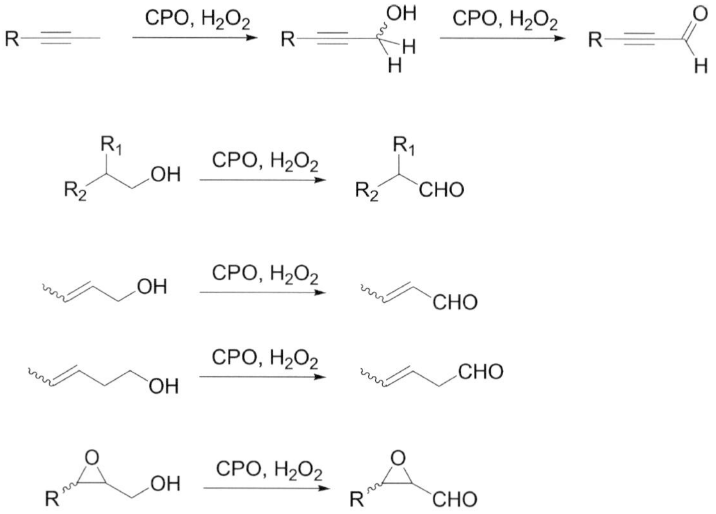

Scheme 2.18 Oxidation of alkynes and primary alcohols by CPO.

Ethylbenzene is transformed into (*R*)-phenethyl alcohol with an *ee* of 97% whereas propylbenzene yields the *S*-configured alcohol in 88% *ee*. The enantioselectivity remains high even with longer-chain substrates, for which the catalytic performance is rather inefficient [249].

Scheme 2.19 summarizes results obtained with indoles and benzofurans as substrates for H_2O_2 and halide-dependent CPO-catalyzed oxidations [250–252]. Indole is usually oxygenated to the corresponding lactone. An unusual double

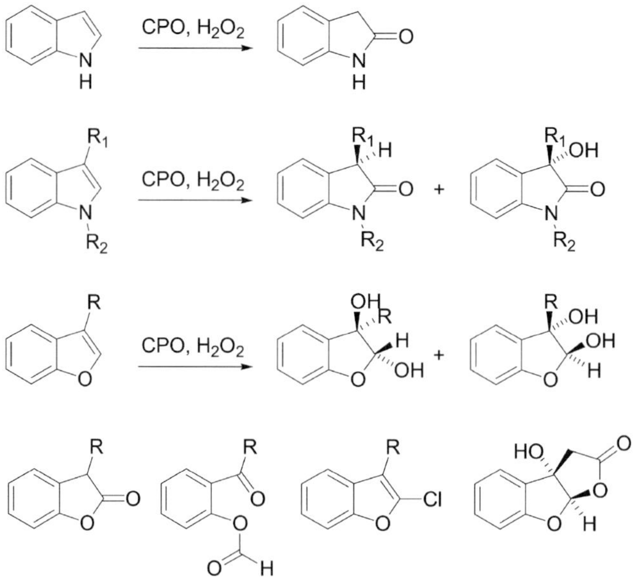

Scheme 2.19 CPO-catalyzed oxidation of indoles and benzofurans.

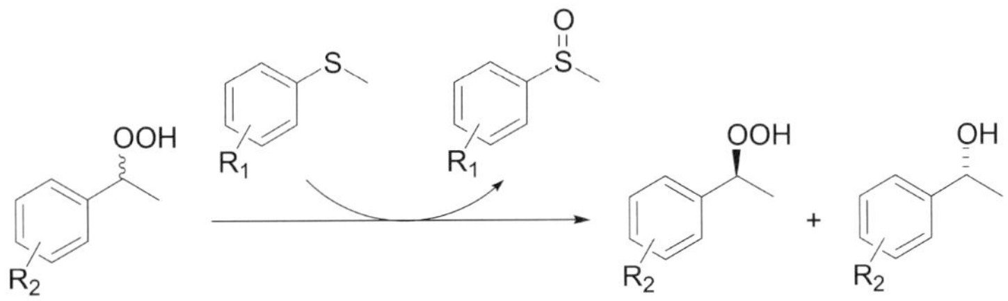

sulfide	hydroperoxide	(*R*)-sulfoxide ee /%	(*S*)-peroxide ee /%	(*R*)-alcohol ee /%
R_1 = H	R_2 = H	70	89	50
R_1 = *p*-MeO	R_2 = H	61	91	38
R_1 = H	R_2 = *p*-F	58	56	68
R_1 = H	R_2 = *o*-OMe	13	n.d.	17

Scheme 2.20 Kinetic resolution of hydroperoxides with CPO in the presence of sulfides.

oxidation has been observed to occur for alkylindoles and also for ergot alkaloids as substrates. 3-Alkylbenzofurans gave predominantly *trans*-2,3-diols as the initial products, which were sufficiently stable for isolation. Upon changing the reaction and/or workup conditions, it is also possible to obtain a lactone or ring-cleaved compounds as major products. Small quantities of halogenated material were isolated as by-products and polycyclic material could be obtained from substrates with suitably functionalized residues R by intramolecular condensation.

Also of significant preparative potential is the kinetic resolution of chiral hydroperoxides in the presence of sulfides or guaiacol [253–261]. The reaction has been shown to occur with CPO, HRP and CiP and provides good to excellent results for a multitude of different substrates. Whereas usually the back-reaction of Compound I to the resting state is used for organic synthesis, Compound I is formed upon the stereoselective decomposition of alkylhydroperoxides here. Scheme 2.20 illustrates the first example described in the literature [253] where CPO and different aryl methyl sulfides have been employed and where it has been found that mainly the *R*-form of the chiral hydroperoxide is reduced to the corresponding alcohol.

Systematic investigations into this reaction have been undertaken and showed that for straight-chain aralkyl hydroperoxides and their cyclic analogues the (*R*)-alcohol forms, whereas the stereoselection is the opposite for branched hydroperoxides. The reaction could be applied to functionalized hydroperoxides such as α- and β-hydroperoxy esters or hydroperoxy alcohols with good to excellent diastereo- and enantioselectivities. Up to 99% *ee*s were obtained for small, sterically non-hindered substrates, whereas for tertiary hydroperoxides and for substrates with substituents at the ω-position neither good *ee*s nor useful turnover numbers have been reported. HRP reacts very sluggishly also with sterically demanding silyl-substituted allyl hydroperoxides.

2.2.3.4 **Halogenation ("Haloperoxidase Reactivity")**

If the substrate for oxygen transfer step in the back-reaction of Compound I to the resting state is a halogen atom, the overall process is denoted "haloperoxidase

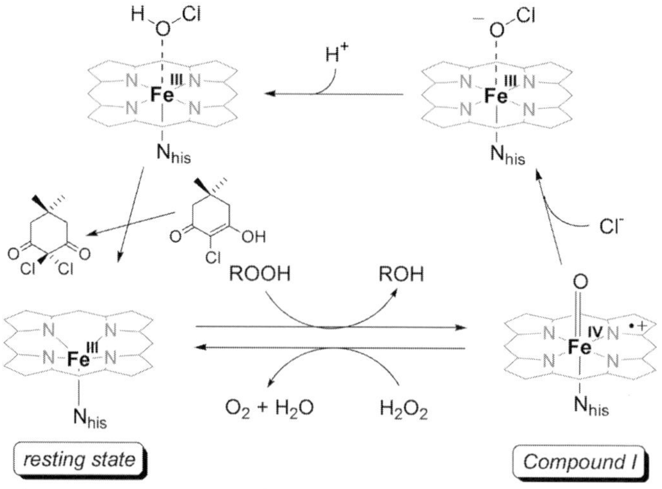

Scheme 2.21 Catalytic cycle of the haloperoxidase reaction.

reactivity" and is catalyzed by haloperoxidases. CPO is by far the most intensely studied heme protein from this enzyme group [63, 64]. The putative mechanism is illustrated in Scheme 2.21 [262].

The reaction has found little application so far; however, one of the first examples of an asymmetric synthesis using peroxidases was the highly regio- and diastereo-selective halohydration of glycals to 2-deoxy-2-halo sugars in the presence of CPO [263]. Some examples of selective halogenation reactions of aromatic compounds using CPO optimized by directed evolution approaches are also known [264].

2.2.3.5 **Epoxidations ("Monoxygenase Activity")**

The selective catalytic epoxidation of alkenes has become the most important reaction catalyzed by heme proteins in organic synthesis. As described above, the monoxygenase activity of a heme peroxidase is restricted to CPO due to the open substrate access of the ferryl subunit for this enzyme. HRP catalyzes epoxidations only after mutagenetic variations, as shown for the substrate *trans*-β-methylstyrene [234]. An exception of this rule is the regioselective epoxidation of (*E*,*E*)-piperylpiperidide, which is successfully catalyzed by native HRP [265].

Excellent enantioselectivity is observed in the CPO/H_2O_2-catalyzed epoxidation of short-chain (*Z*)-alkenes with a chain length of nine of fewer carbon atoms, except for monosubstituted alkenes, which often function as reversible suicide inhibitors of the enzyme [266–271]. (*E*)-Alkenes are highly unreactive substrates and are converted to epoxides in yields below 5%. A number of functionalized (*Z*)-2-alkenes have been successfully epoxidized by CPO using *tert*-butyl hydroperoxide as the terminal oxidant [272]. This procedure appears to be more effective, especially in large-scale reactions, due to the fairly high sensitivity of CPO to hydrogen peroxide.

Terminal alkenes can be suitable substrates if the chain is branched at the 2- or 3-position [249]. The oxidation of 3-hydroxy-1,4-pentadiene proceeds with 98% *de* and 65% *ee*, yielding predominantly (2*S*,3*R*)-1,2-epoxy-4-penten-3-ol, which is the

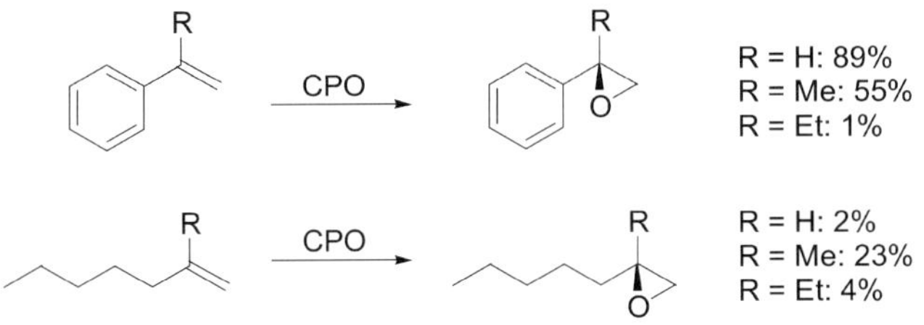

R = H: 89%
R = Me: 55%
R = Et: 1%

R = H: 2%
R = Me: 23%
R = Et: 4%

Scheme 2.22 Selected examples for the epoxidation of 2-substituted 1-alkenes by CPO.

enantiomer of the epoxy alcohol produced via Sharpless epoxidation of the corresponding divinylcarbinol [273]. Both the conversion and the stereoselectivity of the reaction are diminished if the hydroxy group is exchanged for a methyl substituent due to solubility and enzyme recognition issues. 2-Substituted 1-alkenes have been studied in much detail. As a typical result, 2-methyl-1-alkenes are epoxidized efficiently and with good *ee* whereas the 2-H-derivatives lead to low turnover numbers, low *ees* and rapid catalyst deactivation. An exception is styrene, which is epoxidized in 89% yield and with moderate selectivity. Larger 2-ethyl-substituted derivatives seem to suffer from limited access to the active ferryl unit of the enzyme and thus produce only small amounts of products (Scheme 2.22).

The effect of chain length on the catalytic performance was investigated using a series of ω-bromo-2-methylalkenes. In all cases the predominant enantiomer produced had the *R*-configuration except for 3-bromo-2-methylpropene oxide, which was predominantly in the *S*-form due to the priority switch [274]. The short propene and butene derivatives were converted quantitatively whereas the longer pentene, hexene and heptene substrates failed to convert completely. Many other functional groups such as carboxylic ester, methoxy, acetoxy and carbonic ester are accepted by the system. The epoxidation fails, however, for 4-hydroxy-2-methyl-1-butene as substrate [270].

The selective epoxidation of dienes by CPO from *C. fumago* has been reported (Scheme 2.23) [275]. The methacrylate was a good substrate which showed two types of selectivity: only the isolated double bond was epoxidized to produce the monoepoxide in 73% yield and the conjugated α,β-unsaturated bond of the methacrylic acid moiety was untouched. It was suggested that conjugated terminal alkenes might

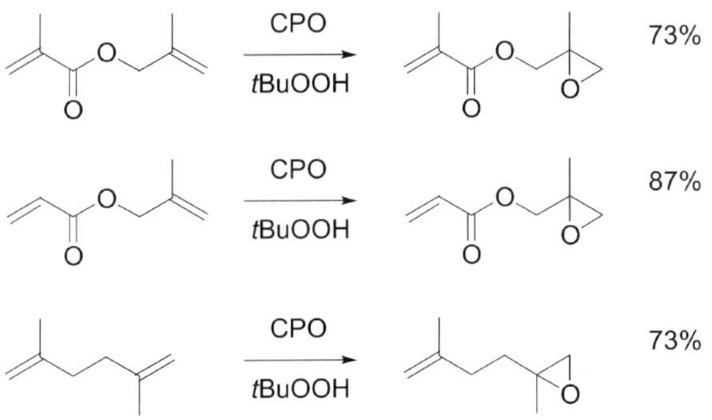

CPO, *t*BuOOH: 73%

CPO, *t*BuOOH: 87%

CPO, *t*BuOOH: 73%

Scheme 2.23 Monoepoxidation of dienes.

have a small effect on the inhibition of CPO activity compared with other aliphatic terminal alkenes [271]. Indeed, the related acrylate was again an excellent substrate for CPO epoxidation and selectively afforded the monoepoxide in high yield and excellent enantioselectivity. This is complementary to the epoxidation of the α,β-unsaturated double bond in enones using synzymes, *viz.* polyleucine, where the epoxidation takes place exclusively at the α,β-unsaturated double bond [276–281]. It has further been proposed that CPO-catalyzed epoxidations should produce only monoepoxides from symmetrical dienes and this was indeed the case. When 2,5-dimethyl-1,5-hexadiene was used as a model substrate, biocatalytic epoxidation afforded exclusively the monoepoxide.

The stereochemistry of the CPO-catalyzed epoxidation of indene has been reported [279]. In aqueous solution the initially formed epoxide is not stable and opens to form the *cis*-diols. When the reaction is carried out in the absence of water, the epoxide enantiomers were isolated, with the 1*S*,2*R*-enantiomer being formed in 30% *ee* (Scheme 2.24).

(1*S*,2*R*)-Indene epoxide is the precursor of *cis*-(1*S*,2*R*)-aminoindanol, a key intermediate of the Merck HIV-1 protease inhibitor Crixivant [280, 281]. As an alternative to the challenging chemical synthesis of this chiral epoxide from indene, the biotransformation route using an enzyme catalyst has been reported [282]. The products were generally racemic *trans*-bromoindanols, which upon basification yielded racemic epoxides (Scheme 2.25). It was found that a crude enzyme preparation from the fungal culture *Curvularia protuberata* MF5400 converted indene to the chiral (1*S*,2*S*)-bromoindanol, which could be chemically converted to the desired (1*S*,2*R*)-epoxide through basification or used directly in the asymmetric synthesis of *cis*-(1*S*, 2*R*)-aminoindanol. The bioconversion rate and the *ee* achieved with this cell-free system were heavily pH dependent. An initial reaction at pH 7.0 gave only a 10% yield of the chiral bromoindanol or epoxide from indene, but was rapidly

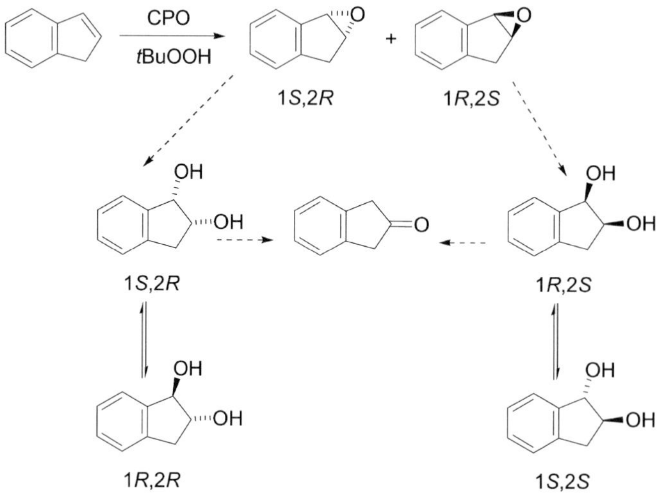

Scheme 2.24 Epoxidation of indene.

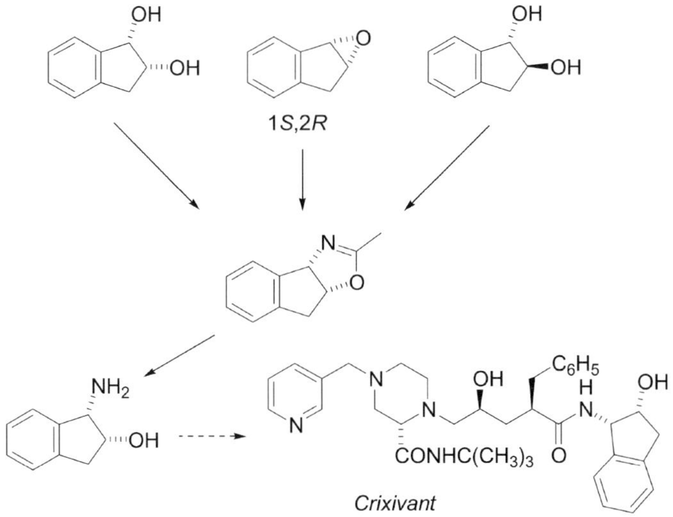

Scheme 2.25 Preparation of the HIV-1 protease inhibitor Crixivant.

improved to 30% of *trans*-(1*S*,2*S*)-bromoindanol with an *ee* of 80%. Reaction mechanistic studies revealed that the stereoselectivity observed was apparently due to a specific dehydrogenase activity present in MF5400, which was also found to resolve chemically synthesized racemic *trans*-2-bromoindanols.

CPO has been used occasionally in complex syntheses. An important application of CPO as an enantioselective epoxidation catalyst is the efficient synthesis of (*R*)-2-mevalonolactone (Scheme 2.26a) [270]. A survey of the literature revealed that the previous methods required many steps to produce the lactone, in low overall yield, with moderate *ee*, in addition to expensive starting materials. Meanwhile, a retrosynthetic analysis starting with an appropriately functionalized epoxide provided confidence that CPO could rescue the situation if used in the key stereogenic step. Another completed synthesis is depicted in Scheme 2.26b. Again, the epoxide is generated in high yield with conversion to (*R*)-dimethyl-2-methylaziridine-1,2-dicarboxylate, which may serve as a synthon for β-methylamino acids [283].

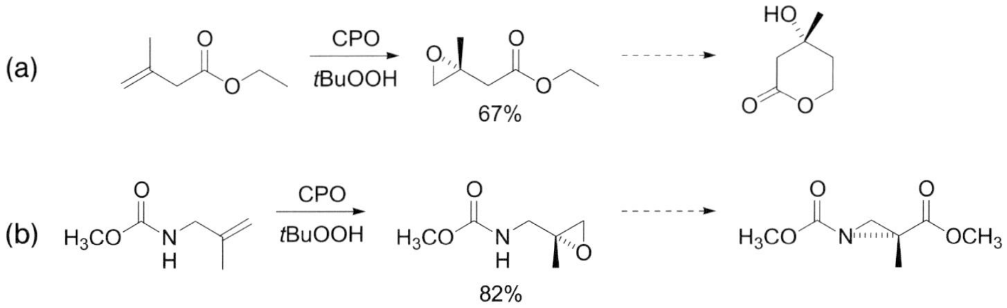

Scheme 2.26 CPO-catalyzed epoxidations as key steps in the syntheses of (*R*)-2-mevalonolactone (a) and (*R*)-dimethyl-2-methylaziridine-1,2-dicarboxylate (b).

References

61 M. Sono, M. P. Roach, E. D. Coulter, J. H. Dawson, *Chem. Rev.* 1996, **96**, 2841–2887.

62 P. Ortiz de Montellano, *Cytochrome P450 – Structure, Mechanism and Biochemistry,* 3rd edn., Kluwer Academic/ Plenum Publishers, New York, 2005.

63 M. P. J. van Deurzen, F. van Rantwijk, R. A. Sheldon, *Tetrahedron* 1997, **53**, 13183–13220.

64 L. F. Zhi, Y. Jiang, M. Hu, S. Li, *Prog. Chem.* 2006, **18**, 1150–1156.

65 Z. Fang, R. Breslow, *Org. Lett.* 2006, **8**, 251–254.

66 K. J. McLean, M. Sabri, K. R. Marshall, R. J. Lawson, D. G. Lewis, D. Clift, P. R. Balding, A. J. Dunford, A. J. Warman, J. P. McVey, A. M. Quinn, M. J. Sutcliffe, N. S. Scrutton, A. W. Munro, *Biochem. Soc. Trans.* 2005, **33**, 796–801.

67 T. L. Poulos, *Biochem. Biophys. Res. Commun.* 2005, **338**, 337–345.

68 M. Newcomb, R. E. Chandrasena, *Biochem. Biophys. Res. Commun.* 2005, **338**, 394–403.

69 M. J. Coon, *Annu. Rev. Pharmacol. Toxicol.* 2005, **45**, 1–25.

70 B. Meunier, S. P. de Visser, S. Shaik, *Chem. Rev.* 2004, **104**, 3947–3980.

71 R. Ullrich, M. Hofrichter, *Cell. Mol. Life Sci.* 2007, **64**, 271–293.

72 J. T. Groves, *J. Inorg. Biochem.* 2006, **100**, 434–447.

73 Z. Li, D. Chang, *Curr. Org. Chem.* 2004, **8**, 1647–1658.

74 W. Adam, M. Lazarus, C. R. Saha-Möller, O. Weichold, U. Hoch, D. Häring, P. Schreier, in *Advances in Biochemical Engineering/ Biotechnology,* Vol. 63, ed. K. Faber, Springer, Heidelberg, 1998, pp. 73–108.

75 J. H. Dawson, *Science* 1988, **240**, 433–439.

76 H. L. Holland, *Organic Synthesis with Oxidative Enzymes,* VCH, New York, 1992, p. 55.

77 H. L. Holland, *Steroids* 1999, **64**, 178–186.

78 I. G. Collado, J. Aleu, *J. Mol. Catal. B* 2001, **13**, 77–93.

79 R. Azerad, in *Stereoselective Biocatalysis,* ed. R. N. Patel, Marcel Dekker, New York, 2000, pp. 153–180.

80 B. M. Fraga, M. G. Hernandez, P. Gonzalez, M. Lopez, S. Suarez, *Tetrahedron* 2001, **57**, 761–770.

81 G. Aranda, L. Moreno, M. Cortes, T. Prange, M. Maurs, R. Azerad, *Tetrahedron* 2001, **57**, 6051–6056.

82 R. Agarwal, N. Deepika, R. Joseph, *Biotechnol. Bioeng.* 1999, **63**, 249–252.

83 G. O. Buchanan, R. B. Reese, *Phytochemistry* 2001, **56**, 141–151.

84 R. A. Johnson, in *Oxidation in Organic Chemistry,* ed. W. S. Trahanovsky, Academic Press, New York, 1978, part C, p. 131.

85 K. Kieslich, *Microbial Transformation of Non-steroid Cyclic Compounds,* Georg Thieme, Stuttgart, 1976.

86 R. Furstoss, A. Archelas, J. D. Fourneron, B. Vigne, in *Organic Synthesis: an Interdisciplinary Challenge,* ed. J. Streith, H. Prinzbach, G. Schill, Blackwell Scientific, Oxford, 1985, pp. 215–226.

87 R. L. Hanson, J. A. Matson, D. B. Brzozowski, T. L. LaPorte, D. M. Springer, R. N. Patel, *Org. Process. Res. Dev.* 2002, **6**, 482–487.

88 R. Okazaki, T. Oritani, Y. Hara, H. Yamamoto, *Biosci. Biotechnol. Biochem.* 2001, **65**, 943–946.

89 W. Abraham, H. Arfmann, K. Kieslich, G. Haufe, *Biocatal. Biotransf.* 2000, **18**, 283.

90 T. Shibasaki, H. Mori, A. Ozaki, *Biosci. Biotechnol. Biochem.* 2000, **64**, 746–750.

91 A. Kraemer-Schafhalter, S. Domenek, H. Boehling, S. Feichtenhofer, H. Griengl, H. Voss, *Appl. Microbiol. Biotechnol.* 2000, **53**, 266–271.

92 M. S. Hemenway, H. F. Olivo, *J. Org. Chem.* 1999, **64**, 6312–6318.

93 S. L. Flitsch, S. J. Aitken, C. S.-Y. Chow, G. Grogan, A. Staines, *Bioorg. Chem.* 1999, **27**, 81–90.

94 T. A. Holton, F. Brugliera, Y. Ranaka, *Plant J.* 1993, **4**, 1003–1010.

95 J. Ogata, Y. Kanno, Y. Ithoh, H. Tsugawa, M. Suzuki, *Nature* 2005, **435**, 757–758.

96 S. Picataggio, T. Rohrer, K. Deanda, D. Lanning, R. Reynolds, J. Mielenz, L. D. Eirich, *Biotechnology* 1992, **10**, 894–898.

97 P. Fickers, P. H. Benetti, Y. Wache, A. Marty, S. Mauersberger, M. S. Smit, J. M. Nicaud, *FEMS Yeast Res.* 2005, **5**, 527–543.

98 C. Dingler, W. Ladner, G. A. Krei, B. S. Cooper, B. Hauer, *Pestic. Sci.* 1996, **46**, 33–53.

99 K. Petzoldt, K. Annen, H. Laurent, R. Wiechert, *US Patent 4 353 985*, 1982.

100 C. Duport, R. Spagnoli, E. Degryse, D. Pompon, *Nat. Biotechnol.* 1998, **16**, 186–189.

101 R. Bernhardt, *J. Biotechnol.* 2006, **124**, 128–145.

102 V. B. Urlacher, S. Lutz-Wahl, R. D. Schmid, *Appl. Microb. Biotechnol.* 2004, **64**, 317–325.

103 U. Schwaneberg, A. Sprauer, C. Schmidt-Dannert, R. D. Schmid, *J. Chromatogr. A* 1999, **848**, 149–159.

104 E. M. Gillam, F. P. Guengerich, *IUBMB Life* 2001, **52**, 271–277.

105 F. P. Guengerich, *Arch. Biochem. Biophys.* 2003, **409**, 59–71.

106 K. Buchholz, V. Kasche, U. T. Bornscheuer, *Biocatalysts and Enzyme Technology*, Wiley-VCH, Weinheim, 2005.

107 A. W. Munro, H. M. Girvan, K. J. McLean, *Nat. Prod. Rep.* 2007, **24**, 585–609.

108 E. G. Funhoff, J. Salzmann, U. Bauer, B. Witholt, J. B. van Beilen, *Enzyme Microbiol. Technol.* 2007, **40**, 806–812.

109 A. Chefson, K. Auclair, *Mol. BioSyst.* 2006, **2**, 462–469.

110 M. A. Noble, C. S. Miles, S. K. Chapman, D. A. Lysek, A. C. Mackay, G. A. Reid, R. P. Hanzlik, A. W. Munro, *Biochem. J.* 1999, **339**, 371–379.

111 C. F. Oliver, S. Modi, W. U. Primrose, L. Y. Lian, G. C. K. Roberts, *Biochem. J.* 1997, **327**, 537–5.

112 T. W. B. Ost, C. S. Miles, J. Murcoch, Y. F. Cheung, G. A. Reid, S. K. Chapman, A. W. Munro, *FEBS Lett.* 2000, **1186**, 173–177.

113 C. F. Oliver, M. Sandeep, M. J. Sutcliffe, W. U. Primrose, L. Y. Lian, G. C. K. Roberts, *Biochemistry* 1997, **36**, 1567–1572.

114 D. Appel, S. Lutz-Wahl, P. Fischer, U. Schwaneberg, R. D. Schmid, *J. Biotechnol.* 2001, **88**, 167–171.

115 Q. S. Li, U. Schwaneberg, M. Fischer, J. Schmitt, J. Pleiss, S. Lutz-Wahl, R. D. Schmid, *Biochim. Biophys. Acta* 2001, **1545**, 114–121.

116 O. Lentz, Q. S. Li, U. Schwaneberg, S. Lutz-Wahl, P. Fischer, R. D. Schmid, *J. Mol. Catal. B* 2001, **15**, 123–133.

117 H. Joo, Z. Lin, F. H. Arnold, *Nature* 1999, **399**, 670–673.

118 E. T. Farinas, U. Schwaneberg, A. Glieder, F. H. Arnold, *Adv. Synth. Catal.* 2001, **343**, 601–606.

119 S. Eiben, L. Kaysser, S. Maurer, K. Kühnel, V. B. Urlacher, R. D. Schmid, *J. Biotechnol.* 2006, **124**, 662–669.

120 T. S. Wong, F. H. Arnold, U. Schwaneberg, *Biotechnol. Bioeng.* 2004, **85**, 351–358.

121 P. Meinhold, M. W. Peters, M. M. Y. Chen, K. Takahashi, F. H. Arnold, *Chembiochem* 2005, **6**, 1765–1768.

122 M. W. Peters, P. Meinhold, A. Glieder, F. H. Arnold, *J. Am. Chem. Soc.* 2003, **125**, 13442–13450.

123 P. Meinhold, M. W. Peters, A. Hartwick, A. R. Hernandez, F. H. Arnold, *Adv. Synth. Catal.* 2006, **348**, 763–772.

124 O. Lentz, A. Feenstra, T. Habicher, B. Hauer, R. D. Schmid, V. B. Urlacher, *Chembiochem* 2005, **7**, 345–350.

125 J. B. vanBeilen, R. Holtackers, D. Luscher, U. Bauer, B. Witholt, W. A. Duetz, *Appl. Environ. Microbiol.* 2005, **71**, 1737–1744.

126 I. Axarli, A. Prigipaki, N. E. Labrou, *Biomol. Eng.* 2005, **22**, 81–88.

127 P. Meinhold, M. W. Peters, A. Hartwick, A. R. Hernandez, F. H. Arnold, *Adv. Synth. Catal.* 2006, **348**, 763–772.

128 M. Landwehr, L. Hochrein, C. R. Otey, A. Kasravan, J.-E. Backwall, F. H. Arnold, *J. Am. Chem. Soc.* 2006, **128**, 6058–6059.

129 J.-A. Stevenson, J. P. Jones, L.-L. Wong, *Isr. J. Chem.* 2000, **40**, 55–62.

130 S. Kumar, J. R. Halpert, *Biochem. Biophys. Res. Commun.* 2005, **338**, 456–464.

131 S. G. Bell, R. J. Sowden, L.-L. Wong, *Chem. Commun.* 2001, 635–636.

132 S. G. Bell, X. Chen, R. J. Sowden, F. Xu, J. N. Williams, L.-L. Wong, Z. Rao, *J. Am. Chem. Soc.* 2003, **125**, 705–714.

133 S. G. Bell, D. A. Rouch, L.-L. Wong, *J. Mol. Catal. B* 1997, **3**, 293–302.

134 J.-A. Stevenson, A. C. G. Westlake, C. Whittock, L.-L. Wong, *J. Am. Chem. Soc.* 1996, **118**, 12846–12847.

135 J.-A. Stevenson, J. L. Bearpark, L.-L. Wong, *New. J. Chem.* 1998, **22**, 551–552.

136 S. G. Bell, J.-A. Stevenson, H. D. Boyd, S. Campbell, A. D. Riddle, E. L. Orton, L.-L. Wong, *Chem. Commun.* 2002, 490–491.

137 F. Xu, S. G. Bell, J. Lednek, A. Insley, Z. Rao, L.-L. Wong, *Angew. Chem. Int. Ed.* 2005, **44**, 4029–4032.

138 C. Colas, P. R. Ortiz de Montellano, *Chem. Rev.* 2003, **103**, 2305–2332.

139 S. G. Bell, E. Orton, H. Boyd, J.-A. Stevenson, A. Riddle, S. Campbell, L.-L. Wong, *Dalton Trans.* 2003, **11**, 2133–2140.

140 J. Limburg, L. A. Bebrun, P. R. Ortiz de Montellano, *Biochemistry* 2005, **44**, 4091–4099.

141 A. Parikh, P. D. Josephy, F. P. Guengerich, *Biochemistry* 1999, **38**, 5283–5289.

142 C.-H. Yun, G. P. Miller, F. P. Guengerich, *Biochemistry* 2000, **39**, 11319–11329.

143 D. Kim, F. P. Guengerich, *Biochemistry* 2004, **43**, 981–988.

144 D. Kim, F. P. Guengerich, *Arch. Biochem. Biophys.* 2004, **432**, 102–108.

145 K. Nakamura, M. V. Martin, F. P. Guengerich, *Arch. Biochem. Biophys.* 2001, **395**, 25–31.

146 Z.-L. Wu, L. M. Podust, F. P. Guengerich, *J. Biol. Chem.* 2005, **280**, 41090–41100.

147 S. Kumar, E. E. Scott, H. Liu, J. R. Halpert, *J. Biol. Chem.* 2003, **278**, 17178–17184.

148 S. Kumar, C. S. Chen, D. J. Waxman, J. R. Halpert, *J. Biol. Chem.* 2005, **280**, 19569–19575.

149 W. Tischer, F. Wedekind, *Top. Curr. Chem.* 1999, **200**, 95–126.

150 D. L. King, M. R. Azari, A. Wiseman, *Methods Enzymol.* 1988, **137**, 675–686.

151 E. I. Karaseva, A. N. Eremin, D. I. Metelitsa, *Vestn. Akad. Nauk SSSR* 1986, **4**, 76–80.

152 K. Seelbach, B. Riebel, W. Hummel, M.-R. Kula, V. I. Tishkov, A. M. Egorov, C. Wandrey, U. Kragl, *Tetrahedron Lett.* 1996, **37**, 1377–1380.

153 F. Hollmann, K. Hofstetter, A. Schmid, *Trends Biotechnol.* 2006, **24**, 163–171.

154 K. Vuorilehto, S. Lutz, C. Wandrey, *Bioelectrochemistry* 2004, **65**, 1–7.

155 F. Hollmann, B. Witholt, A. Schmid, *J. Mol. Catal. B* 2002, **19–20**, 167–176.

156 K. M. Faulkner, M. S. Shet, C. W. Fisher, R. W. Estabrook, *Proc. Natl. Acad. Sci. USA* 1995, **92**, 7705–7709.

157 R. W. Estabrook, K. M. Faulkner, M. S. Shet, C. W. Fisher, *Methods Enzymol.* 1996, **272**, 44–51.

158 U. Schwaneberg, D. Appel, J. Schmitt, R. D. Schmid, *J. Biotechnol.* 2000, **84**, 249–257.

159 J. Nazor, U. Schwaneberg, *Chembiochem* 2006, **7**, 638–644.

160 A. K. Udit, F. H. Arnold, H. B. Gray, *J. Inorg. Biochem.* 2004, **98**, 1547–1550.

161 V. V. Shumyantseva, T. V. Bulko, A. I. Archakov, *J. Inorg. Biochem.* 2005, **99**, 1051–1063.

162 J. Kazlauskaite, A. C. G. Westlake, L.-L. Wong, H. Hill, O. Allen, *Chem. Commun.* 1996, 2189–2190.

163 K. Di Gleria, H. Hill, O. Allen, L.-L. Wong, *FEBS Lett.* 1996, **390**, 142–144.

164 K. Di Gleria, D. P. Nickerson, H. Hill, O. Allen, L.-L. Wong, V. Fueloep, *J. Am. Chem. Soc.* 1998, **120**, 46–52.

165 K. K.-W. Lo, L.-L. Wong, H. Hill, O. Allen, *FEBS Lett.* 1999, **451**, 342–346.

166 Z. Zhang, A.-E. F. Nassar, Z. Lu, J. B. Schenkman, J. F. Rusling, *J. Chem. Soc., Faraday Trans.* 1997, **93**, 1769–1774.

167 Y. M. Lvov, Z. Lu, J. B. Schenkman, X. Zu, J. F. Rusling, *J. Am. Chem. Soc.* 1998, **120**, 4073–4080.

168 X. Zu, Z. Lu, Z. Zhang, J. B. Schenkman, J. F. Rusling, *Langmuir* 1999, **15**, 7372–7377.

169 B. Munge, C. Estavillo, J. B. Schenkman, J. F. Rusling, *Chembiochem* 2003, **4**, 82–89.

170 C. Lei, U. Wollenberger, C. Jung, F. W. Scheller, *Biochem. Biophys. Res. Commun.* 2000, **268**, 740–744.

171 K.-F. Aguey-Zinsou, P. V. Bernhardt, J. J. De Voss, K. E. Slessor, *Chem. Commun.* 2003, 418–419.

172 B. D. Fleming, Y. Tian, S. G. Bell, L.-L. Wong, V. Urlacher, H. Hill, O. Allen, *Eur. J. Biochem.* 2003, **270**, 4082–4088.

173 A. K. Udit, K. D. Hagen, P. J. Goldman, A. Star, J. M. Gillan, H. B. Gray, M. G. Hill, *J. Am. Chem. Soc.* 2006, **128**, 10320–10325.

174 V. V. Shumyantseva, T. V. Bulko, T. T. Bachmann, U. Bilitewski, R. D. Schmid, A. I. Archakov, *Arch. Biochem. Biophys.* 2000, **377**, 43–48.

175 V. V. Shumyantseva, T. V. Bulko, S. A. Usanov, R. D. Schmid, C. Nicolini, A. I. Archakov, *J. Inorg. Biochem.* 2001, **87**, 185–190.

176 V. V. Shumyantseva, Y. D. Ivanov, N. Bistolas, F. W. Scheller, A. I. Archakov, U. Wollenberger, *Anal. Chem.* 2004, **76**, 6046–6052.

177 C. Estavillo, Z. Lu, I. Jansson, J. B. Schenkman, J. F. Rusling, *Biophys. Chem.* 2003, **104**, 291–296.

178 S. Joseph, J. F. Rusling, Y. M. Lvov, T. Friedberg, U. Fuhr, *Biochem. Pharmacol.* 2003, **65**, 1817–1826.

179 A. Fantuzzi, M. Fairhead, G. Gilardi, *J. Am. Chem. Soc.* 2004, **126**, 5040–5041.

180 A. Shukla, E. M. Gillam, D. J. Mitchell, P. V. Bernhardt, *Electrochem. Commun.* 2005, **7**, 437–442.

181 V. V. Shumyantseva, G. Deluca, T. Bulko, S. Carrara, C. Nicolini, S. A. Usanov, A. I. Archakov, *Biosens. Bioelectron.* 2004, **19**, 971–976.

182 V. V. Shumyantseva, S. Carrara, V. Bavastrello, D. Jason Riley, T. V. Bulko, K. G. Skryabin, A. I. Archakov, C. Nicolini, *Biosens. Bioelectron.* 2005, **21**, 217–222.

183 Y.-S. Kim, M. Hara, K. Ikebukuro, J. Miyake, H. Ohkawa, I. Karube, *Biotechnol. Tech.* 1996, **10**, 717–720.

184 M. Hara, H. Ohkawa, M. Narato, M. Shirai, Y. Asada, I. Karube, J. Miyake, *J. Ferment. Bioeng.* 1997, **84**, 324–329.

185 M. Hara, S. Iazvovskaia, H. Ohkawa, Y. Asada, J. Miyake, *J. Biosci. Bioeng.* 1999, **87**, 793–797.

186 V. V. Shumyantseva, V. Uvarov, V. Yu, O. E. Byakova, A. I. Archakov, *Biochem. Mol. Biol. Int.* 1996, **38**, 829–838.

187 V. V. Shumyantseva, V. Uvarov, V. Yu, O. E. Byakova, A. I. Archakov, *Arch. Biochem. Biophys.* 1998, **354**, 133–138.

188 V. V. Shumyantseva, T. V. Bulko, R. D. Schmid, A. I. Archakov, *Biosens. Bioelectron.* 2002, **17**, 233–238.

189 V. Ullrich, H. J. Staudinger, Model systems in studies of the chemistry and the enzymatic activation of oxygen, in *Handbuch der Experimentellen Pharmakologie*, Vol. 28/2, ed. B. B. Brodie, J. R. Gilette, H. S. Ackermann, Springer, Berlin, 1971, pp. 251–263.

190 F. F. Kadlubar, K. C. Morton, D. M. Ziegler, *Biochem. Biophys. Res. Commun.* 1973, **54**, 1255–1261.

191 A. D. Rahimtula, P. J. O'Brien, *Biochem. Biophys. Res. Commun.* 1974, **60**, 440–447.

192 A. D. Rahimtula, P. J. O'Brien, *Biochem. Biophys. Res. Commun.* 1975, **62**, 268–275.

193 A. Ellin, S. Orrenius, *FEBS Lett.* 1975, **50**, 378–381.

194 H. Joo, Z. Lin, F. H. Arnold, *Nature* 1999, **399**, 670–673.

195 K. Matsuura, T. Tosha, S. Yoshioka, S. Takahashi, K. Ishimori, I. Morishima, *Biochem. Biophys. Res. Commun.* 2004, **323**, 1209–1215.

196 P. C. Cirino, F. H. Arnold, *Angew. Chem. Int. Ed.* 2003, **115**, 3421–3423.

197 A. Chefson, J. Zhao, K. Auclair, *Chembiochem* 2006, **7**, 916–919.

198 E. G. Hrycay, J. A. Gustafsson, M. Ingelman-Sundberg, L. Ernster, *FEBS Lett.* 1975, **56**, 161–165.
199 J. A. Gustafsson, E. G. Hrycay, L. Ernster, *Arch. Biochem. Biophys.* 1976, **174**, 438–451.
200 E. G. Hrycay, J. A. Gustafsson, M. Ingelman-Sundberg, L. Ernster, *Eur. J. Biochem.* 1976, **61**, 43–52.
201 G. D. Nordblom, R. E. White, M. J. Coon, *Arch. Biochem. Biophys.* 1976, **175**, 524–533.
202 X. Fang, J. R. Halpert, *Drug Metab. Dispos.* 1996, **24**, 1282–1285.
203 L. Y. Xie, D. Dolphin, in *Metalloporphyrin Catalyzed Oxidations*, ed. F. Montanari, L. Casella, Kluwer, Dordrecht, 1994, pp. 269–306.
204 H. B. Dunford,in *Peroxidases in Chemistry and Biology*, Vol. II, ed. J. Everse, K. E. Everse, M. B. Grisham, CRC Press, Boca Raton. FL, 1991, pp. 1–24.
205 B. Meunier,in *Peroxidases in Chemistry and Biology*, Vol. II, ed. J. Everse, K. E. Everse, M. B. Grisham, CRC Press, Boca Raton, FL, 1991, pp. 201–217.
206 H. Zemel, J. F. Quinn,*PCT Int. Appl. WO 9410327*, 1994.
207 J. A. Akkara, K. J. Senecal, D. L. Kaplan, *J. Polym. Sci., Part A* 1991, **29**, 1561–1574.
208 M. Akita, D. Tsutsumi, M. Kobayashi, H. Kise, *Biosci. Biotechnol. Biochem.* 2001, **65**, 1581–1588.
209 M. S. Ayyagari, K. A. Marx, S. K. Tripathy, J. A. Akkara, D. L. Kaplan, *Macromolecules* 1995, **28**, 5192–5197.
210 Y.-P. Xu, G.-L. Huang, Y.-T. Yu, *Biotechnol. Bioeng.* 1995, **47**, 117–119.
211 H. Kurioka, I. Komatsu, H. Uyama, S. Kobayashi, *Macromol. Rapid Commun.* 1994, **15**, 507–510.
212 H. Ritter, *Trends Polym. Sci.* 1993, **1**, 171–173.
213 J. S. Dordick, M. A. Marletta, A. M. Klibanov, *Biotechnol. Bioeng.* 1987, **30**, 31–36.
214 A. F. Naves, A. M. Carmona-Ribeiro, D. F. S. Petri, *Langmuir* 2007, **23**, 1981–1987.
215 M. Reihmann, H. Ritter, *Adv. Polym. Sci.* 2006, **194**, 1–49.
216 M. Akita, D. Tsutsumi, M. Kobayashi, H. Kise, *Biotechnol. Lett.* 2001, **23**, 1827–1831.
217 J. A. Nicell, J. K. Bewtra, N. Biswas, C. C. St. Pierre, K. E. Taylor, *Can. J. Civ. Eng.* 1993, **20**, 725–735.
218 C. Flock, A. Bassi, M. Gijzen, *J. Chem. Technol. Biotechnol.* 1999, **74**, 303–309.
219 L. M. Colosi, D. J. Burlingame, Q. Huang, W. J. Weber, Jr., *Environ. Sci. Technol.* 2007, **41**, 891–896.
220 Y. Wu, K. E. Taylor, N. Biswas, J. K. Bewtra, *Enzyme Microbial Technol.* 1998, **22**, 315–322.
221 K. Woithe, N. Geib, K. Zerbe, D. B. Li, M. Heck, S. Fournier-Rousset, O. Meyer, F. Vitali, N. Matoba, K. Abou-Hadeed, J. A. Robinson, *J. Am. Chem. Soc.* 2007, **129**, 6887–6895.
222 E. Baciocchi, O. Lanzalunga, S. Malandrucco, M. Ioele, S. Steenken, *J. Am. Chem. Soc.* 1996, **118**, 8973–8974.
223 A. Tuynman, M. K. S. Vink, H. L. Dekker, H. E. Schoemaker, R. Wever, *Eur. J. Biochem.* 1998, **258**, 906–913.
224 P. R. Ortiz de Montellano, *J. Biol. Chem.* 1993, **268**, 1637–1645.
225 U. Pérez, H. B. Dunford, *Biochim. Biophys. Acta* 1990, **1038**, 98–104.
226 S. Kobayashi, M. Nakano, T. Kimura, A. P. Schaap, *Biochemistry* 1987, **26**, 5019–5022.
227 M. P. J. van Deurzen, I. J. Remkes, F. van Rantwejk, R. A. Scheldon, *J. Mol. Catal. A* 1997, **117**, 329–337.
228 S. Colonna, N. Gaggero, L. Casella, G. Carrea, P. Pasta, *Tetrahedron: Asymmetry* 1992, **3**, 93–106.
229 H. Fu, H. Kondo, Y. Ichikawa, G. C. Look, C. H. Wong, *J. Org. Chem.* 1992, **57**, 7265–7270.
230 S. Colonna, N. Gaggero, G. Carrea, P. Pasta, *J. Chem. Soc., Chem. Commun.* 1992, 357–358.
231 S. Colonna, N. Gaggero, C. Richelmi, G. Carrea, P. Pasta, *Gazz. Chim. Ital.* 1995, **125**, 9103–9104.
232 K. Lee, J. M. Brans, D. T. Gibson, *Biochem. Biophys. Res. Commun.* 1995, **212**, 9–15.

233 S.-I. Ozaki, P. R. Ortiz de Montellano, *J. Am. Chem. Soc.* 1994, **116**, 4487–4488.

234 S.-I. Ozaki, P. R. Ortiz de Montellano, *J. Am. Chem. Soc.* 1995, **117**, 7056–7064.

235 S. Colonna, N. Gaggero, A. Manfredi, L. Casella, M. Gullotti, G. Carrea, P. Pasta, *Biochemistry* 1990, **29**, 10465–10468.

236 S. G. Allenmark, M. Andersson, *Tetrahedron: Asymmetry* 1996, **7**, 1089–1094.

237 V. M. Dembitsky, *Tetrahedron* 2003, **59**, 4701–4720.

238 R. R. Vargas, E. J. H. Bechara, L. Marzorati, B. Wladislaw, *Tetrahedron: Asymmetry* 1999, **10**, 3219–3227.

239 M. Sundaramoorthy, J. Terner, T. L. Poulos, *Structure* 1995, **3**, 1367–1375.

240 S. G. Allenmark, M. A. Andersson, *Chirality* 1998, **10**, 146–153.

241 I. Fita, M. G. Rossmann, *J. Mol. Biol.* 1985, **185**, 21–37.

242 K. Matsummura, S. Hashiguchi, T. Ikariya, R. Noyori, *J. Am. Chem. Soc.* 1997, **119**, 8738–8739.

243 S. Hu, L. P. Hager, *Biochem. Biophys. Res. Commun.* 1998, **253**, 544–546.

244 S. Hu, L. P. Hager, *J. Am. Chem. Soc.* 1999, **121**, 872–873.

245 E. Kiljunen, L. T. Kanerva, *Tetrahedron: Asymmetry* 1999, **10**, 3529–3535.

246 E. Kiljunen, L. T. Kanerva, *J. Mol. Catal. B* 2000, **9**, 163–172.

247 P. R. Ortiz de Montellano, Y. S. Choe, G. DePillis, E. E. Catalano, *J. Biol. Chem.* 1987, **262**, 11641–11646.

248 J. A. Morgan, Z. Lu, D. S. Clark, *J. Mol. Catal. B* 2002, **18**, 147–154.

249 A. F. Dexter, R. A. Lakner, J. Campbell, L. P. Hager, *J. Am. Chem. Soc.* 1995, **117**, 6412–6414.

250 F. van de Velde, M. Bakker, F. van Rantwijk, F. P. Rai, L. P. Hager, R. A. Sheldon, *J. Mol. Catal. B* 2001, **11**, 765–769.

251 V. Kren, L. Jawulokova, P. Sedmera, M. Polasek, T. K. Lindhorst, K. H. van Pee, *Liebigs Ann. Chem.* 1997, 2379–2383.

252 R. G. Alvarez, I. S. Hunter, C. J. Suckling, M. Thomas, U. Vitinius, *Tetrahedron* 2001, **57**, 8581–8587.

253 H. Fu, H. Kondo, Y. Ichikawa, G. C. Look, C.-H. Wong, *J. Org. Chem.* 1992, **57**, 7265–7270.

254 A. K. Abelskov, A. T. Smith, C. B. Rasmussen, H. B. Dunford, K. G. Welinder, *Biochemistry* 1997, **36**, 9453–9463.

255 W. Adam, U. Hoch, M. Lazarus, C. R. Saha-Möller, P. Schreier, *J. Am. Chem. Soc.* 1995, **117**, 11898–11901.

256 W. Adam, R. T. Fell, C. R. Saha-Möller, P. Schreier, *Tetrahedron: Asymmetry* 1995, **6**, 1047–1050.

257 W. Adam, U. Hoch, C. R. Saha-Möller, P. Schreier, *Angew. Chem.* 1993, **105**, 1800–1801.

258 W. Adam, C. Mock-Knoblauch, C. R. Saha-Möller, *Tetrahedron: Asymmetry* 1997, **8**, 1947–1950.

259 W. Adam, U. Hoch, H. U. Humpf, C. R. Saha-Möller, P. Schreier, *J. Chem. Soc., Chem. Commun.* 1996, 2701–2702.

260 W. Adam, C. R. Saha-Möller, K. Schmidt, *J. Org. Chem.* 2000, **65**, 1431–1433.

261 W. Adam, C. R. Saha-Möller, K. Mock-Knoblauch, *J. Org. Chem.* 1999, **64**, 4834–4839.

262 R. D. Libby, A. L. Shedd, A. Phipps, T. M. Beachy, S. M. Gerstberger, *J. Biol. Chem.* 1992, **267**, 1769–1775.

263 K. K. C. Liu, C.-H. Wong, *J. Org. Chem.* 1992, **57**, 3748–3750.

264 G. P. Rai, S. Sakai, A. M. Florez, L. Mogollon, L. P. Hager, *Adv. Synth. Catal.* 2001, **343**, 638–645.

265 A. B. Rao, M. V. Rao, A. Kumar, G. L. D. Krupadanam, G. Srimannarayana, *Tetrahedron Lett.* 1994, **35**, 279–280.

266 S. Colonna, N. Gaggero, L. Casella, G. Carrea, P. Pasta, *Tetrahedron: Asymmetry* 1993, **4**, 1325–1330.

267 E. J. Allain, L. P. Hager, L. Deng, E. N. Jacobsen, *J. Am. Chem. Soc.* 1993, **115**, 4415–4416.

268 A. Zaks, D. R. Dodds, *J. Am. Chem. Soc.* 1995, **117**, 10419–10424.

269 R. R. Everett, H. S. Soedjak, A. Butler, *J. Biol. Chem.* 1990, **265**, 15671–15677.

270 F. J. Lakner, L. P. Hager, *J. Org. Chem.* 1996, **61**, 3923–3925.

271 A. F. Dexter, L. P. Hager, *J. Am. Chem. Soc.* 1995, **117**, 817–818.

272 S. Hu, L. P. Hager, *Tetrahedron Lett.* 1999, **40**, 1641–1644.

273 S. Colonna, N. Gaggero, C. Richelmi, G. Carrea, P. Past, *Gazz. Chim. Ital.* 1995, **125**, 479–482.

274 F. J. Lakner, K. P. Cain, P. L. Hager, *J. Am. Chem. Soc.* 1997, **119**, 443–444.

275 S. Hu, P. Gupta, A. K. Prasad, R. A. Gross, V. S. Parmar, *Tetrahedron Lett.* 2002, **43**, 6763–6766.

276 S. Julia, J. Masana, J. C. Vega, *Angew. Chem., Int. Ed. Engl.* 1980, **19**, 929–934.

277 P. A. Bentley, S. Bergeron, M. W. Cappi, D. E. Hibbs, M. B. Hursthouse, T. C. Nugent, R. Pulido, S. M. Roberts, L. E. Wu, *Chem. Commun.* 1997, 739–740.

278 P. A. Bentley, W. Kroutil, J. A. Littlechild, S. M. Roberts, *Chirality* 1997, **9**, 198–207.

279 K. M. Manoj, F. J. Lakner, L. P. Hager, *J. Mol. Catal. B* 2000, **9**, 107–111.

280 B. C. Buckland, W. D. Stephen, N. Connors, M. Cartrain, C. Lee, P. Salmon, K. Gbewonyo, P. Gailliot, R. Singhvi, R. Olewinski, J. Reddy, J. Zhang, K. Gklen, B. Junker, R. Greasham, *Metabol. Eng.* 1999, **1**, 63–74.

281 R. Bishop, *Comprehensive Org. Synth.* 1991, **6**, 261–300.

282 J. Zhang, C. Roberge, J. Reddy, N. Connors, M. Chartrain, B. Buckland, R. Greasham, *Enzyme Microbiol. Technol.* 1999, **24**, 86–95.

283 F. J. Lakner, L. P. Hager, *FASEB J. Suppl.* 1997, **31**, P9.

3
Iron-catalyzed Oxidation Reactions

3.1
Oxidations of C–H and C=C Bonds

Agathe Christine Mayer and Carsten Bolm

3.1.1
Gif Chemistry

This section deals with Gif and GoAgg systems that were discovered by Barton and coworkers in the 1980s. After the presentation of the various systems, we will focus on the mechanism of the reaction. The last section will focus on the latest applications of Gif and GoAgg type systems to alkane oxidation.

In 1983, Barton *et al.* described "a new procedure for the oxidation of saturated hydrocarbons" [1] by the use of metallic iron, adamantane, hydrogen sulfide or sodium sulfide, pyridine, acetic acid and a small amount of water. The reactions were run under air and afforded adamant-1-ol together with a mixture of adamant-2-ol and adamantanone. The substrate scope could be extended to other hydrocarbons. However, all of them led to product mixtures of the corresponding alcohol(s) and ketone (Scheme 3.1).

Interestingly, the presence of hydrogen sulfide, which is much more easily oxidized than the saturated hydrocarbon, did not inhibit the reaction but, in contrast, was an essential additive. Later, it was shown that the same holds true for other easy-to-oxidize compounds such as Ph_2S, PPh_3 and $P(OMe)_3$. It was assumed that the iron species that attacks the alkane is only activated by contact with it (called the "Sleeping Beauty Effect" by Barton). This observation is known as the "Gif paradox" [2].

Subsequently, the original Gif systems (described above, Gif^{I} and Gif^{II}, respectively) [3] were subjected to various alterations named after the places of their discovery (Table 3.1).

In all systems, the precatalyst is derived from readily available metal source, i.e. iron or copper. In addition to pyridine, acetonitrile proved to be a possible solvent, although a minimum amount of a suitable unhindered pyridine base was found necessary for turnover [4]. In addition to molecular oxygen, H_2O_2 and *t*-BuOOH are

Iron Catalysis in Organic Chemistry. Edited by Bernd Plietker

ISBN: 978-3-527-31927-5

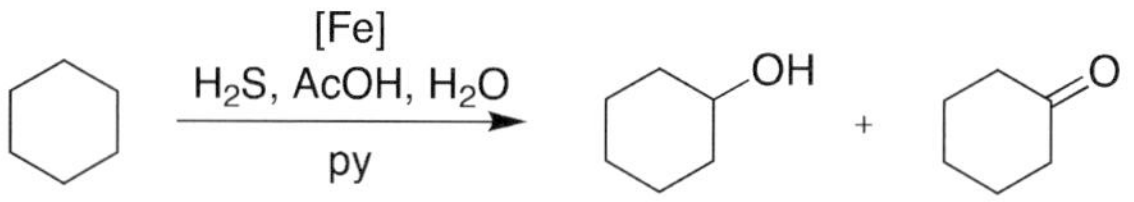

[Fe]
H_2S, AcOH, H_2O
py

OH
+
O

Scheme 3.1 Gif oxygenation.

suitable oxidants (for mechanistic details, see below). The heterogeneous mixture Fe–O_2–Zn (Gif^{IV}) deserves special attention as it allows for turnover numbers (TONs) above 2000 for adamantane oxygenation. Hence the reaction can be run catalytic in iron with zinc as the source of electrons [5]. Replacement of acetic acid by 2-picolinic acid (PicH) results in significantly enhanced reaction rates. This beneficial aspect was realized in the GoAgg^{III} system consisting of a homogeneous mixture of $FeCl_3$, H_2O_2 and PicH [6]. As early as 1983, Barton *et al.* proposed that the iron powder not only adopted the role of a reductant but should also be considered as a precursor to a cationic iron complex. Furthermore, the organic carboxylic acid was assumed to be more than just a source of protons, namely (in its deprotonated form) a ligand for iron [5]. In the absence of the acid, disproportionation of H_2O_2 is exclusively observed [7].

Much effort has been devoted to the elucidation of the oxygenation mechanism (see reviews [8, 9]). Extensive studies by Barton and coworkers led to the assumption that Gif chemistry was not a radical process. This assumption was based on two observations. First, in radical chemistry, the order of reactivity is C_{tert}-H > C_{sec}-H ≫ C_{prim}-H, whereas in Gif systems, a significant preference for oxidation in the secondary position of the hydrocarbon was observed [10, 11] (C_{sec}-H > C_{tert}-H ≫ C_{prim}-H). A second hint was the low kinetic isotope effect (about 2 for the ketone). Both findings were combined to the mechanism depicted in Scheme 3.2.

It consists of two manifolds [7, 12]. The first is the "non-radical Fe^{III}/Fe^{V} manifold". A high-valent Fe^{V}=O species undergoes a [2 + 2]-type addition to a C–H bond followed by conversion to an alkyl hydroperoxide and eventual formation of the ketone and the alcohol with the ketone being the major product. However, both Knight and Perkins [13] and Barton *et al.* [14] found that the oxygen content of the alkyl

Table 3.1 Some Gif oxygenation systems.

Entry	System	Catalyst	Oxidant	Reductant	Solvent[a]
1	Gif^{I}	Fe	O_2	Fe/Na_2S	py/AcOH
2	Gif^{II}	Fe	O_2	Fe/H_2S	py/AcOH/H_2O
3	Gif^{IV}	Fe$^{II/III}$	O_2	Zn	py/AcOH/H_2O[b]
4	GoAgg^{II}	Fe^{III}	H_2O_2	–	py/AcOH
5	GoAgg^{III}	Fe^{III}/PicH	H_2O_2	–	py/AcOH or py
6	GoAgg^{IVc}	Fe^{III}	*t*-BuOOH	–	py/AcOH, 60 °C
7	GoAgg^{Vc}	Fe^{III}/PicH	*t*-BuOOH	–	py/AcOH, 60 °C

[a] Reactions were performed at room temperature, except where noted.
[b] Addition of H_2O is optional.
[c] This system was later expelled from the Gif family due to a different reaction mechanism (see below).

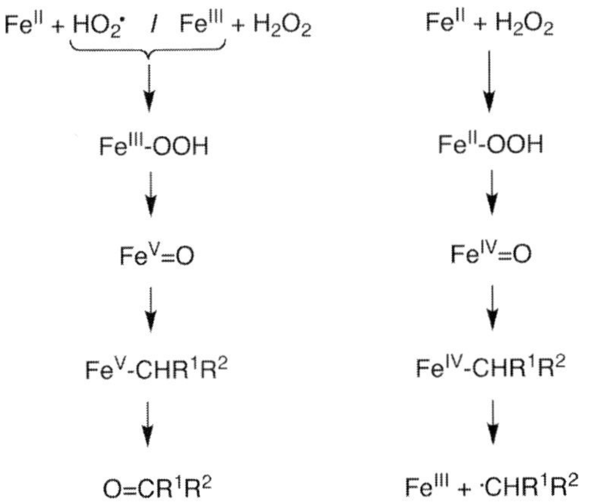

$Fe^{II} + HO_2^{\bullet}$ / $Fe^{III} + H_2O_2$ → Fe^{III}-OOH → Fe^{V}=O → Fe^{V}-CHR^1R^2 → O=CR^1R^2

$Fe^{II} + H_2O_2$ → Fe^{II}-OOH → Fe^{IV}=O → Fe^{IV}-CHR^1R^2 → Fe^{III} + $\cdot CHR^1R^2$

Scheme 3.2 Barton's manifolds for alkane oxygenation.

hydroperoxides of cyclodecane and cyclohexane stemmed from dioxygen (which in fact is a strong indication for a radical mechanism). The second manifold ("radical Fe^{II}/Fe^{IV} manifold") was assumed to be due to accidental Fe^{II}–H_2O_2 interactions, producing a high-valent Fe^{IV}=O oxidant that behaves similarly to the Fe^{V}=O species present in the Fe^{III}/Fe^{V} manifold. The only difference between them was assumed to be that the Fe^{IV}–C group would collapse to afford alkyl radicals whereas this would not happen with the Fe^{V}–C system except for special cases (for example, when R = *tert*-adamantyl). The systems were believed to interconvert via reversible O_2 activation.

Despite the enormous efforts by Barton and coworkers to gain insight into the oxygenation mechanism (or, better, to support their hypothesis that it was not a radical reaction), major aspects remained unclear. Furthermore, the choice of oxidant turned out to be crucial for the reaction mechanism and this led to further elaboration by various groups, including elegant studies of the GoAgg^{III} system by Newcomb *et al.* using radical clocks [15]. Knight and Perkins suggested a radical autoxidation pathway [16] with hydroxyl and alkoxy radicals playing a major role (Scheme 3.3). They investigated the oxidation of cyclohexane with H_2O_2 and were able to detect both cyclohexyl and hydroxyl radicals [13].

According to these assumptions, Fenton-type chemistry [17] would generate hydroxyl radicals from H_2O_2 and Fe^{2+}. Subsequent alkane attack would lead to alkyl radicals that could then be trapped by dioxygen and later abstract hydrogen from H_2O_2 to yield alkyl hydroperoxides. These would, upon reaction with Fe^{2+}, decompose into alkoxyl radicals and, in turn, abstract hydrogen from the hydrocarbon to form an alkyl radical and alcohol. Although it is well known that both types of radicals can abstract hydrogen from alkanes, it should be noted that hydroxy radicals exhibit almost no selectivity whereas alkoxy radicals show a reactivity ratio of 4–5 : 1 for tertiary to secondary positions [18]. The values observed in Gif chemistry lie somewhere between those extremes. Some further mechanistic suggestions by Knight and Perkins [16] concerning the influence of $P(OMe)_3$ or PPh_3 and also on halogenation reactions were almost immediately rejected by Barton [19].

$$Fe^{3+} + H_2O_2 \xrightarrow{\text{slow}} Fe^{2+} + H^+ + HOO\bullet$$

$$Fe^{2+} + H_2O_2 \xrightarrow{\text{very fast}} Fe^{3+} + HO^- + HO\bullet$$

$$HO\bullet + RH \longrightarrow R\bullet + H_2O$$

$$R\bullet + O_2 \longrightarrow ROO\bullet$$

$$ROO\bullet + H_2O_2 \longrightarrow ROOH$$

$$Fe^{2+} + ROOH \longrightarrow Fe^{3+} + HO^- + RO\bullet$$

$$RO\bullet + RH \longrightarrow ROH + R\bullet$$

Scheme 3.3 Perkins' radical autoxidation pathway.

In 2001, Stavropoulos *et al.* pointed out some inconsistencies within Barton's arguments and presented new experimental evidence for the involvement of radical chemistry in Gif systems that use O_2 as oxidant [9]. Barton *et al.* had recognized earlier that *tert*-adamantyl pyridines were formed from coupling of *tert*-adamantyl radicals and protonated pyridine (positions 2 and 4) [20]. However, as coupling to secondary positions was not observed, they claimed that no secondary radicals were involved in the course of the reaction [14]. This was shown not to be the case [21]. Both *tert*- and *sec*-adamantyl radicals are being formed and dioxygen and pyridine compete for the radicals. The addition of pyridinium to the radicals is reversible but the rate coefficients for the tertiary radicals are much higher (two orders of magnitude). Notably, increasing partial pressure of O_2 leads to a preferential formation of oxo products and this happens at the expense of pyridine-coupled products.

Although the mechanism of Gif chemistry still gives reason for debate, experimental results provide evidence for the existence of a radical pathway. Four hints on a radical mechanism involving hydroxyl radicals are:

1. The ratio of alcohol to ketone products (A/K) is low (about 1).
2. The product distribution depends on the presence of O_2 and the alcohol oxygen comes from O_2.
3. The selectivity in C–H oxidation is low.
4. The reaction proceeds with little or no stereoselectivity due to long-lived tertiary alkyl radicals that are prone to epimerization.

On the other hand, the Gif–*tert*-butyl hydroperoxide (TBHP) systems seem to be much less complicated. These have been extensively studied by many groups and all workers agree that the reaction proceeds via radical pathways based on the reactivity of *tert*-butylperoxy and *tert*-butyloxy radicals [22–24]. Minisci *et al.* [22] suggested a Haber–Weiss radical chain mechanism [25] that accounts for the observed selectivities.

The oxidation of activated methylenes to ketones under GoAgg^V-type conditions was reported by Kim *et al.* in 2002 [24] and by Nakanishi and Bolm in 2007 [26]. In the latter case, no acid was needed and, as pyridine was used as a solvent, an additional ligand was unnecessary. Interestingly, *p*-methoxytoluene was converted to the corresponding carboxylic acid, albeit in moderate yield (53%). The oxidation of triphenylmethane did not lead to the expected trityl alcohol but gave the corresponding hydroperoxide in excellent yield (91%). This protocol was a major improvement on the outcome of the oxidation reaction compared with an earlier report by the same group [27]. Noteworthy also is that these conditions gave much better results than the parent $GoAgg^V$ system.

Hydrocarbon oxidation is a major field of research. Generally, the most challenging aspect is not the oxidation itself but to achieve selectivity towards alcohols and aldehydes/ketones. Hence a large number of reports on iron-catalyzed procedures have appeared in recent years. The major drawback of the present systems is their low activity, presumably due to decomposition of H_2O_2 in the presence of transition metals (catalase activity) and oxidative degradation of the catalyst.

In this chapter, we will focus only on the latest reports, first dealing with a ligand-free system and later with multidentate *N,N*- and *N,O*-ligand systems. As a general rule of thumb, non-heme iron catalysts work best if they have exchangeable ligands [28].

In a comparison of the catalytic performances of various simple iron compounds with H_2O_2, remarkable differences were observed [29]. They originated from the variable ligand environment of the iron ion. In all cases, the primarily formed product was the alkyl hydroperoxide, which decomposed with time to form alcohol and ketone. Compounds with weakly bound ligands such as perchlorate react via a hydroxyl radical mechanism, whereas ferryl species are involved when more strongly bound ligands are present, as is the case with $FeCl_3$. In the $Fe(ClO_4)_3$-catalyzed oxidation of ethane, a TON of 68 was reported [29]. Shul'pin *et al.* described TONs as high as 205 for cyclohexane oxidation with $FeCl_3$ and bipyridine as a bidentate co-catalyst [30]. This combination facilitated the transformation of the primarily formed cyclohexyl hydroperoxide into cyclohexanone.

Tridentate ligands (Figure 3.1) have been reported by various groups [31–34] but the oxidation performance was moderate. One has to keep in mind that two tridentate ligands (or one hexadentate ligand) on one iron core will occupy all coordination sites, thereby hampering the reaction [32]. That means that oxidation reactions can only occur if one of the iron–ligand bonds dissociates beforehand. The catalytic activity remains low or even absent if this requirement is not met.

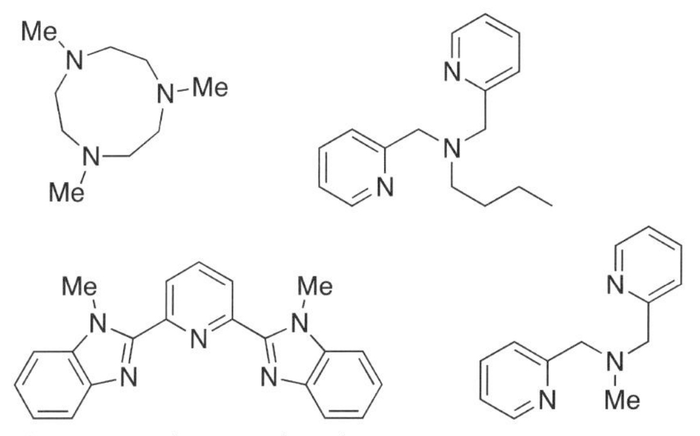

Figure 3.1 Tridentate *N*-ligands.

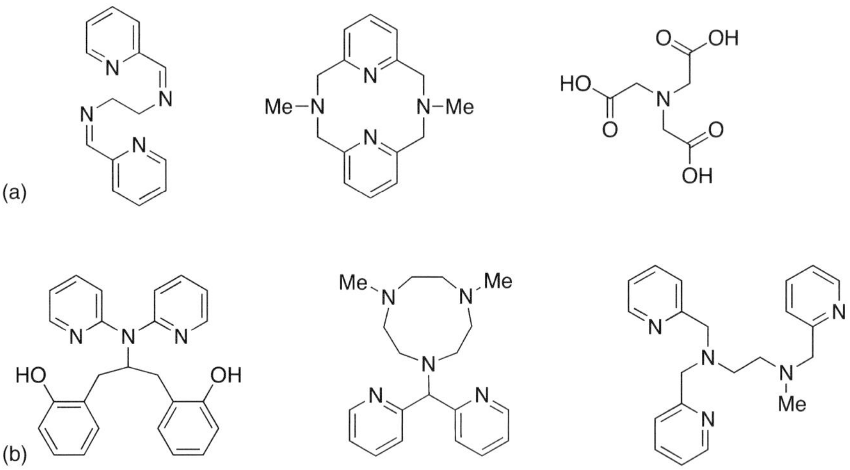

Figure 3.2 Tetra- (a) and pentadentate (b) *N*-ligands.

Better results were obtained with tetra- and pentadentate ligands [35a, 36–38] (Figure 3.2). In this case, two or one coordination sites, respectively, are open for binding of the oxidant (O_2 or H_2O_2) [39].

A linear tetradentate ligand, bpmen [*N,N'*-dimethyl-*N,N'*-bis(2-pyridylmethyl)-1,2-diaminoethane] was reported to allow stereospecific alkane hydroxylation [36, 40]. The same holds true for the $[Fe^{II}(TPA)(CH_3CN)_2]^{2+}$ [TPA = tris(2-pyridylmethyl) amine] catalyst family (Figure 3.3) [35].

Chen and Que were the first to provide experimental evidence on a high-valent $Fe^{V}{=}O$ species in alkane hydroxylation with H_2O_2 using the TPA family [35a]. Such species are normally only observed in alkane oxidation with heme systems [41]. The authors pointed out that the substitution pattern on the TPA ligand strongly influences both the spin state of the Fe^{III}–OOH intermediate and in consequence also the reaction mechanism, as proven by incorporation of isotope-labeled $H_2{}^{18}O$ into the product alcohols. Both the unsubstituted parent TPA ligand and those bearing substituents at the β-pyridyl positions were found to be low spin in iron. The β-substituted ligands are able to catalyze the hydroxylation reaction stereospecifically. In this case, the intermediate radicals are short-lived and no ^{18}O is incorporated into

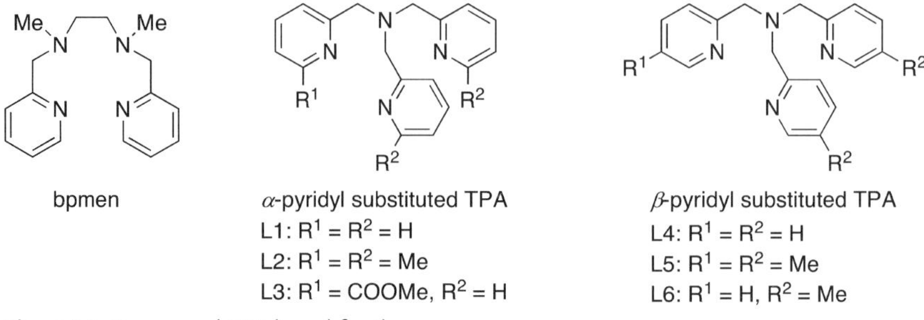

Figure 3.3 Bpmen and TPA ligand families.

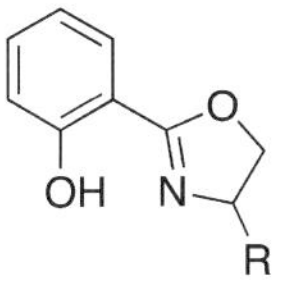

R = COOH, *i*Pr

Hphox

Figure 3.4 Hphox ligand.

the product alcohol [35a, 42]. In contrast, α-pyridyl substituents on the ligands shield the metal center and prevent the pyridyl groups from approaching the metal too closely. Hence these complexes favor large ionic radii as present in Fe^{II} and they have high-spin Fe^{II} centers. It was found that the oxidation of *cis*-1,2-dimethylcyclohexane using α-substituted ligands predominantly gave rise to two epimeric tertiary alcohols. This indicated the intermediate presence of long-lived alkyl radicals.

Pentadentate ligands also work well in alkane oxidations [39, 43, 44] (see Figure 3.2b). Feringa and coworkers reported on the successful use of N_4Py [*N,N*-bis(2-pyridylmethyl)-*N*-bis(2-pyridyl)methylamine] ligands with peracids such as peracetic acid and *m*-chloroperbenzoic acid instead of H_2O_2 [37]. The authors made two remarkable observations. First, the selectivity for alcohol formation was much higher than in H_2O_2 systems. Second, KIE values as high as 4.5–6.0 for cyclohexane oxidation were observed. It was therefore suggested that the reaction did not proceed via hydroxyl radicals and that a high-valent $Fe^{IV}{=}O$ species was involved.

A different approach for hydrocarbon oxidation is the use of *N,O*-ligands. Bouwman and coworkers reported on phenoloxazoline (Hphox) ligands (Figure 3.4) that exhibited moderate activity in the oxidation of activated methylenes such as toluene, ethylbenzene and cumene [45]. Interestingly, application to hydrocarbons with higher C–H bond dissociation energies such as cyclohexane, cyclooctane and adamantane failed. Despite hexacoordination in the complex $(HNEt_3)_2[Fe(phoxCOO)_2](ClO_4)$, some catalytic activity was observed, which was probably due to loss of one ore more ligands as the reaction proceeds (see above).

Li *et al.* designed the N_4O tpoen ligand [46] (Figure 3.5) with the aim of creating a hemi-labile coordination site through the ether oxygen function. However, the catalytic performance of several tpoen complexes in alkane oxidation was moderate. Again, *m*-CPBA drastically improved the selectivity for alcohol formation. Depending on the oxidant, two different mechanisms were proposed, one with a radical-based oxidant in H_2O_2 systems and the other with a metal-based oxidant in *m*-CPBA oxidations.

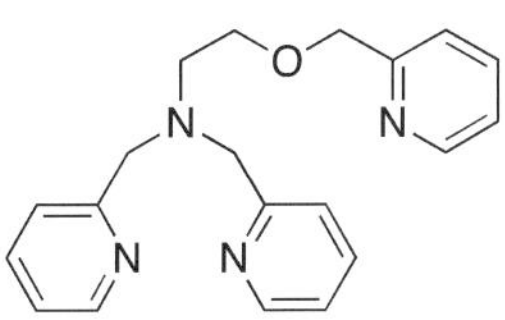

tpoen

Figure 3.5 Tpoen ligand.

Shul'pin and coworkers described the application of a di- and a tetranuclear iron complex with triazacyclononane acetate ligands in the oxidation of alkanes and alcohols with H_2O_2 [47]. A highly complex tetranuclear iron complex with octadentate pyridine carboxylate ligands was described by Gutkina *et al.* [48]. However, the TONs for cyclohexane oxidation did not exceed 5.0 in the latter case.

In summary, the oxidation of alkanes with iron-based catalysts remains a challenging task. Much progress has been made but the field is still far from mature. Higher efficiencies in the transformation of hydrocarbons to alcohols and ketones are desirable.

3.1.2
Alkene Epoxidation

The functionalization of C=C double bonds to furnish epoxides is a challenging field of research. Ideally, environmentally friendly oxidants such as molecular oxygen or hydrogen peroxide should be used in combination with cheap and non-toxic metal catalysts (Scheme 3.4) (iodosobenzene can also be used as oxidant; the disadvantage is the formation of one equivalent of iodobenzene as waste. For an example, see [49]).

Since H_2O_2 is easier to handle than O_2, we will focus on the use of the former. Many metals can be used for this transformation [50]. Among them, iron compounds are of interest as mimics of naturally occurring non-heme catalysts such as methane monooxygenase (MMO) [51a] or the non-heme anti-tumor drug bleomycin [51b]. Epoxidation catalysts should meet several requirements in order to be suitable for this transformation [50]. Most importantly, they must activate the oxidant without formation of radicals as this would lead to Fenton-type chemistry and catalyst decomposition. Instead, heterolytic cleavage of the O–O bond is desired. In some cases, alkene oxidation furnishes not only epoxides but also diols. The latter transformation will be the topic of the following section.

Non-heme iron catalysts containing multidentate nitrogen ligands such as pyridines and amines have been studied by various groups [42a, 52–54]. Jacobsen and coworkers presented an MMO mimic system for the epoxidation of aliphatic alkenes in which the catalyst self-assembles to form the active species [54] (Scheme 3.5). Interestingly, small amounts of an additive (one equivalent of acetic acid) increased the catalytic performance, presumably due to the intermediate formation of peroxyacetic acid [55, 56]. The reactions proceeded quickly even with terminal aliphatic alkenes, which are generally considered difficult substrates. Another catalyst system available for the epoxidation of terminal alkenes uses phenanthroline as ligand [57].

In a seminal contribution [42a], Que and coworkers revealed insight into the reaction mechanisms of epoxidation and dihydroxylation using the pentadentate TPA and bpmen ligands. Some of those compounds had earlier shown remarkable

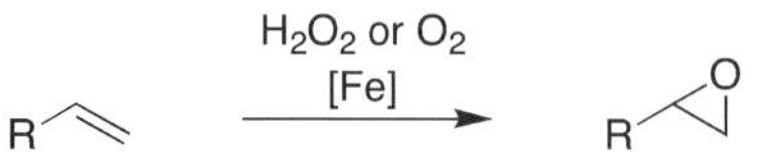

H_2O_2 or O_2
[Fe]
R ⟶ R (epoxide, O)

Scheme 3.4 Alkene epoxidation.

H_2O_2 (1.5 equiv.),
$[LFe^{II}(MeCN)_2](SbF_6)_2$ (3 mol %),
AcOH (30 mol %)
MeCN,
4 °C, 5 min

nC_8H_{17} → nC_8H_{17} (epoxide) 85%

L = Me, N, N, Me, N, N

Scheme 3.5 Jacobsen's epoxidation.

efficiency in the stereoselective hydroxylation of alkanes [35a, 58]. The authors reported that the substitution pattern on the TPA ligands strongly influenced the outcome of the reaction and that the presence of two labile *cis* coordination sites on iron was crucial. Both epoxide and diol were assumed to be formed from a common Fe^{III}–OOH intermediate. The two ligand families, TPA and bpmen, furnished mixtures of epoxide and diol and the ratio of the two products varied significantly. The bpmen ligands predominantly formed the epoxide (epoxide : diol ratio = 6–8) and TPA ligands with two or three 6-methyl substituents led mainly to the *cis*-diol (see the next section). TPA ligands bearing no or only one 6-methyl substituent gave mixtures. Reactions catalyzed by other N_2Py_2 ligands such as bpmcn and bpmpn (Figure 3.6) were highly sensitive to the ligand topology [59]. Cyclooctene epoxidation with iron–bispidine complexes is also possible, albeit with low turnovers [60].

A pentadentate ligand consisting of a rigid pyridine-containing triazamacrocycle and a flexible aminopropyl pendant arm was recently reported [61]. The complex has a pseudo-square pyramidal geometry in which the aminopropyl pendant arm adopts the axial position and can be reversibly protonated. The activity of the complexes in alkene epoxidation can be varied through this feature (Scheme 3.6). The presence of a non-coordinating acid such as triflic acid was found to be beneficial for the catalytic performance. These ligands were compared with structural analogs lacking the pendant arm. Overall, the pentadentate complexes were superior to the tetradentate complexes in terms of epoxide yield (up to 89%) and selectivity (Scheme 3.6).

Diastereopure complexes of iron(II) with pentadentate pyridine/pyrrolidine moieties performed well in sulfide oxidation but were inefficient in alkene epoxidation [62].

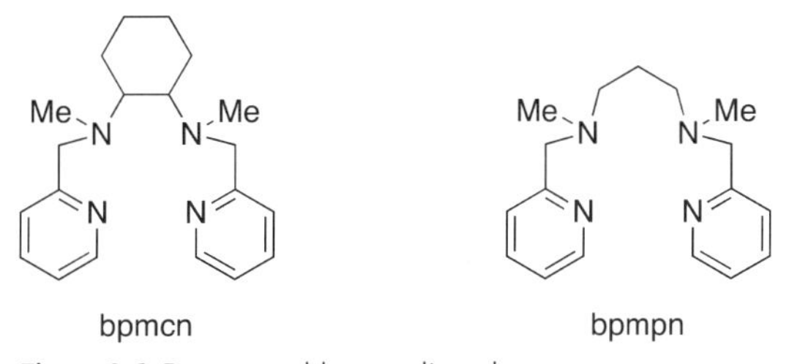

Figure 3.6 Bpmcn and bpmpn ligands.

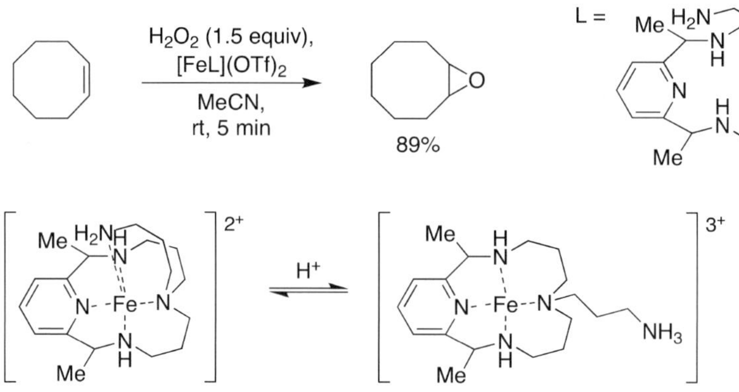

H_2O_2 (1.5 equiv), $[FeL](OTf)_2$, MeCN, rt, 5 min — 89%

Scheme 3.6 Alkene epoxidation with a pyridine-containing macrocycle bearing a pendant arm.

A tridentate 2,6-bis(*N*-methylbenzimidazol-2-yl)pyridine catalyst (Figure 3.1) gave moderate to high yields for epoxide formation [33].

A very simple yet elegant method for efficient epoxidation of aromatic and aliphatic alkenes was presented by Beller and coworkers [63, 64]. $FeCl_3$ hexahydrate in combination with 2,6-pyridinedicarboxylic acid and various organic amines gave a highly reactive and selective catalyst system. An asymmetric variant (for epoxidations of *trans*-stilbene and related aromatic alkenes) was published recently [65] using *N*-monosulfonylated diamines as chiral ligands (Scheme 3.7).

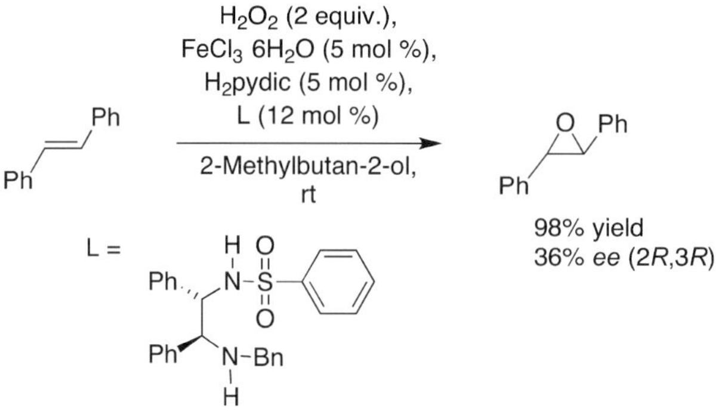

H_2O_2 (2 equiv.), $FeCl_3$ $6H_2O$ (5 mol %), H_2pydic (5 mol %), L (12 mol %), 2-Methylbutan-2-ol, rt

98% yield
36% *ee* (2*R*,3*R*)

Scheme 3.7 Simple alkene epoxidation reported by Beller and coworkers.

3.1.3 Alkene Dihydroxylation

For alkene dihydroxylations, heavy metal oxides such as OsO_4 and RuO_4^- can be applied. They are efficient catalysts but their toxicity makes their use less desirable and there is a clear need for non-toxic metal catalysts. Nevertheless, only a few reports have focused on the use of iron catalysts for alkene dihydroxylations. All systems described so far try to model the naturally occurring Rieske dioxygenase, an enzyme responsible for the biodegradation of arenes via *cis*-dihydroxylation by soil bacteria [66].

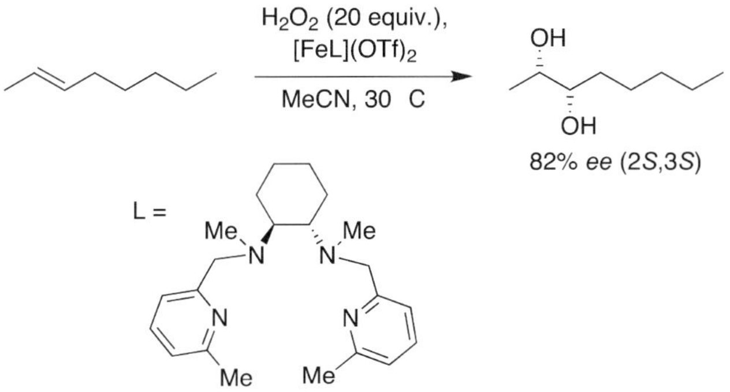

Scheme 3.8 Bmpcn ligand for alkene dihydroxylation.

Chen and Que presented an excellent model for Rieske oxygenase in 1999 [67] using 6-Me_3-TPA ligands (see L2 in Figure 3.3). As mentioned in the previous section, the TPA derivatives were suitable ligands for iron-catalyzed alkane and alkene oxidation [42, 68]. In the case of two or three 6-methyl substituents on the TPA pyridine rings, the selectivity switches to the predominant formation of *cis*-diols. This is due to a change in the spin state of iron depending on the ligand used [42, 69]. Isotope labeling experiments revealed that in the 6-Me_3-TPA systems, both oxygen atoms that are incorporated into the diol stem from one molecule of H_2O_2. No oxygen comes from either O_2 or H_2O. The product is thought to be formed via a cyclic intermediate, similar to oxidations with $KMnO_4$ or OsO_4. Furthermore, two labile *cis* coordination sites on iron seem to be required for dihydroxylation [67], which is in contrast to iron-catalyzed epoxidation reactions.

Later, ligands with *trans*-cyclohexane-1,2-diamine backbones were applied in catalysis (bpmcn ligands) (Figure 3.6) [36, 58, 70]. In this context, Que and coworkers also described the first asymmetric *cis*-hydroxylation of alkenes in 2001 using such bmpcn ligands [70]. In the *cis*-dihydroxylation of *trans*-2-octene with 6-Me_2-bpmcn as ligand, an *ee* of 82% was obtained (Scheme 3.8). A drawback of this method was the need for a large excess of the substrate.

Que and coworkers also presented the first *N,N,O*-ligand to mimic the Rieske dioxygenase. It is the most reactive catalyst for *cis*-dihydroxylation found to date (Figure 3.7) [71].

A solvent-dependent product distribution for iron-catalyzed dihydroxylation with N_3Py-derived ligands was reported by Feringa and coworkers [72]. In acetonitrile, stereoselective *cis*-dihydroxylation was observed. On the other hand, acetone gave rise to *trans*-diols, thereby indicating that the choice of solvent determines the mechanism and hence the outcome of this reaction [72].

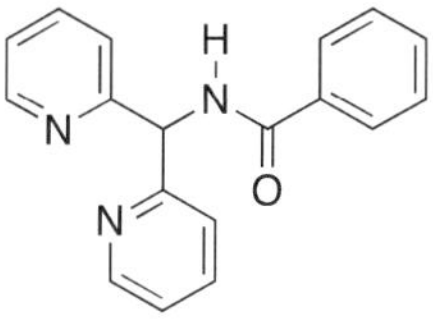

Figure 3.7 First *N,N,O*-ligand described by Que and coworkers.

3.1.4
The Kharasch Reaction and Related Reactions

Halocarbons are known to add to alkenes under transition metal catalysis. The reaction proceeds via radicals and was first described by Kharasch and coworkers in 1945 (Scheme 3.9) [73–76].

Although many transition metals can be used, iron compounds were found to be remarkably efficient, especially in the presence of co-catalysts or ligands [77]. The presence of iron strikingly enhances the reactivity of CCl_4 due to an electron transfer [78]. In addition to CCl_4, many other polyhalogenated compounds can be used. They are equally or even more reactive than CCl_4 itself. The weak C–Br bond in CCl_3Br accounts for the increased reactivity and usually gives high yields.

Other polyhalogenated compounds can be used with similar success. $CpFe(CO)_2$ dimer leads to the formation of mixtures of lactones and esters when reacted with an alkene and methyl trichloroacetate [79] with the lactone being the major product (Scheme 3.10). Similar results were reported earlier by Freidlina and Velichko [80]. $FeCl_3$ and $Fe(CO)_5$ are both suitable for catalyzing the addition of methyl dibromoacetate to electron-deficient alkenes such as methyl propenoate. It was observed that the ratio of products (acyclic vs. lactone) could be tuned by varying the reaction conditions. In all cases, the acyclic product is predominantly formed. Only in the presence of a co-catalyst such as *N,N*-dimethylaniline are small amounts of lactone observed. Noteworthy, elevated temperatures (above 100 °C) are necessary for this transformation.

The use of iron carbonyls such as $Fe_2(CO)_9$ and $Me_3NFe(CO)_4$ was described by Elzinga and Hogeveen [81]. As these compounds are more reactive than $FeCl_3$, the reaction temperature could be lowered to ambient. $Me_3NFe(CO)_4$, which is easily accessible from $Fe(CO)_5$ and Me_3NO, turned out to be best suited. For the addition reactions, optimal results were obtained with stoichiometric amounts of the iron compounds. Terminal alkenes gave better results than their internal counterparts. This observation was probably due to steric hindrance. Strained ring systems such as norbornadiene gave rise to rearranged products in high yields [81]. When terminal

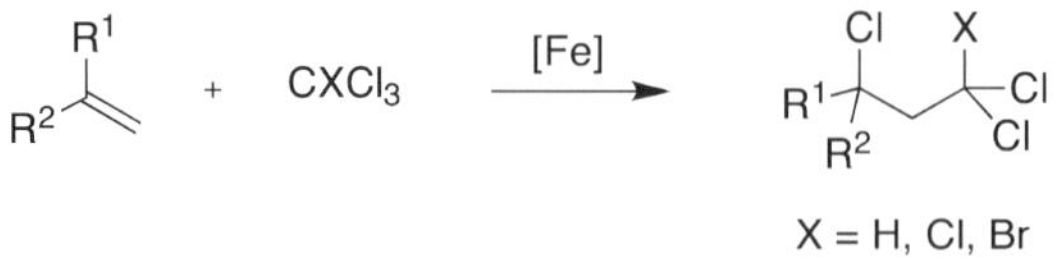

R1, R2 + CXCl3 —[Fe]→ product; X = H, Cl, Br

Scheme 3.9 Kharasch reaction.

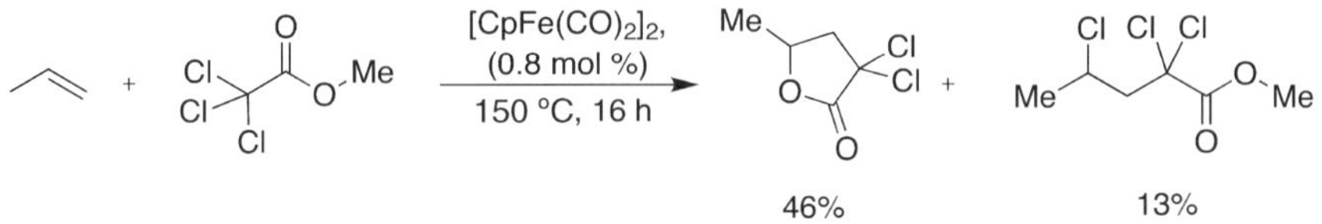

[CpFe(CO)2]2, (0.8 mol %), 150 °C, 16 h; 46%; 13%

Scheme 3.10 Lactonization with methyl trichloroacetate.

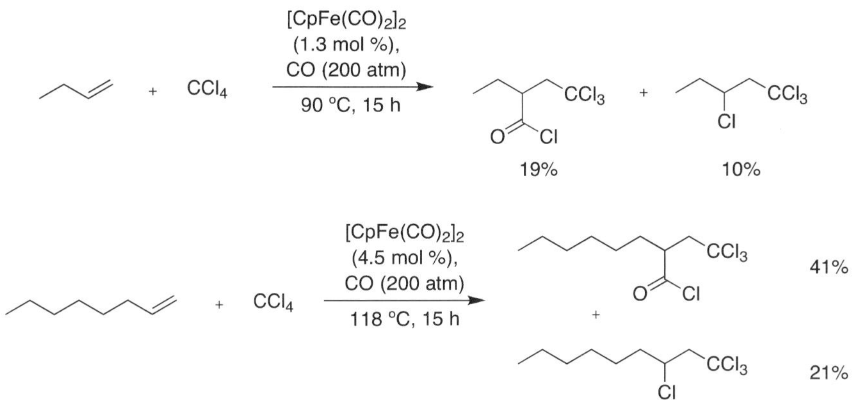

Scheme 3.11 Acyl chlorides via the Kharasch reaction.

alkenes were treated with $CpFe(CO)_2$ dimer, CCl_4 and high pressures of carbon monoxide, not only the expected Kharasch addition product but also the corresponding acyl chlorides were formed (Scheme 3.11) [82].

Intramolecular variants of the Kharasch reaction were described by Weinreb and coworkers in a short series of papers [83, 84]. In an *exo*-cyclization of 7,7,7-trichlorohept-1-ene with $FeCl_2[P(OEt_3)]_3$, 1,1-dichloro-2-chloromethylcyclohexane and 1,1,3-trichlorocycloheptane were obtained in 75 and 13% yield, respectively (Scheme 3.12). Interestingly, α-keto radicals carrying a carbonyl group endocyclic to the ring that is formed cyclize in an *endo* mode. In this way, cyclohexanones are formed from α-keto 5-hexenyl radicals and cycloheptanones are formed from α-keto 6-heptenyl radicals. Furthermore, it is possible to cyclize alkeneic α,α-dichloro esters, acids and nitriles as described earlier by the same authors [83].

An intramolecular lactonization reaction with extremely low catalyst loadings was reported in 1998 [77]. This development led to milder reaction conditions without concomitant loss of product yield. Thus, medium-sized lactones were accessible with a combination of as little as 0.03 mol% of $FeCl_2$ and 0.03 mol% of a multidentate nitrogen-based ligand (Scheme 3.13).

Recently, immobilized metal ion-containing ionic liquids were presented for the Kharasch reaction [85]. Whereas copper salts proved to be suitable catalysts in the addition of CCl_4 to styrene, $FeCl_2$ gave poor results (12% product yield).

Another interesting functionalization of carbon–carbon double bonds is the aminochlorination reaction. This transformation has been known for a long time [86]. Not only iron but also chromium, palladium and copper compounds can be used [87].

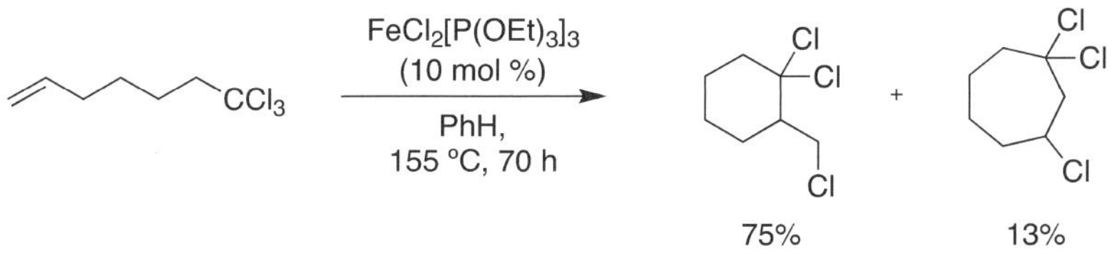

Scheme 3.12 Intramolecular Kharasch reaction.

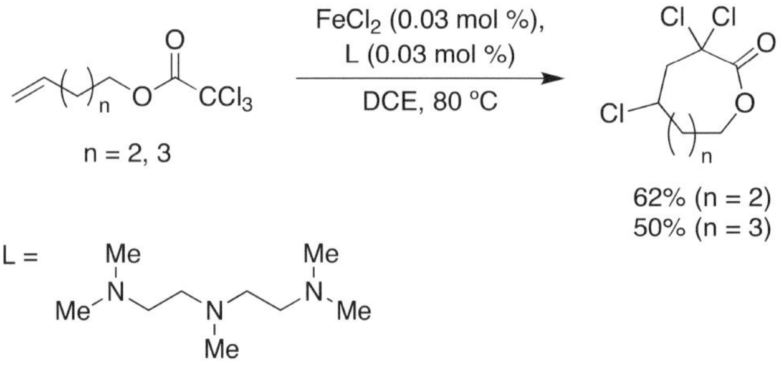

$FeCl_2$ (0.03 mol %), L (0.03 mol %)
DCE, 80 °C

n = 2, 3

62% (n = 2)
50% (n = 3)

L =

Scheme 3.13 Intramolecular lactonization.

Bach and coworkers [88] reported on intramolecular chloroamination reactions with $FeCl_2$ as catalyst starting from 2-alkenyloxycarbonyl azides. TMSCl was found to be an essential additive for the efficient addition to double and triple bonds. The corresponding oxazolidinones were obtained in moderate to high yields and with high diastereoselectivity (Scheme 3.14).

Interestingly, the catalytic reaction yielded predominantly the *threo*-4-(chloromethyl)oxazolidinones, whereas under thermal conditions in 1,1,2,2-tetrachloroethane (TCE) the *erythro* isomer was formed exclusively. If the reaction proceeded via aziridine formation followed by nucleophilic ring opening, the product stereochemistry would be *erythro*. As this was not the case, the involvement of an *N*-centered radical species was suggested (Figure 3.8).

A different finding was made by Li *et al.* in the aminochlorination of arylmethylenecyclopropanes [89]. In this case an aziridinium intermediate accounts for the high regio- and stereoselectivity. Interestingly, screening of various metal catalysts revealed that $FeCl_3$ gave the best results (73% yield), whereas $FeCl_2$ did not show any conversion. Hence the oxidation state of the metal plays an important role. The products were formed in moderate to high yields. Arylmethylenecyclopropanes with

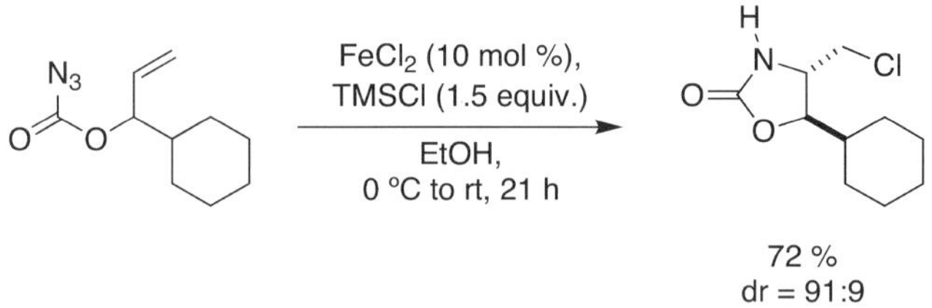

$FeCl_2$ (10 mol %), TMSCl (1.5 equiv.)
EtOH, 0 °C to rt, 21 h

72 %
dr = 91:9

Scheme 3.14 Bach's aminochlorination.

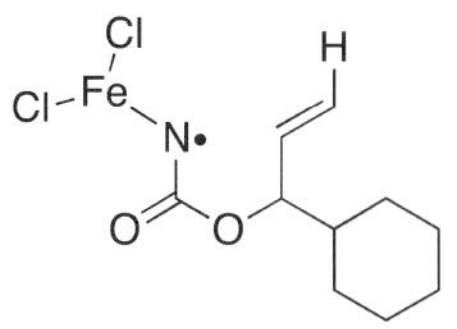

Figure 3.8 *N* centered radical species in aminochlorination reactions.

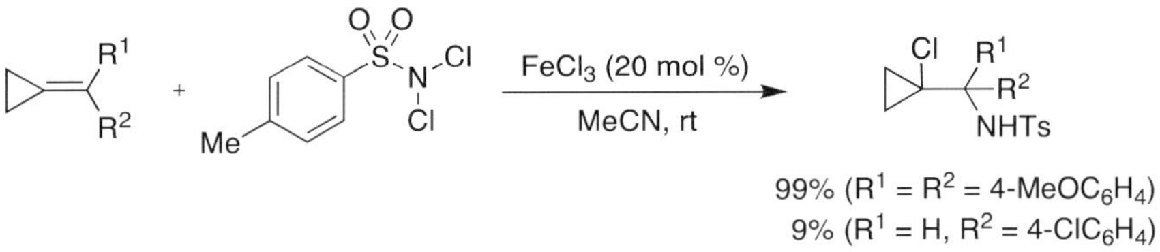

99% ($R^1 = R^2$ = 4-MeOC_6H_4)
9% (R^1 = H, R^2 = 4-ClC_6H_4)

Scheme 3.15 Li's aminochlorination.

electron-donating substituents on the aromatic ring performed best, whereas those with electron-withdrawing substituents gave only sluggish reactions (Scheme 3.15).

3.1.5 Aziridination and Diamination

The first reports on iron-catalyzed aziridinations date back to 1984, when Mansuy *et al.* reported that iron and manganese porphyrin catalysts were able to transfer a nitrene moiety on to alkenes [90]. They used iminoiodinanes PhIN=R (R = tosyl) as the nitrene source. However, yields remained low (up to 55% for styrene aziridination). It was suggested that the active intermediate formed during the reaction was an Fe^{V}=NTs complex and that this complex would transfer the NTs moiety to the alkene [91–93]. However, the catalytic performance was hampered by the rapid iron-catalyzed decomposition of PhI=NTs into iodobenzene and sulfonamide. Other reports on aziridination reactions with iron porphyrins or corroles and nitrene sources such as bromamine-T or chloramine-T have been published [94]. An asymmetric variant was presented by Marchon and coworkers [95]. Biomimetic systems such as those mentioned above will be dealt with elsewhere.

An interesting new catalytic system was described by Hossain and coworkers [96]. Using the relatively mild Lewis acid iron catalyst $[CpFe(CO)_2(THF)]^+[BF_4]^-$, moderate to high yields were obtained in the aziridination of aromatic alkenes with preformed PhI=NTs (Scheme 3.16) [recently it was found that simple $Fe(acac)_3$ can also be used as catalyst for the aziridination of alkenes. In this case, the nitrogen transfer reagent can even be generated *in situ* starting from $PhI(OAc)_2$ and aromatic sulfonamides] [97]. The reaction occurs stereospecifically and yields *cis*-aziridines from *cis*-alkenes and *trans*-aziridines from *trans*-alkenes. Noteworthily, however, the latter transformations do not proceed well, as indicated by that fact that the aziridination of *trans*-β-methylstyrene yields a meager 7% of the product. Furthermore, *trans*-stilbene does not react at all.

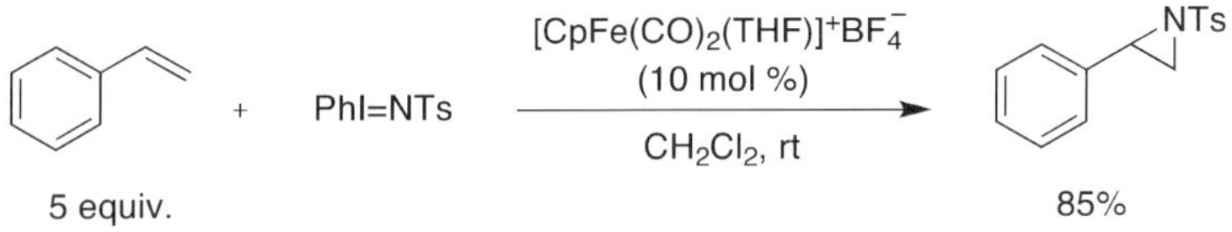

5 equiv. + PhI=NTs → $[CpFe(CO)_2(THF)]^+BF_4^-$ (10 mol %), CH_2Cl_2, rt → 85%

Scheme 3.16 Hossain's aziridination.

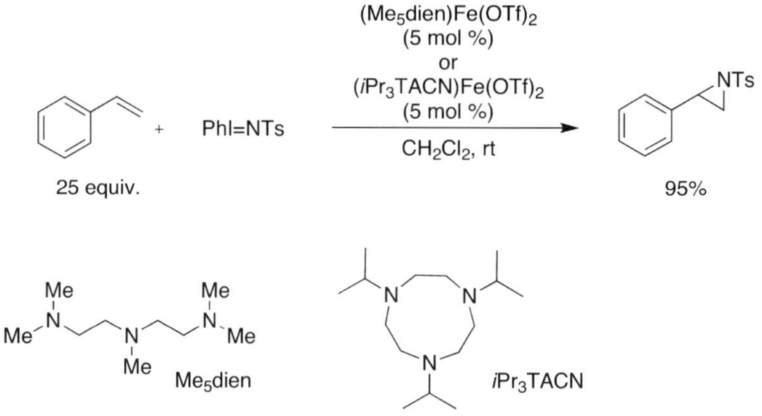

(Me$_5$dien)Fe(OTf)$_2$ (5 mol %) or (*i*Pr$_3$TACN)Fe(OTf)$_2$ (5 mol %)

PhI=NTs, 25 equiv. alkene, CH_2Cl_2, rt → NTs aziridine, 95%

Me$_5$dien   *i*Pr$_3$TACN

Scheme 3.17 Halfen's aziridination.

Another non-heme system made use of hexadentate phenol ligands [98]. However, the catalytically active species was only formed upon ligand oxidation by excess of PhI=NTs. Furthermore, a large excess of the alkene was required (2000 equiv. vs. PhI=NTs). The reaction of cyclooctene and 1-hexene gave yields of about 50%, which represents a significant improvement over the earlier described copper systems [99].

Halfen and coworkers found that simple non-heme iron(II) complexes with Me$_5$dien and TACN ligands performed well in alkene aziridination (Scheme 3.17) [100]. With a comparably small excess of alkene (25 equiv. vs. PhI=NTs), almost quantitative yields of the corresponding aziridine were obtained. Lowering the substrate loading to 5–10 equiv. resulted in decreased yields (68%). Interestingly, when Fe(OTf)$_2$·2MeCN was used as the catalyst, almost no product was formed. A recent study revealed that Fe (OTf)$_2$ can be used as catalyst for aziridinations of silyl enol ethers and simple alkenes; with chiral catalysts up to 40% *ee* have been achieved [101]. A comparison with other multidentate nitrogen ligands revealed that the presence of at least one pair of *cis* labile coordination sites was required for efficient catalysis. This parallels the findings for alkene *cis*-dihydroxylation with H_2O_2 and iron(II) catalysts [42a].

Attempts to aziridinate alkenes with iron catalysts in an asymmetric manner have met with only limited success to date [101]. In an early report on the use of various chiral metal salen complexes, it was found that only the Mn complex catalyzed the reaction whereas all other metals investigated (Cr, Fe, Co, Ni etc.) gave only unwanted hydrolysis of the iminoiodinane to the corresponding sulfonamide and iodobenzene [102]. Later, Jacobsen and coworkers and Evans *et al.* achieved good results with chiral copper complexes [103].

Literature reports on iron-catalyzed alkene diamination are scarce. Li *et al.* described the synthesis of imidazolidine derivatives with an $FeCl_3$–PPh_3 complex. As substrates, α,β-unsaturated ketones and α,β-unsaturated esters were used. The products were obtained in good to high yields and with excellent stereoselectivity (Scheme 3.18). Interestingly, the iron catalyst system worked much better than a previously described rhodium catalyst. Furthermore, the iron catalyst is inexpensive and easier to handle because it is less hygroscopic [104].

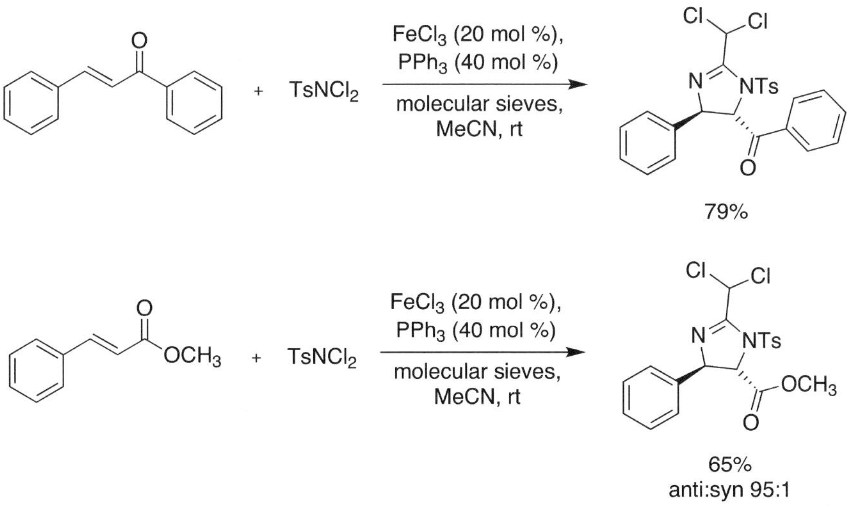

Scheme 3.18 Li's diamination of alkenes.

The iron-catalyzed nitrene transfer to alkenes is a challenging field of research. Remarkable results have been reported but much more effort has to be made to develop further this area of catalysis.

References

1 D. H. R. Barton, M. J. Gastiger, W. B. Motherwell, *J. Chem. Soc., Chem. Commun.* 1983, 41.

2 D. H. R. Barton, D. Doller, *Acc. Chem. Res.* 1992, **25**, 504, and references cited therein.

3 Named after Gif-sur-Yvette, France, where it was developed. D. H. R. Barton, A. K. Göktürk, J. W. Morzycki, W. B. Motherwell, *J. Chem. Soc., Perkin Trans.* 1985, **1**, 583.

4 D. H. R. Barton, T. Li, *Tetrahedron* 1998, **54**, 1735.

5 D. H. R. Barton, M. J. Gastiger, W. B. Motherwell, *J. Chem. Soc., Chem. Commun.* 1983, 731.

6 E. About-Jaudet, D. H. R. Barton, E. Csuhai, N. Ozbalik, *Tetrahedron Lett.* 1990, **31**, 1657.

7 D. H. R. Barton, B. Hu, D. K. Taylor, R. U. Rojas Wahl, *J. Chem. Soc., Perkin Trans.* 1996, **2**, 1031.

8 D. H. R. Barton, *Tetrahedron* 1998, **54**, 5805.

9 P. Stavropoulos, R. Celenligil-Cetin, A. E. Tapper, *Acc. Chem. Res.* 2001, **34**, 745.

10 D. H. R. Barton, F. Halley, N. Ozbalik, M. Schmitt, E. Young, G. Balavoine, *J. Am. Chem. Soc.* 1989, **111**, 7144.

11 D. H. R. Barton, W. Chavasiri, *Tetrahedron* 1997, **53**, 2997.

12 C. Bardin, D. H. R. Barton, B. Hu, R. Rojas Wahl, D. K. Taylor, *Tetrahedron Lett.* 1994, **35**, 5805.

13 C. Knight, M. J. Perkins, *J. Chem. Soc., Chem. Commun.* 1991, 925.

14 D. H. R. Barton, S. D. Beviere, W. Chavasiri, E. Csuhai, D. Doller, W.-G. Liu, *J. Am. Chem. Soc.* 1992, **114**, 2147.

15 M. Newcomb, P. A. Simakov, S.-U. Park, *Tetrahedron Lett.* 1996, **37**, 819.

16 (a) C. Knight, M. J. Perkins, *J. Chem. Soc., Chem. Commun.* 1991, 925;

(b) M. J. Perkins, *Chem. Soc. Rev.* 1996, **25**, 229.

17 (a) H. J. H. Fenton, *Chem. News* 1876, **33**, 190; (b) H. J. H. Fenton, *Proc. Chem. Soc.* 1894, **10**, 157.

18 (a) G. V. Buxton, C. L. Greenstock, W. P. Helman, A. B. Ross, *J. Phys. Chem. Ref. Data* 1988, **17**, 513; (b) C. Walling, B. B. Jacknow, *J. Am. Chem. Soc.* 1960, **82**, 6108.

19 D. H. R. Barton, *Chem. Soc. Rev.* 1996, **25**, 237.

20 D. H. R. Barton, J. Boivin, N. Ozbalik, K. M. Schwarzentruber, *Tetrahedron Lett.* 1985, **26**, 447.

21 F. Recupero, A. Bravo, H.-R. Bjorsvik, F. Fontana, F. Minisci, M. Piredda, *J. Chem. Soc., Perkin Trans.* **2**, 1997, 2399.

22 F. Minisci, F. Fontana, S. Araneo, F. Recupero, L. Zhao, *Synlett* 1996, 119.

23 D. W. Snelgrove, P. A. MacFaul, K. U. Ingold, D. D. M. Wayner, *Tetrahedron Lett.* 1996, **37**, 823.

24 S. S. Kim, S. K. Sar, P. Tamrakar, *Bull. Korean Chem. Soc.* 2002, **23**, 937.

25 (a) F. Haber, R. Willstätter, *Ber. Dtsch. Chem. Ges.* 1931, **64**, 2844; (b) F. Haber, J. Weiss, *Naturwissenschaften* 1932, **20**, 948.

26 M. Nakanishi, C. Bolm, *Adv. Synth. Catal.* 2007, **349**, 861.

27 C. Pavan, J. Legros, C. Bolm, *Adv. Synth. Catal.* 2005, **347**, 703.

28 M. Costas, K. Chen, L. Que, *Coord. Chem. Rev.* 2000, **200–202**, 517.

29 G. B. Shul'pin, G. V. Nizova, Y. N. Kozlov, L. Gonzalez Cuervo, G. Süss-Fink, *Adv. Synth. Catal.* 2004, **346**, 317.

30 G. B. Shul'pin, C. C. Golfeto, G. Süss-Fink, L. S. Shul'pina, D. Mandelli, *Tetrahedron Lett.* 2005, **46**, 4563.

31 N. Kitajima, H. Fukui, Y. Moro-oka, *J. Chem. Soc. Chem. Commun.* 1988, 485.

32 M. Klopstra, R. Hage, R. M. Kellogg, B. L. Feringa, *Tetrahedron Lett.* 2003, **44**, 4581.

33 X. Wang, S. Wang, L. Li, E. B. Sundberg, G. P. Gacho, *Inorg. Chem.* 2003, **42**, 7799.

34 G. J. P. Britovsek, J. England, S. K. Spitzmesser, A. J. P. White, D. J. Williams, *Dalton Trans.* 2005, 945.

35 (a) K. Chen, L. Que, *J. Am. Chem. Soc.* 2001, **123**, 6327; (b) C. Kim, K. Chen, J. Kim, L. Que, *J. Am. Chem. Soc.* 1997, **119**, 5964.

36 M. Costas, L. Que, *Angew. Chem.* 2002, **114**, 2283; *Angew. Chem. Int. Ed.* 2002, **41**, 2179.

37 T. A. van den Berg, J. W. de Boer, W. R. Browne, G. Roelfes, B. L. Feringa, *Chem. Commun.* 2004, 2550.

38 J. England, G. J. P. Britovsek, N. Rabadia, A. J. P. White, *Inorg. Chem.* 2007, **46**, 3752.

39 G. Roelfes, M. Lubben, R. Hage, L. Que, B. L. Feringa, *Chem. Eur. J.* 2000, **6**, 2152.

40 K. Chen, L. Que, *Chem. Commun.* 1999, 1375.

41 (a) J. T. Groves, Y.-Z. Han, in *Cytochrome P-450. Structure, Mechanism and Biochemistry*, 2nd edn., ed. P. R. Ortiz de Montellano, Plenum Press, New York, 1995, pp. 3–48; (b) J. T. Groves, R. C. Haushalter, M. Nakamura, T. E. Nemo, B. J. Evans, *J. Am. Chem. Soc.* 1981, **103**, 2884.

42 (a) K. Chen, M. Costas, J. Kim, A. K. Tipton, L. Que, *J. Am. Chem. Soc.* 2002, **124**, 3026; (b) K. Chen, M. Costas, L. Que, *J. Chem. Soc., Dalton Trans.* 2002, 672.

43 C. Nguyen, R. J. Guajardo, P. K. Mascharak, *Inorg. Chem.* 1996, **35**, 6273.

44 J. M. Rowland, M. Olmstead, P. K. Mascharak, *Inorg. Chem.* 2001, **40**, 2810.

45 M. D. Godbole, M. P. Puig, S. Tanase, H. Kooijman, A. L. Spek, E. Bouwman, *Inorg. Chim. Acta* 2007, **360**, 1954.

46 F. Li, M. Wang, C. Ma, A. Gao, H. Chen, L. Sun, *Dalton Trans.* 2006, 2427.

47 V. B. Romakh, B. Therrien, G. Süss-Fink, G. B. Shul'pin, *Inorg. Chem.* 2007, **46**, 3166.

48 E. A. Gutkina, V. M. Trukhan, C. P. Pierpont, S. Mkoyan, V. V. Strelets, E. Nordlander, A. A. Shteinman, *Dalton Trans.* 2006, 492.

49 Y. Suh, M. S. Seo, K. M. Kim, Y. S. Kim, H. G. Jang, T. Tosha, T. Kitagawa, J. Kim, W. Nam, *J. Inorg. Biochem.* 2006, **100**, 627.

50 B. S. Lane, K. Burgess, *Chem. Rev.* 2003, **103**, 2457.

51 (a) M. Ono, I. Okura, *J. Mol. Catal.* 1990, **61**, 113; (b) R. M. Burger, *Chem. Rev.* 1998, **98**, 1153.
52 W. Nam, R. Ho, J. S. Valentine, *J. Am. Chem. Soc.* 1991, **113**, 7052.
53 R. J. Guajardo, S. E. Hudson, S. J. Brown, P. K. Mascharak, *J. Am. Chem. Soc.* 1993, **115**, 7971.
54 M. C. White, A. G. Doyle, E. Jacobsen, *J. Am. Chem. Soc.* 2001, **123**, 7194.
55 E. A. Duban, K. P. Bryliakov, E. P. Talsi, *Eur. J. Inorg. Chem.* 2007, 852.
56 M. Fujita, L. Que, *Adv. Synth. Catal.* 2004, **346**, 190.
57 G. Dubois, A. Murphy, T. D. P. Stack, *Org. Lett.* 2003, **5**, 2469.
58 For mechanistic studies, see: D. Quinonero, D. G. Musaev, K. Morokuma, *Inorg. Chem.* 2003, **42**, 8449.
59 R. Mas-Ballesté, M. Costas, T. van den Berg, L. Que, *Chem. Eur. J.* 2006, **12**, 7489.
60 M. R. Bukowski, P. Comba, A. Lienke, C. Limberg, C. Lopez de Laorden, R. Mas-Ballesté, M. Merz, L. Que, *Angew. Chem.* 2006, **118**, 3524; *Angew. Chem. Int. Ed.* 2006, **45**, 3446.
61 S. Taktak, W. Ye, A. M. Herrera, E. V. Rybak-Akimova, *Inorg. Chem.* 2007, **46**, 2929.
62 S. Gosiewska, M. Lutz, A. L. Spek, R. J. M. Klein Gebbink, *Inorg. Chim. Acta* 2007, **360**, 405.
63 B. Bitterlich, G. Anilkumar, F. G. Gelalcha, B. Spilker, A. Grotevendt, R. Jackstell, M. K. Tse, M. Beller, *Chem. Asian J.* 2007, **2**, 521.
64 G. Anilkumar, B. Bitterlich, F. G. Gelalcha, M. K. Tse, M. Beller, *Chem. Commun.* 2007, 289.
65 F. G. Gelalcha, B. Bitterlich, G. Anilkumar, M. K. Tse, M. Beller, *Angew. Chem.* 2007, **119**, 7431; *Angew. Chem. Int. Ed.* 2007, **46**, 7293.
66 D. T. Gibson, R. E. Parales, *Curr. Opin. Biotechnol.* 2000, **11**, 236.
67 K. Chen, L. Que, *Angew. Chem.* 1999, **111**, 2365; *Angew. Chem. Int. Ed.* 1999, **38**, 2227
68 J. Y. Ryu, J. Kim, M. Costas, K. Chen, W. Nam, L. Que, *Chem. Commun.* 2002, 1288.
69 M. Fujita, M. Costas, L. Que, *J. Am. Chem. Soc.* 2003, **125**, 9912.
70 M. Costas, A. K. Tipton, K. Chen, D.-H. Jo, L. Que, *J. Am. Chem. Soc.* 2001, **123**, 6722.
71 P. D. Oldenburg, A. A. Shteinman, L. Que, *J. Am. Chem. Soc.* 2005, **127**, 15672.
72 M. Klopstra, G. Roelfes, R. Hage, R. M. Kellogg, B. L. Feringa, *Eur. J. Inorg. Chem.* 2004, 846.
73 M. S. Kharasch, E. V. Jensen, W. H. Urry, *Science* 1945, **102**, 128.
74 M. S. Kharasch, E. V. Jensen, W. H. Urry, *J. Am. Chem. Soc.* 1947, **69**, 1100.
75 For a mechanistic study, see: V. I. Tararov, T. F. Saveleva, Y. T. Struchkov, A. P. Pisarevskii, N. I. Raevskii, Y. N. Belokon, *Russ. Chem. Bull.* 1996, **45**, 600.
76 C. Bolm, J. Legros, J. Le Paih, L. Zani, *Chem. Rev.* 2004, **104**, 6217.
77 F. de Campo, D. Lastécouères, J.-B. Verlhac, *Chem. Commun.* 1998, 2117.
78 M. Asscher, D. Vofsi, *J. Chem. Soc.* 1963, 3921.
79 Y. Mori, J. Tsuji, *Tetrahedron* 1972, **28**, 29.
80 R. K. Freidlina, F. K. Velichko, *Synthesis* 1977, 145, and references cited therein.
81 J. Elzinga, H. Hogeveen, *J. Org. Chem.* 1980, **45**, 3957.
82 T. Susuki, J. Tsuji, *J. Org. Chem.* 1970, **35**, 2982.
83 (a) T. K. Hayes, A. J. Freyer, M. Parvez, S. M. Weinreb, *J. Org. Chem.* 1986, **51**, 5501; (b) T. K. Hayes, R. Villani, S. M. Weinreb, *J. Am. Chem. Soc.* 1988, **110**, 5533; (c) G. M. Lee, M. Parvez, S. M. Weinreb, *Tetrahedron* 1988, **44**, 4671.
84 G. M. Lee, S. M. Weinreb, *J. Org. Chem.* 1990, **55**, 1281.
85 T. Sasaki, C. Zhong, M. Tada, Y. Iwasawa, *Chem. Commun.* 2005, 2506.
86 F. Minisci, *Acc. Chem. Res.* 1975, **8**, 165.
87 (a) J. Lessard, R. Cote, P. Mackiewicz, R. Furstoss, B. Waegell, *J. Org. Chem.* 1978, **43**, 3750; (b) H.-X. Wei, S. H. Kim, G. Li, *Tetrahedron* 2001, **57**, 3869; (c) R. Göttlich, *Synthesis* 2000, 1561.
88 (a) T. Bach, B. Schlummer, K. Harms, *Chem. Commun.* 2000, 287; (b) H.

Danielec, J. Klügge, B. Schlummer, T. Bach, *Synthesis* 2006, 551.
89 Q. Li, M. Shi, C. Timmons, G. Li, *Org. Lett.* 2006, **8**, 625.
90 D. Mansuy, J.-P. Mahy, A. Dureault, G. Bedi, P. Battioni, *J. Chem. Soc., Chem. Commun.* 1984, 1161.
91 J.-P. Mahy, P. Battioni, D. Mansuy, *J. Am. Chem. Soc.* 1986, **108**, 1079.
92 J.-P. Mahy, G. Bedi, P. Battioni, D. Mansuy, *J. Chem. Soc., Perkin Trans. 2* 1988, 1517.
93 For a recent theoretical study, see Y. Moreau, H. Chen, E. Derat, H. Hirao, C. Bolm, S. Shaik, *J. Phys. Chem. B* 2007, **111**, 10288.
94 (a) R. Vyas, G.-Y. Gao, J. D. Harden, X. P. Zhang, *Org. Lett.* 2004, **6**, 1907; (b) L. Simkhovich, Z. Gross, *Tetrahedron Lett.* 2001, **42**, 8089.
95 J.-P. Simonato, J. Pécaut, W. R. Scheidt, J.-C. Marchon, *Chem. Commun.* 1999, 989.
96 B. D. Heuss, M. F. Mayer, S. Dennis, M. M. Hossain, *Inorg. Chim. Acta* 2003, **342**, 301.
97 A. Mayer, C. Bolm to be published.
98 F. Avenier, J.-M. Latour, *Chem. Commun.* 2004, 1544.
99 (a) J. A. Halfen, J. M. Uhan, D. C. Fox, M. P. Mehn, L. Que, *Inorg. Chem.* 2000, **39**, 4913; (b) P. Dauban, L. Saniere, A. Tarrade, R. H. Dodd, *J. Am. Chem. Soc.* 2001, **123**, 7707.
100 K. L. Klotz, L. M. Slominski, A. V. Hull, V. M. Gottsacker, R. Mas-Ballesté, L. Que, J. A. Halfen, *Chem. Commun.* 2007, 2063.
101 M. Nakanishi, C. Bolm, to be published.
102 K. J. O'Connor, S.-J. Wey, C. J. Burrows, *Tetrahedron Lett.* 1992, **33**, 1001.
103 For early asymmetric aziridinations using chiral copper catalysts, see (a) Z. Li, K. R. Conser, E. N. Jacobsen, *J. Am. Chem. Soc.* 1993, **115**, 5326; (b) D. A. Evans, M. M. Faul, M. T. Bilodeau, B. A. Anderson, D. M. Barnes, *J. Am. Chem. Soc.* 1993, **115**, 5328.
104 H.-X. Wei, S. H. Kim, G. Li, *J. Org. Chem.* 2002, **67**, 4777.

3.2
Oxidative Allylic Oxygenation and Amination

Sabine Laschat, Volker Rabe, and Angelika Baro

3.2.1
Introduction

Allylic alcohols and allylic amines are important building blocks in organic chemistry, which can be further functionalized, for example, by Sharpless epoxidation, asymmetric dihydroxylation, rearrangements or cross-metathesis to give access to a broad variety of synthetic intermediates for natural products and fine chemicals. In principle, allylic alcohols and amines are available by two synthetic strategies. The first, the introduction of the hydroxy or amino group via nucleophilic displacement of a suitable leaving group at the allyl system, will be discussed in Chapter 7. This chapter will focus on the second route, namely the direct insertion of oxygen or nitrogen into the allylic C–H bond via a transition metal-catalyzed oxidative process.

As shown in Scheme 3.19, two competing pathways are possible with regard to allylic oxidation. The alkene **1** can either undergo abstraction of an allylic hydrogen and subsequent formation of the allylic alcohol **2** and the enone **3** (path A), respectively, or alternatively epoxidation of the C=C double bond occurs to give derivative **4** (path B). In order to develop a suitable catalytic system for path A, it is of utmost importance to achieve high chemoselectivity in addition to high catalytic

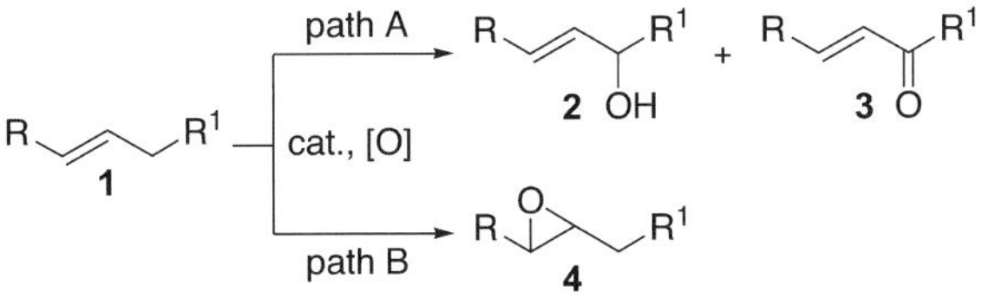

Scheme 3.19

activity, long-term stability of the catalyst under oxidative conditions and the use of non-toxic oxidants without any safety hazards.

Most work in the field of iron-catalyzed allylic oxidations has been devoted to the following classes of catalysts: (1) simple Fe^{3+} (or Fe^{2+}) salts, (2) Fe complexes with bidentate ligands such as acetylacetonate or picolinate, (3) porphyrin and phthalocyanine Fe complexes, (4) salen Fe complexes and (5) non-heme Fe complexes with tetra- or pentadentate ligands. It should be noted that in many iron-catalyzed oxidations a direct comparison of different catalysts is not possible, simply because yields of isolated pure products and detailed experimental procedures are lacking. This has to be considered as a major drawback concerning synthetic applications. In order to guide readers who are searching for the most suitable catalyst for a particular alkene substrate, the yields given in the schemes below refer to isolated yields, whereas conversions were usually determined by gas chromatography. In some examples, just turnover numbers (TON = moles of product per mole of catalyst) are given.

3.2.2 Iron-catalyzed Allylic Oxidations

3.2.2.1 Simple Iron Salts

Simple Fe^{3+} salts have rarely been used for catalytic allylic oxidations. Covalent metal nitrates are well known to be strong oxidants which undergo dissociation of the bidentate metal nitrate bond resulting in the formation of the NO_3 radical as reactive species [105]. However, Sahle-Demessie and coworkers were the first who showed the utility of even commercially available $Fe(NO_3)_3{\cdot}9H_2O$ as an oxidation catalyst [106]. Turnover and chemoselectivity turned out to be strongly dependent on the alkene substrate and the partial pressure (Scheme 3.20).

Whereas under ambient oxygen pressure only 20% of cyclohexene **1a** was converted to an equimolar mixture of 2-cyclohexen-1-ol (**2a**) and 2-cyclohexen-1-one (**3a**), a mixture of oxygen and nitrogen at elevated pressures (3.4 bar/10.3 bar) improved both conversion and chemoselectivity, favoring the enone **3a** (70%) as compared with the allylic alcohol **2a** (14%). In addition, cyclohexanediol **5** was produced in 16% yield. In contrast, the product distribution of 1,5-cyclooctadiene **1b** clearly varied. Under aerobic conditions, **1b** gave exclusively the epoxide **4b**; under pressurized conditions the allylic oxidation products **2b** and **3b** were also formed in minor amounts (Scheme 3.20). When the reaction was carried out with $FeCl_3{\cdot}2H_2O$ instead of $Fe(NO_3)_3{\cdot}9H_2O$ as catalyst, no oxidation products could be detected, thus indicating that nitrate must be directly involved in the catalytic cycle.

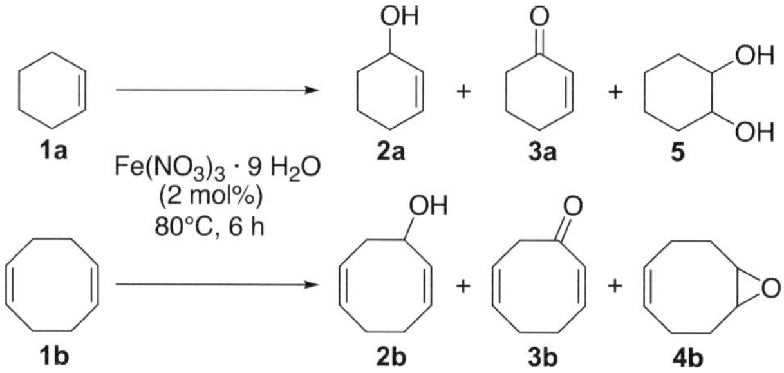

$Fe(NO_3)_3 \cdot 9\,H_2O$ (2 mol%), 80°C, 6 h

1a → 2a + 3a + 5

1b → 2b + 3b + 4b

O_2 (1 bar)				O_2/N_2 (3.4/10.3 bar)			
[%] conv.	**2a**	**3a**	**5**	[%] conv.	**2a**	**3a**	**5**
20	45	55	–	62	14	70	16
[%] conv.	**2b**	**3b**	**4b**	[%] conv.	**2b**	**3b**	**4b**
8	–	–	>95	18	16	16	68

Scheme 3.20

3.2.2.2 Fe(III) Complexes with Bidentate Ligands

Jiang *et al.* studied homogeneous mono- and bimetallic catalysts derived from first-row transition metal acetylacetonate complexes and $RuCl_2(PPh_3)_3$ in the oxidation of cyclohexene **1a** under atmospheric oxygen pressure [107]. With catalysts $Fe(acac)_3$ and $RuCl_2(PPh_3)_3$, neat **1a** was oxidized to allylic alcohol **2a** and enone **3a** in ratios **2a** : **3a** = 38 : 62 (Fe) and 42 : 58 (Ru) with a TON of 198 and 140, respectively. For a 1:1 mixture of both catalysts, however, a synergistic effect was observed, resulting in an improved TON (294) [107].

Owing to their relevance in steroid chemistry Okamoto *et al.* investigated aerobic allylic hydroxylations of octahydronaphthalene derivatives such as **1c** in the presence of Fe(III) picolinate complexes $Fe(PA)_3{\cdot}H_2O$ (Scheme 3.21) [108]. The combination of electrolysis and the $Fe(PA)_3$–O_2–MeCN system suppressed epoxidation almost completely, leading exclusively to the oxidation products **2c** and **3c**, albeit with low yields. In contrast, when alkene **1c** was submitted to chemical oxidation using the

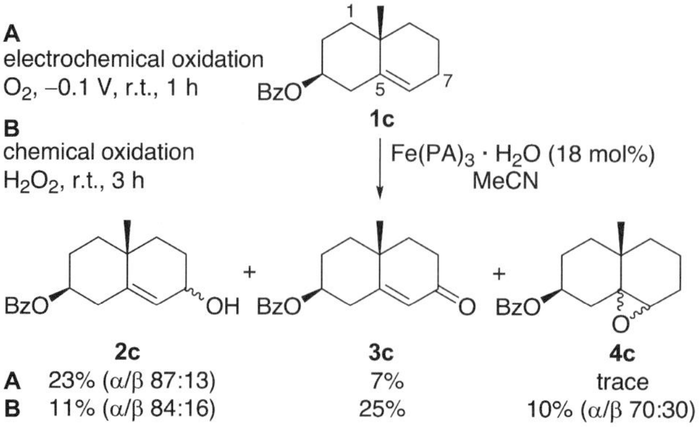

A electrochemical oxidation O_2, −0.1 V, r.t., 1 h

B chemical oxidation H_2O_2, r.t., 3 h

1c → $Fe(PA)_3 \cdot H_2O$ (18 mol%), MeCN → **2c** + **3c** + **4c**

	2c	**3c**	**4c**
A	23% (α/β 87:13)	7%	trace
B	11% (α/β 84:16)	25%	10% (α/β 70:30)

Scheme 3.21

system $Fe(PA)_3{\cdot}H_2O$–35% H_2O_2–MeCN, a mixture of alcohol **2c**, enone **3c** and epoxide **4c** was obtained. It should be emphasized that in both cases the allylic oxidation took place only at the 7-H with a slight preference for the 7α-hydrogen abstraction. However, the electrochemical method is limited to trisubstituted bridgehead double bonds.

3.2.2.3 Fe^{3+}/Fe^{2+} Porphyrin and Phthalocyanine Complexes

Iron porphyrin complexes were of tremendous interest because they can serve as chemical models for cytochrome P450 monooxygenase [109]. Allylic hydroxylation by heme iron complexes follows the "rebound mechanism" depicted in Scheme 3.22. Starting from iron(III) porphyrin complex **6**, the high-valent iron(IV)-oxo porphyrin π-cation radical **7** is formed with various oxidants such as single oxygen atom donors (i.e. iodosobenzene), hydroperoxides (i.e. alkyl hydroperoxides, H_2O_2, peracids) or molecular oxygen. Iron(IV)-oxo species **7**, which corresponds to Compound I in CYP 450 monooxygenases [109c], induces hydrogen abstraction from the alkene substrate **1**, resulting in the formation of $Fe^{3+}(OH)$ complex **8** and allylic radical **9**. In a "rebound process" and subsequent cleavage, the allylic alcohol **2** is produced and the precursor catalyst **6** is regenerated by replacement of the axial ligand with water or solvent.

Porphyrin complexes, however, are prone to oxidative decomposition and therefore synthetic applications are hampered by rapid catalyst deactivation. This problem can be overcome by attaching electron-withdrawing groups to the periphery of the porphyrin system. Another problem is the poor chemoselectivity. In many cases, addition to the C=C double bond and formation of the epoxide are much faster than the corresponding hydrogen abstraction, which leads to the allylic alcohols. This is

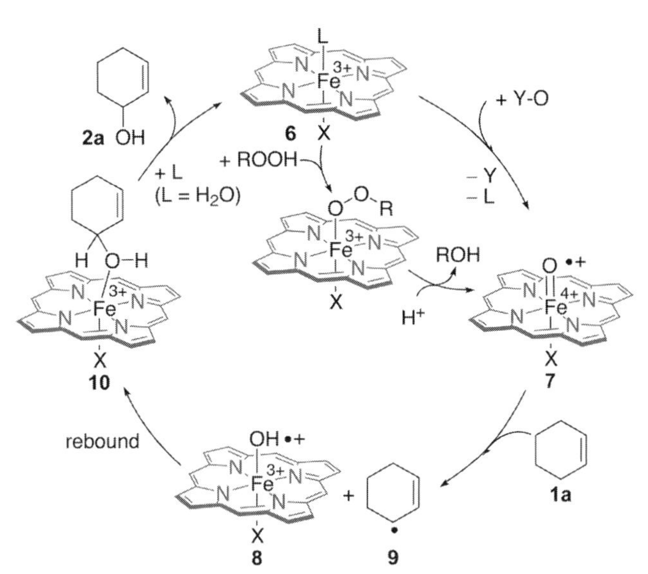

Scheme 3.22 Proposed rebound mechanism shown for the hydroxylation of cyclohexene **1a** [109].

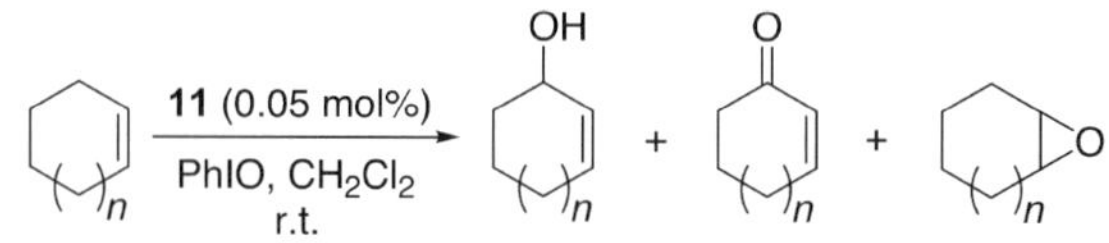

11 (0.05 mol%), PhIO, CH_2Cl_2, r.t.

1a (n = 1)
1d (n = 0)
1e (n = 2)
1f (n = 3)

2a,d–f + **3a,d–f** + **4a,d–f**

	conv.	2	3	4
a	90%	10%	4%	86%
d	92%	22%	2%	76%
e	91%	–	–	100%
f	89%	–	–	100%

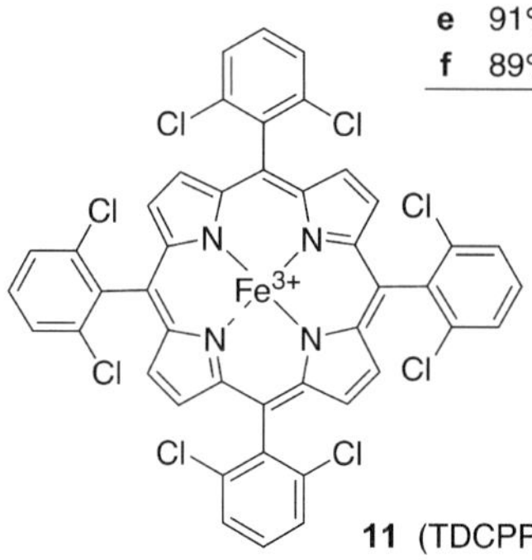

11 (TDCPP)Fe

Scheme 3.23

illustrated by a study by Appleton *et al.* [110], who also observed that the chemoselectivity depended on the ring size of the alkene (Scheme 3.23). In all cases, the formation of epoxide **4** dominated, but cyclopentene **1d** and cyclohexene **1a** gave up to 24% of the corresponding enones **3a,d** and allylic alcohols **2a, d**. However, virtually no trace of allylic oxidation was found for cycloheptene **1e** and cyclooctene **1f**.

The prevalence for epoxidation seems to be independent of the electronic structure of the porphyrin or the type of oxidant. For example, the catalytic oxidation of cyclohexene **1a** with molecular oxygen and isobutyraldehyde in the presence of Fe^{3+}(TPP)Cl (TPP = tetraphenylporphyrin) in 1,2-dichloroethane yielded exclusively the epoxide **4a** and only trace amounts of the allylic alcohol **2a** or ketone **3a**, as demonstrated by Nam *et al.* [111].

Similar results were reported by Niño *et al.* for FeTPPCl intercalated in α-zirconium phosphate and isobutyraldehyde (α-ZrP-Imi-Fe(TPPCl)] [112]. Several strategies have been developed to overcome the epoxidation problem. O'Shea *et al.* used iron(III) porphyrin nitrite ($PFeNO_2$) which was generated *in situ* from octaethylporphyrin in $FeCl_3$ with potassium crown ether nitrite in NMP–1% HOAc for the oxidation of cyclohexene **1a** (Scheme 3.24, **A**) [113]. Allylic alcohol **2a** was formed with exceptional selectivity (90%) together with enone **3a** as a minor byproduct. By using ^{15}N-labeled nitrite salts, it was demonstrated that $^{15}NO_2^-$ is bound directly to the iron center and oxygen atom transfer resulted in the formation of $PFe^{15}NO$.

A remarkable solvent effect on the chemoselectivity was discovered by Agarwala and Bandyopadhyay (Scheme 3.24, **B**) [114]. When cyclohexene **1a** was oxidized with *t*BuOOH in the presence of an electronegative substituted iron(III) porphyrin complex in CH_2Cl_2–MeOH, epoxide **4a** was the predominant product (69% yield) in addition to alcohol **2a** and ketone **3a** as byproducts in 20% and 11% yields,

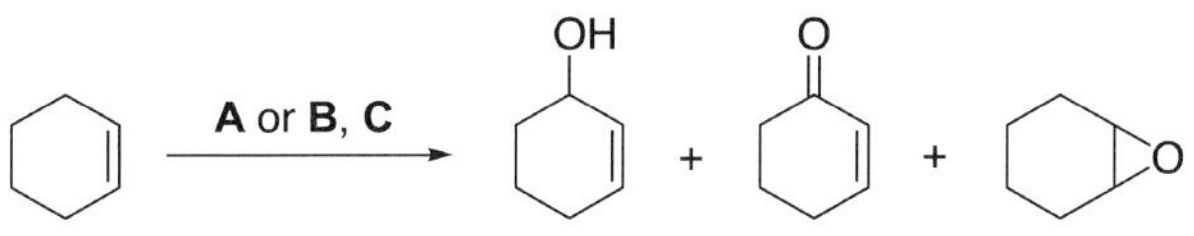

		1a → 2a (OH)	3a (O)	4a (epoxide)
A		90%	10%	–
B	72% conv.	20%	11%	69%
C	97% conv.	100%	–	–

A PFeCl, K(18-crown-6)NO_2,NMP, HOAc (1%), Ar, r.t., 1 h
B F_{20}TPPFeCl, *t*BuOOH, CH_2Cl_2–MeOH, 30 min
C F_{20}TPPFeCl, *t*BuOOH, CH_3CN–H_2O, 10 min

P = octaethylporphyrin
F_{20}TPP = *meso*-tetrakis(pentafluorophenyl)porphinato dianion

Scheme 3.24

respectively. However, on replacing the solvent mixture with MeCN–H_2O the reaction was much faster, yielding entirely 2-cyclohexen-1-ol (**2a**) (Scheme 3.24, **C**). In a competing experiment, a 1:1 mixture of cyclohexane and cyclohexene **1a** was converted to cyclohexanol (17%) and 2-cyclohexen-1-ol (**2a**) (76%), indicating a much faster allylic hydroxylation than alkane hydroxylation. *In situ* UV/Vis studies revealed the presence of the iron(IV)-oxo porphyrin π-cation radical **7**.

Another approach to improved chemoselectivity utilizes sterically hindered alkenes, as reported by Konoike *et al.* (Scheme 3.25) [115]. Ursolic acid **12**, a steroid with a highly congested trisubstituted double bond, undergoes allylic hydroxylation at the C-11 position with MCPBA and tetrakis(pentafluorophenyl)porphyrin iron chloride [Fe(PFPP)Cl] as a catalyst to give a single diastereomer **13** in 91% yield. With sterically less encumbered systems only epoxidation was observed.

Exclusive enone formation could be achieved by electrocatalytic oxygenation of 2-cyclopentene-1-acetic acid in the presence of a water-soluble iron(III) porphyrin (2-TMPyP)Fe [2-TMPyP = tetrakis(*N*-methyl-2-pyridyl)porphyrin]. Unfortunately, neither yields nor TONs are given [116].

A recent contribution to the chemoselectivity problem is the composite photocatalysis developed by Maldotti *et al.* [117]. Cyclohexene **1a** was subjected to photoinitiated oxidation in Nafion membranes containing Pd(II) porphyrin Pd(4-TMPyP) [4-TMPyP = *meso*-tetrakis(*N*-methyl-4-pyridyl)porphyrin] and Fe(III) porphyrin **11** to

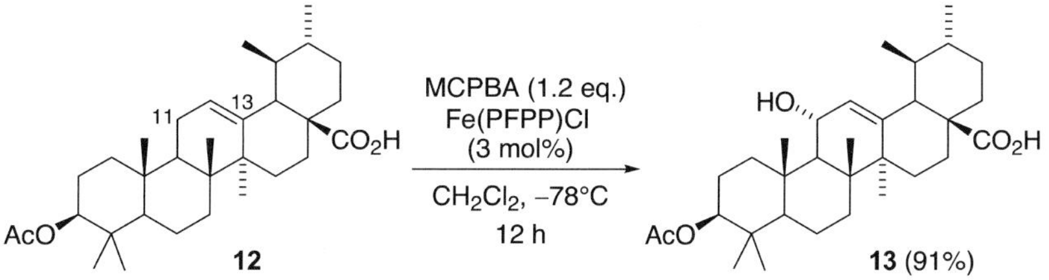

12 (AcO, 11, 13, CO_2H) → MCPBA (1.2 eq.), Fe(PFPP)Cl (3 mol%), CH_2Cl_2, −78°C, 12 h → **13** (91%) (HO, AcO, CO_2H)

Scheme 3.25

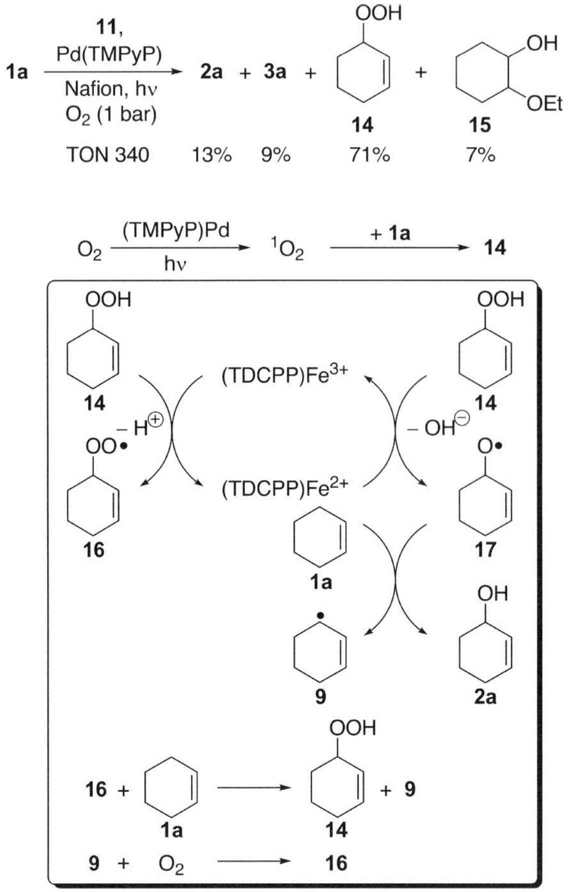

Scheme 3.26

give a mixture of the allylic hydroperoxide **14** (71%), alcohol **2a**, enone **3a** and additionally the ethyl ether **15** (29% overall yield) (Scheme 3.26).

As illustrated in Scheme 3.26, the role of the Pd(II) porphyrin complex is the photochemically sensitized generation of singlet oxygen ($^1O_2$), which forms the allylic hydroperoxide **14**. Subsequent Fe^{3+} porphyrin-catalyzed hydrogen abstraction from **14** yields the peroxy radical **16**, while the catalytic Fe^{3+} species is recycled by homolytic O–O bond cleavage of **14** to give the alkoxy radical **17**. Both radicals **16** and **17** are involved in the dark reaction where the remaining cyclohexene is consumed.

In comparison with metal porphyrins, the corresponding metal phthalocyanines are much more stable against oxidative decomposition. Murahashi *et al.* reported that chlorinated Fe(II) phthalocyanine is particularly well suited for aerobic allylic oxidation employing acetic aldehyde as a cofactor (Scheme 3.27) [118]. Under these conditions, cyclohexene **1a** is converted to a mixture of **2a** and **3a** in 70% overall yield and the epoxide **4a** as byproduct (30%). Acetic aldehyde is proposed to autoxidize by

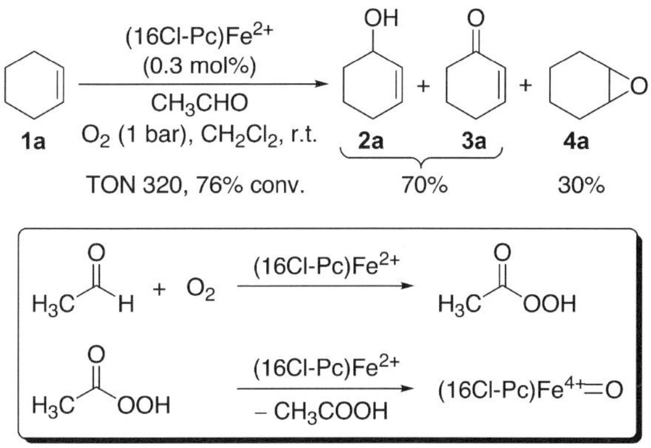

Scheme 3.27

the Fe catalyst to acetic peracid, which then transfers an oxygen atom to the Fe phthalocyanine complex leading to an iron(IV) species as reactive intermediate.

Weber *et al.* utilized dinuclear μ-oxo iron(III) phthalocyanines **18** as catalysts in the aerobic oxidation of α-pinene **1g**, resulting in the formation of *trans*-verbenol **2g**, verbenone **3g** and α-pinene oxide **4g** in almost equimolar amounts. Additionally, 3-pinen-2-ol **19** was obtained (Scheme 3.28) [119].

Applying the same conditions to cyclohexene **1a** gave preferably enone **3a** (39%) and cyclohexenol **2a** (48%), while epoxide **4a** was detected in only 12% yield.

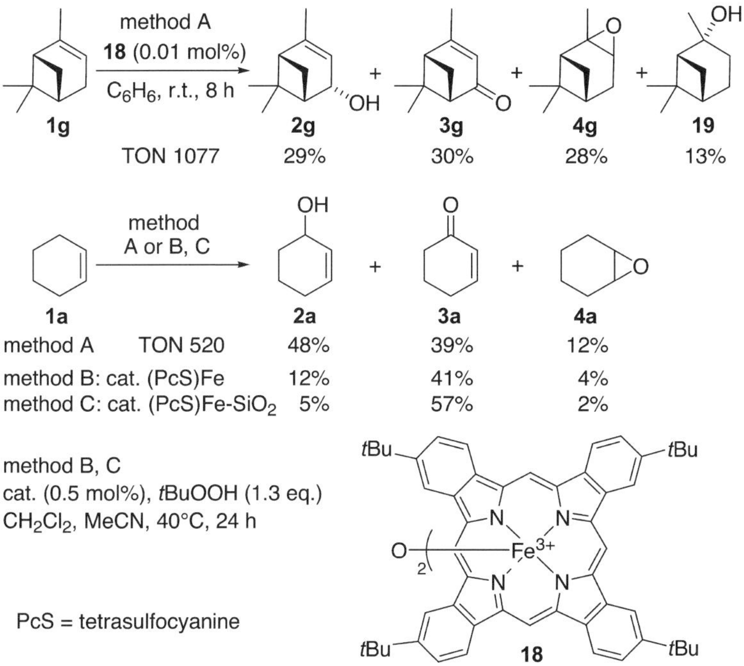

Scheme 3.28

The strong preference towards enone formation, which was realized by González *et al.* employing iron(III) tetrasulfophthalocyanine and *t*BuOOH (Scheme 3.28, method B) [120], could be further improved by immobilization of the phthalocyanine complex on silica (method C). Under these conditions, μ-oxo dimeric species are suggested to be the active catalysts.

3.2.2.4 Iron(III) Salen Complexes

Whereas metal porphyrins suffer from catalyst deactivation, phthalocyanines are often poorly soluble, hampering the synthesis of tailor-made metal phthalocyanines. In contrast, the salen complexes provide an attractive alternative as potential ligands for oxidation catalysis due to their convergent synthesis from easily available building blocks and their pronounced stability towards oxidative degradation. Thus, Böttcher *et al.* designed a series of iron complexes bearing achiral and chiral salen ligands with different electronic and steric properties [121]. The most promising catalyst was **20**, which converted cyclohexene **1a** under aerobic conditions into a mixture of **2a** and **3a** in about a 1 : 2 ratio (Scheme 3.29). Cyclovoltammetric experiments revealed that unlike porphyrin complexes, the catalytic activity is not correlated with the redox potential of the salen complexes.

3.2.2.5 Non-heme Iron Complexes with Tetra- and Pentadentate Ligands

Heme and non-heme iron complexes have special relevance to biological systems. The former represent model compounds for cytochrome P450 monooxygenase and the latter those for both mononuclear iron enzymes such as catechol dioxygenase, lipoxygenases, isopenicillin N-synthase (IPNS), 1-aminocyclopropane-1-carboxylic acid oxidase, pterin-dependent hydroxylases, Rieske dioxygenases [122] and dinuclear iron enzymes such as methane monooxygenase or ribonucleotide reductase [122, 123]. Mukerjee *et al.* developed a tetradentate ligand **21** structurally related to the salen ligands previously reported [124]. Upon deprotonation, ligand **21** forms the dinuclear Fe complex **22** in the presence of *trans*-Fe^{2+}(*N*-methylimidazole)$_2$$Cl_2$(MeOH)$_2$ in anhydrous MeOH–MeCN (1:99) under anaerobic conditions (Scheme 3.30). Interestingly, the turnover depended strongly on the oxidation state of the iron centers, with

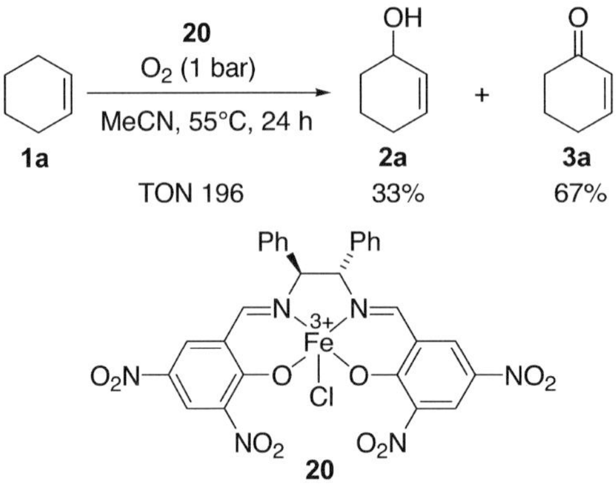

20
O_2 (1 bar)
MeCN, 55°C, 24 h
1a
TON 196
2a 33%
3a 67%
20

Scheme 3.29

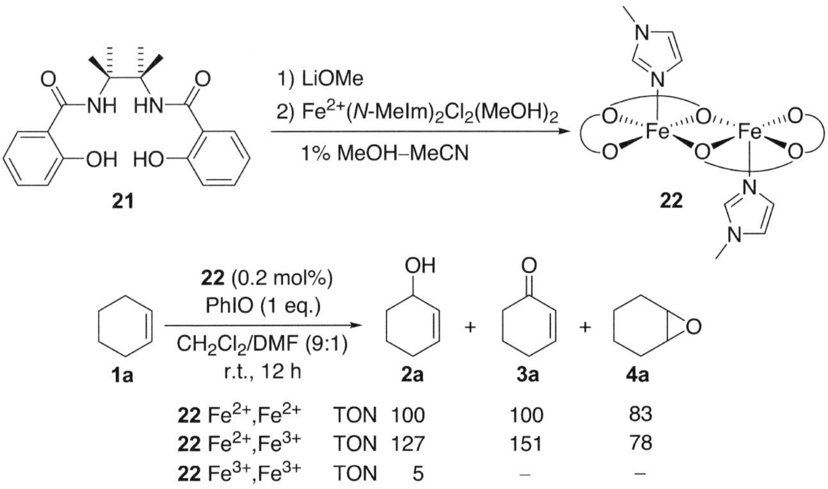

1) LiOMe
2) $Fe^{2+}(N\text{-MeIm})_2Cl_2(MeOH)_2$
1% MeOH–MeCN

21 → 22

22 (0.2 mol%)
PhIO (1 eq.)
CH_2Cl_2/DMF (9:1)
r.t., 12 h

1a → 2a + 3a + 4a

		2a	3a	4a
22 Fe^{2+},Fe^{2+}	TON	100	100	83
22 Fe^{2+},Fe^{3+}	TON	127	151	78
22 Fe^{3+},Fe^{3+}	TON	5	–	–

Scheme 3.30

the mixed valence state giving the highest TONs. The active catalysts **22** provided alcohol **2a** and enone **3a** as the major products. The Fe^{3+},Fe^{3+} catalyst **22**, however, was almost inactive (Scheme 3.30).

Two examples of non-heme iron complexes with tetradentate ligands should be presented, illustrating the problems associated with the use of such complexes for synthetic purposes.

Rowland *et al.* prepared a low-spin Fe(III) complex **24** from *N,N*-bis(2-pyridylmethyl) amine-*N*-ethyl-2-pyridine-2-carboxamide **23** (Scheme 3.31) [125]. When cyclohexene **1a** reacted with complex **24** and a large excess of H_2O_2, among the oxidation products epoxide **4a** predominated (TON 17).

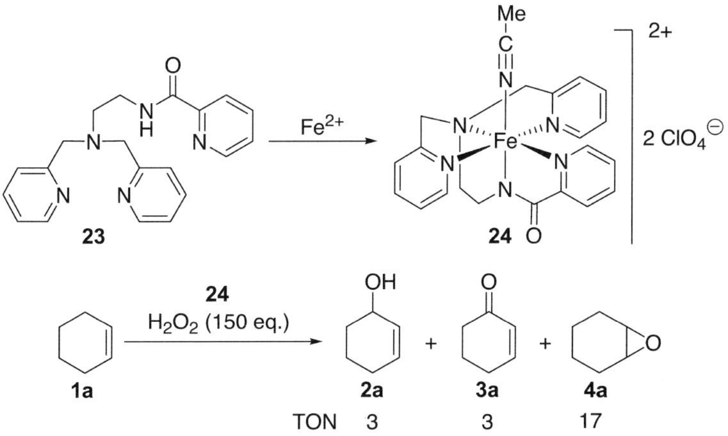

Fe^{2+}

23 → 24 $2+$ $2\ ClO_4^{\ominus}$

24
H_2O_2 (150 eq.)

1a → 2a + 3a + 4a

	2a	3a	4a
TON	3	3	17

Scheme 3.31

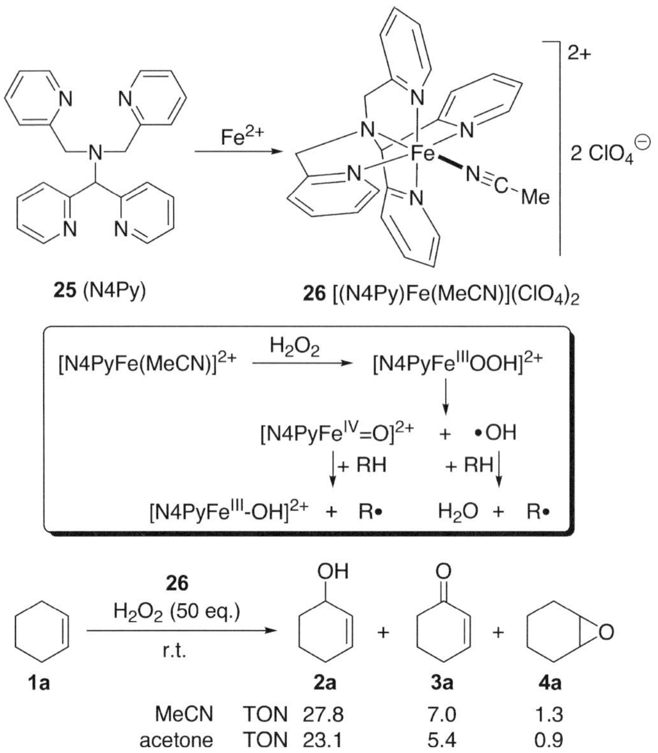

25 (N4Py) $\xrightarrow{Fe^{2+}}$ **26** $[(N4Py)Fe(MeCN)](ClO_4)_2$

$$[N4PyFe(MeCN)]^{2+} \xrightarrow{H_2O_2} [N4PyFe^{III}OOH]^{2+}$$

$$[N4PyFe^{III}OOH]^{2+} \rightarrow [N4PyFe^{IV}{=}O]^{2+} + {\bullet}OH$$

$$[N4PyFe^{IV}{=}O]^{2+} \xrightarrow{+\,RH} [N4PyFe^{III}\text{-}OH]^{2+} + R{\bullet} \qquad {\bullet}OH \xrightarrow{+\,RH} H_2O + R{\bullet}$$

1a $\xrightarrow[\text{r.t.}]{\mathbf{26},\ H_2O_2\ (50\ \text{eq.})}$ **2a** + **3a** + **4a**

		2a	3a	4a
MeCN	TON	27.8	7.0	1.3
acetone	TON	23.1	5.4	0.9

Scheme 3.32

Roelfes *et al.* prepared a non-heme iron(II) complex **26** from pentadentate ligand *N,N*-bis(2-pyridylmethyl)-*N*-bis(2-pyridyl)methylamine **25** (Scheme 3.32) [126]. In the presence of H_2O_2, complex **26** reacted to a low-spin Fe(III)OOH intermediate, which was cleaved homolytically to an oxo Fe(IV) species and a hydroxy radical. Both species are capable of oxidizing various organic substrates via a radical pathway (Scheme 3.32). Under the catalysis of complex **26**, cyclohexene **1a** was oxidized with excess H_2O_2 to a mixture of products **2a**, **3a** and **4a**. The TON was found to be solvent dependent, with higher TON in acetonitrile than in acetone (Scheme 3.32). In no case were isolated yields given and, furthermore, the allylic oxidation is limited to cyclohexene **1a**.

Finally, a biomimetic iron complex should be discussed, which serves as a model for the antitumor antibiotic bleomycin. Activated bleomycin, an oxygenated Fe–bleomycin complex, is probably responsible for the oxidative damage of DNA [127]. Starting from pentadentate ligand **27**, Nguyen *et al.* synthesized the iron(III) complex **28**, which was used to oxidize cyclohexene **1a** (Scheme 3.33) [127d]. Chemoselectivity and catalytic activity depended on the oxidant. Whereas *t*BuOOH produced only allylic oxidation products **2a** and **3a**, H_2O_2 gave an almost equimolar mixture of **2a**, **3a** and **4a** with decreased TONs. Both the TON and chemoselectivity of **3a** versus **2a** were improved by the addition of 10% water to *t*BuOOH, albeit with accompanying epoxide formation.

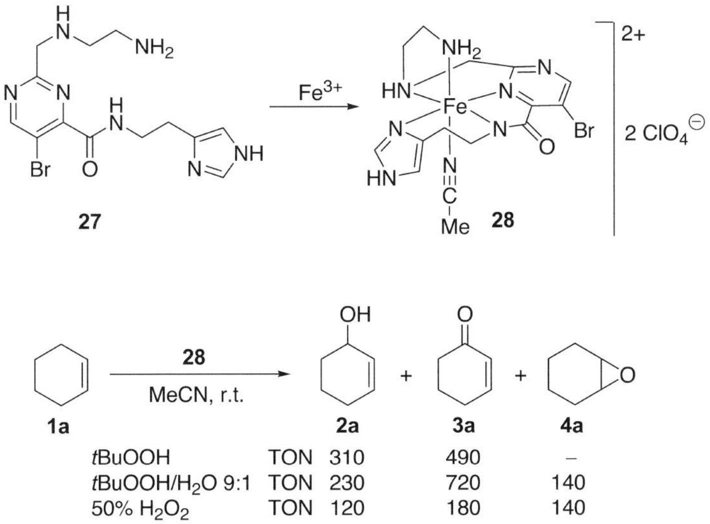

27 → (Fe^{3+}) → 28 ($2+$, $2\ ClO_4^{\ominus}$)

1a → (28, MeCN, r.t.) → 2a + 3a + 4a

		2a	3a	4a
*t*BuOOH	TON	310	490	–
*t*BuOOH/H_2O 9:1	TON	230	720	140
50% H_2O_2	TON	120	180	140

Scheme 3.33

3.2.3 Oxidative Allylic Aminations

Although allylic amines are valuable building blocks for synthetic organic chemistry, the direct catalytic insertion of a nitrogen atom into the allylic C–H bond is difficult [128]. Seminal contributions came from the groups of Jørgensen [129] and Nicholas [130]. They reported independently that iron phthalocyanines and mixtures of Fe^{2+}/Fe^{3+} salts, respectively, catalyzed the allylic amination of alkenes with *N*-phenylhydroxylamine **29a** as the nitrogen transfer reagent.

Upon screening various transition metal catalysts in the reaction of α-methylstyrene **1h** with *N*-phenylhydroxylamine **29a**, Jørgensen and coworkers obtained the best result for 2-phenyl-3-(phenylamino)propene **30a** with iron(III) phthalocyanine [Fe(Pc)]. In this case, the decomposition of **29a** to aniline, azobenzene and azoxybenzene was minimized (Scheme 3.34).

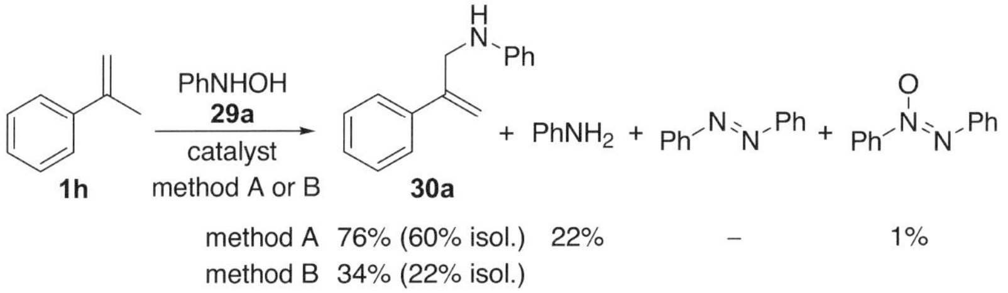

1h → (PhNHOH **29a**, catalyst, method A or B) → 30a + $PhNH_2$ + PhN=NPh + PhN(O)=NPh

	30a	$PhNH_2$	PhN=NPh	PhN(O)=NPh
method A	76% (60% isol.)	22%	–	1%
method B	34% (22% isol.)			

method A: Fe(Pc) (2 mol%), toluene, 110°C, 10 h
method B: $FeCl_2 \cdot 4\ H_2O/FeCl_3 \cdot 6\ H_2O$ 9:1 (10 mol%), dioxane, 70–100°C, 8 h

Scheme 3.34 Allylic amination of α-methylstyrene **1h** with *N*-phenylhydroxylamine **29a** according to the groups of Jørgensen (method A) [129] and Nicholas (method B) [130].

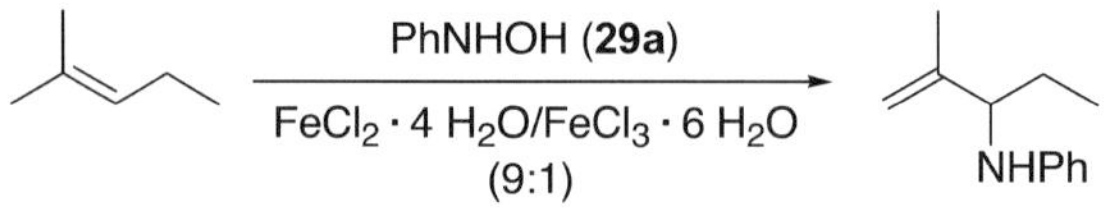

PhNHOH (**29a**)

$FeCl_2 \cdot 4\,H_2O/FeCl_3 \cdot 6\,H_2O$ (9:1)

1i → NHPh **31a** (61%, 48% isol.)

Scheme 3.35

Iron phthalocyanine catalysis is more effective for alkenes conjugated to an aryl ring as compared with the Fe^{2+}/Fe^{3+} salts (Scheme 3.34), whereas the Nicholas method worked much better for acyclic non-conjugated alkenes such as **1i** (Scheme 3.35).

Depending on the type of iron catalyst, the reaction seems to take different mechanistic pathways. According to Johannsen and Jørgensen's results, the catalytic cycle starts with the formation of nitrosobenzene **32** either by disproportionation of hydroxylamine **29a** to **32** and aniline in the presence of oxo iron(IV) phthalocyanine ($PcFe^{4+}{=}O$) or by oxidation of **29a** [131]. The second step, a hetero-ene reaction between the alkene **1** and nitrosobenzene **32**, yields the allylic hydroxylamine **33**, which is subsequently reduced by iron(II) phthalocyanine to afford the desired allylic amine **30** with regeneration of oxo iron(IV) phthalocyanine (Scheme 3.36). That means the nitrogen transfer proceeds as an "off-metal" reaction. The other byproduct, azoxybenzene, is probably formed by reaction of **29a** with nitrosobenzene **32**.

In contrast, Srivastava and Nicholas proved that in the $FeCl_3/FeCl_2$-catalyzed allylic amination an intermediate free nitrosobenzene can be excluded (Scheme 3.37) [130, 132]. For example, when a mixture of α-methylstyrene **1h** and

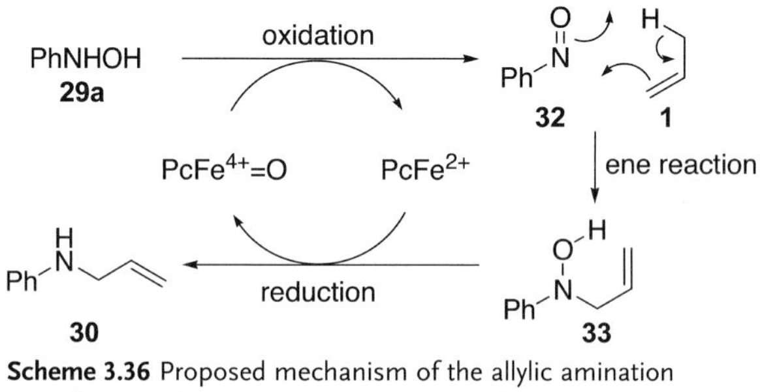

PhNHOH **29a** — oxidation → **32** + **1**

$PcFe^{4+}{=}O$ $PcFe^{2+}$

ene reaction → **33**

30 ← reduction

Scheme 3.36 Proposed mechanism of the allylic amination according to Johannsen and Jørgensen [128b, 131].

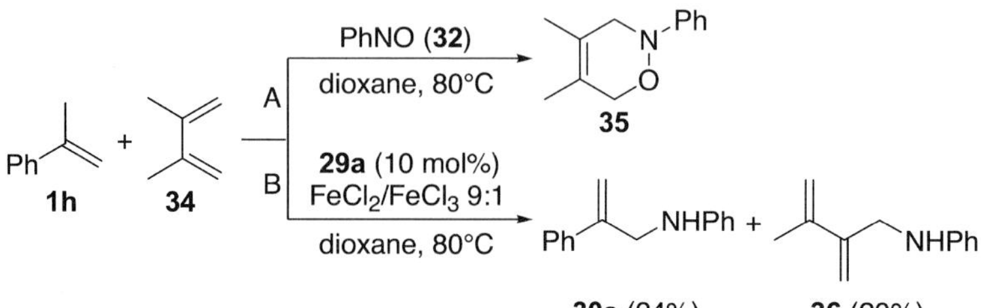

1h + **34**

A: PhNO (**32**), dioxane, 80°C → **35**

B: **29a** (10 mol%), $FeCl_2/FeCl_3$ 9:1, dioxane, 80°C → **30a** (24%) + **36** (20%)

Scheme 3.37 Trapping reaction (A) and amination (B) according to Srivastava and Nicholas [130, 132].

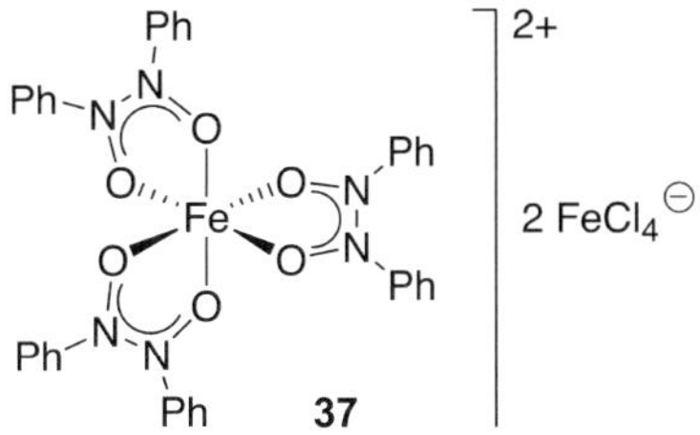

Ph–N(N–Ph)(O)... Fe ... 37, $2+$, $2\ FeCl_4^{\ominus}$

Figure 3.9 Isolated catalytically active intermediate **37** [132, 133].

2,3-dimethylbutadiene **34** was heated with nitrosobenzene **32** in dioxane, a smooth hetero-Diels–Alder reaction to derivative **35** occurred (A). However, when the same mixture was heated with hydroxylamine **29a** in the presence of $FeCl_2/FeCl_3$, only allylic amines **30a** and **36** were obtained in 24% and 20% yield, respectively. No trace of hetero-Diels–Alder product **35** was found.

Furthermore, as shown in Figure 3.9, Nicholas and coworkers isolated azodioxide complex **37** as a reactive intermediate, which could be characterized by X-ray crystal structure analysis [132, 133]. Isotope labeling studies established complex **37** to be indeed the catalytically active species.

More recently, Srivastava and Nicholas discovered the direct allylic amination with nitroaromatics **38** as effective aminating agents under catalysis by the inexpensive $[CpFe(CO)_2]_2$ (Scheme 3.38) [134].

The electronic properties of the nitroarene strongly influence the reactivity. Whereas nitrobenzene **38a** gave the amination product **30a** in high yield, electron-withdrawing substituents decreased the yield. With 4-methoxynitrobenzene **38d**, almost no product **30d** was isolated.

The use of $[Cp^*Fe(CO)_2]_2$ rather than $[CpFe(CO)_2]_2$ had a particular pronounced effect on the reaction with aliphatic alkenes, as depicted in Scheme 3.39 [135]. The yields increased significantly. Again, the reaction proceeded with high regioselectivity in favor of the less substituted carbon of the vinyl unit in starting alkene **1**.

It was originally proposed that under the reaction conditions CO reduces the nitrobenzene **38a** to nitrosobenzene **32**. However, no free nitrosobenzene could be detected by trapping experiments. Instead, the carbamoyl complex **39** was isolated and crystallized (Figure 3.10) [135]. Although complex **39** showed decreased catalytic activity in the allylic amination as compared with $[Cp^*Fe(CO)_2]_2$

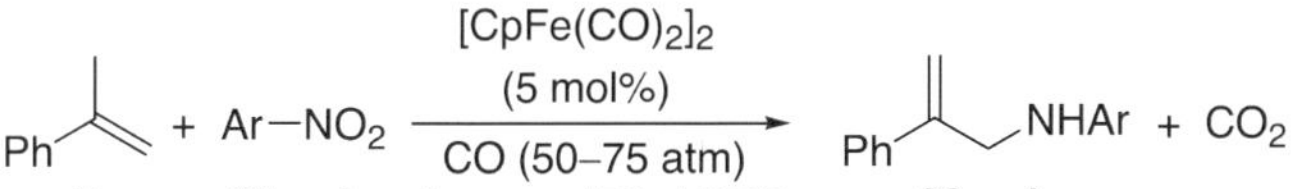

1h + $Ar{-}NO_2$ **38a–d** $\xrightarrow[\text{CO (50–75 atm), dioxane, 150–180°C}]{[CpFe(CO)_2]_2\ (5\ mol\%)}$ **30a–d** (Ph–C(=CH₂)–CH₂–NHAr) + CO_2

	a	b	c	d
Ar	C_6H_5	C_6F_5	$3\text{-}CF_3C_6H_4$	$4\text{-}MeOC_6H_4$
yield	92%	57%	52%	2%

Scheme 3.38

1j + $PhNO_2$ (**38a**), catalyst (4 mol%), CO (50 atm), dioxane, 160°C, 24 h → **31b** (NHPh)

1k → **31c** (Ph, NHPh)

catalyst	31b	31c
$[CpFe(CO)_2]_2$	10%	13%
$[Cp^*Fe(CO)_2]_2$	80%	32%

Scheme 3.39

$\{k_{rel}$ $[Cp^*Fe(CO)_2]_2 : \mathbf{39} \approx 15{:}1\}$, it seems to be a minor but significant contributor to the overall catalytic cycle (e.g. a resting state) with the 17-electron monomer $Cp^*Fe(CO)_2$ being the active catalyst [136]. Two other putative intermediates, **40** and **41**, were isolated and crystallized [137], but in contrast to complex **39** they were completely inactive.

In order to reduce the reaction temperature and CO pressure necessary for the allylic amination with nitrobenzene **38a** in the presence of $[Cp^*Fe(CO)_2]_2$, Nicholas and coworkers developed a photochemically assisted method, that requires only 3–6 atm CO and 80–120 °C to produce the allylic amines in reasonable yields [138].

The above-mentioned catalytic allylic aminations employing *N*-phenylhydroxylamine **29a** and nitroso- (**32**) or nitrobenzene **38a**, respectively, as nitrogen transfer reagent are mainly limited by the rather difficult removal of the phenyl ring from the final *N*-phenyl-*N*-allylamine. As a possible solution, Nicholas's group studied a novel aminating agent, 2,4-dinitrophenylhydroxylamine **29b**, in the $FeCl_2/FeCl_3$-catalyzed amination [139] (Scheme 3.40).

The allylic amines **30e** and **31d** were obtained in 87% and 75% yield, respectively. *N*-Alkylation and subsequent treatment with methylamine–40% water or pure butylamine gave the corresponding *N*-methyl-*N*-allylamine **42** and **43** in good overall yield.

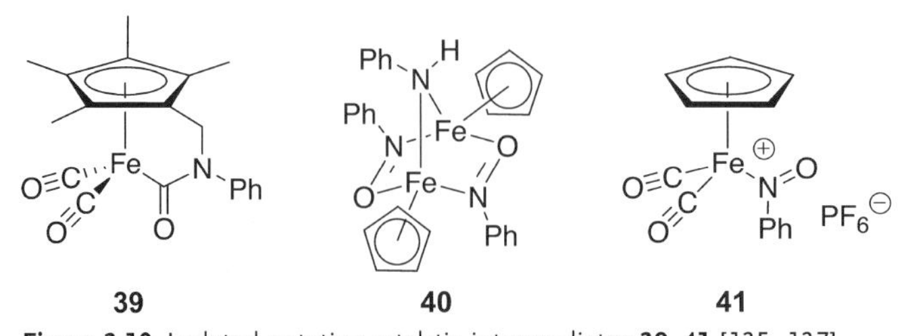

Figure 3.10 Isolated putative catalytic intermediates **39–41** [135, 137].

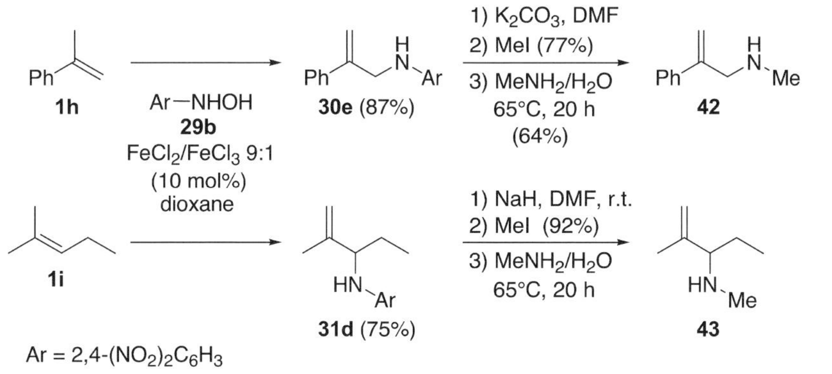

1h

Ar—NHOH
29b
$FeCl_2/FeCl_3$ 9:1
(10 mol%)
dioxane

30e (87%)

1) K_2CO_3, DMF
2) MeI (77%)
3) $MeNH_2/H_2O$
65°C, 20 h
(64%)

42

1i

31d (75%)

1) NaH, DMF, r.t.
2) MeI (92%)
3) $MeNH_2/H_2O$
65°C, 20 h

43

Ar = 2,4-$(NO_2)_2C_6H_3$

Scheme 3.40

3.2.4 Conclusion

From the discussion above, the following conclusions can be drawn. Apart from some selected examples, the issue of chemoselectivity and catalytic activity in iron-catalyzed allylic hydroxylation has not so far been solved. In particular, synthetically useful methods with a broad scope concerning alkene substrates are still lacking. Furthermore, in many cases it seems to be difficult to avoid overoxidation of the allylic alcohol to the corresponding enone. In addition, most published procedures utilize the alkene in a large excess (often as a solvent), thus limiting the use of functionalized alkenes which are not commercially available.

On the other hand, for allylic aminations very promising results have been realized with regard to regioselectivity, catalytic activity and substrate scope. However, further efforts are necessary to find nitrogen transfer reagents which allow convenient deprotection and isolation of the allylic amine.

Finally, the issue of stereoselectivity has not even been touched upon, neither in allylic hydroxylations or aminations. Therefore, much more effort is needed to develop reliable preparative methods.

References

105 C. C. Addison, N. Logan, S. C. Wallwork, C. D. Garner, *Q. Rev.* 1971, **25**, 289–322.

106 U. R. Pillai, E. Sahle-Demessie, V. V. Namboodiri, R. S. Varma, *Green Chem.* 2002, **4**, 495–497.

107 H. Jiang, Y. Xu, S. Liao, D. Yu, *React. Kinet. Catal. Lett.* 1998, **63**, 179–183.

108 I. Okamoto, W. Funaki, S. Nobuchika, M. Sawamura, E. Kotani, T. Takeya, *Chem. Pharm. Bull.* 2005, **53**, 248–252.

109 (a) W. Nam, *Acc. Chem. Res.* 2007, **40**, 522–531; (b) S. Shaik, H. Hirao, D. Kumar, *Acc. Chem. Res.* 2007, **40**, 532–542; (c) S. Shaik, S. Cohen, S. P. de Visser, P. K. Sharma, D. Kumar, S. Kozuch, F. Ogliaro, D. Danovich, *Eur. J. Inorg. Chem.* 2004, 207–226.

110 A. J. Appleton, S. Evans, J. R. Lindsay Smith, *J. Chem. Soc., Perkin Trans. 2* 1996, 281–285.

111 W. Nam, H. J. Kim, S. H. Kim, R. Y. N. Ho, J. Selverstone Valentine, *Inorg. Chem.* 1996, **35**, 1045–1049.

112 M. E. Niño, S. A. Giraldo, E. A. Páez-Mozo, *J. Mol. Catal. A* 2001, **175**, 139–151.

113 S. K. O'Shea, W. Wang, R. S. Wade, C. E. Castro, *J. Org. Chem.* 1996, **61**, 6388–6395.

114 A. Agarwala, D. Bandyopadhyay, *Chem. Commun.* 2006, 4823–4825.

115 T. Konoike, Y. Araki, Y. Kanda, *Tetrahedron Lett.* 1999, **40**, 6971–6974.

116 T. Lee, Y. O. Su, *J. Electroanal. Chem.* 1996, **414**, 69–73.

117 A. Maldotti, L. Andreotti, A. Molinari, S. Borisov, V. Vasil'ev, *Chem. Eur. J.* 2001, **7**, 3564–3571.

118 S.-I. Murahashi, X.-G. Zhou, N. Komiya, *Synlett* 2003, 321–324.

119 L. Weber, M. Grosche, H. Hennig, G. Haufe, *J. Mol. Catal.* 1993, **78**, L9–L13.

120 L. M. González, A. L. Villa de P., C. Montes de C., A. Sorokin, *Tetrahedron Lett.* 2006, **47**, 6465–6468.

121 A. Böttcher, M. W. Grinstaff, J. A. Labinger, H. B. Gray, *J. Mol. Catal. A* 1996, **113**, 191–200.

122 Reviews: (a) E. G. Kovaleva, M. B. Neibergall, S. Chakrabarty, J. D. Lipscomb, *Acc. Chem. Res.* 2007, **40**, 475–483; (b) C. Krebs, D. Galonic Fujimori, C. T. Walsh, J. M. Bollinger, *Acc. Chem. Res.* 2007, **40**, 484–492; (c) L. Que, *Acc. Chem. Res.* 2007, **40**, 493–500; (d) I. V. Korendovych, S. V. Kryatov, E. V. Rybak-Akimova, *Acc. Chem. Res.* 2007, **40**, 510–521; (e) M. Costas, M. P. Mehn, M. P. Jensen, L. Que, *Chem. Rev.* 2004, **104**, 939–986.

123 Review: E. Y. Tshuva, S. J. Lippard, *Chem. Rev.* 2004, **104**, 987–1012.

124 S. Mukerjee, A. Stassinopoulos, J. P. Caradonna, *J. Am. Chem. Soc.* 1997, **119**, 8097–8098.

125 J. M. Rowland, M. Olmstead, P. K. Mascharak, *Inorg. Chem.* 2001, **40**, 2810–2817.

126 G. Roelfes, M. Lubben, R. Hage, L. Que, B. L. Feringa, *Chem. Eur. J.* 2000, **6**, 2152–2159.

127 Reviews: (a) R. M. Burger, *Chem. Rev.* 1998, **98**, 1153–1169; (b) J. Stubbe, J. W. Kozarich, W. Wu, D. E. Vanderwall, *Acc. Chem. Res.* 1996, **29**, 322–330; (c) R. A. Marusak, C. F. Meares, *Structure Energetics and Reactivity in Chemistry Series 3 (Active Oxygen in Biochemistry)*, Chapman and Hall, Glasgow, 1995, pp. 336–400; (d) C. Nguyen, R. J. Guajardo, P. K. Mascharak, *Inorg. Chem.* 1996, **35**, 6273–6281.

128 Reviews: (a) C. Bolm, J. Legros, J. Le Paih, L. Zani, *Chem. Rev.* 2004, **104**, 6217–6254; (b) M. Johannsen, K. A. Jørgensen, *Chem. Rev.* 1998, **98**, 1689–1708; (c) Y. Tamaru, M. Kimura, *Synlett* 1997, 749–757.

129 M. Johannsen, K. A. Jørgensen, *J. Org. Chem.* 1994, **59**, 214–216.

130 R. S. Srivastava, K. M. Nicholas, *Tetrahedron Lett.* 1994, **35**, 8739–8742.

131 M. Johannsen, K. A. Jørgensen, *J. Org. Chem.* 1995, **60**, 5979–5982.

132 R. S. Srivastava, K. M. Nicholas, *J. Am. Chem. Soc.* 1997, **119**, 3302–3310.

133 R. S. Srivastava, M. A. Khan, K. M. Nicholas, *J. Am. Chem. Soc.* 1996, **118**, 3311–3312.

134 R. S. Srivastava, K. M. Nicholas, *Chem. Commun.* 1998, 2705–2706.

135 M. K. Kolel-Veetil, M. A. Khan, K. M. Nicholas, *Organometallics* 2000, **19**, 3754–3756.

136 R. S. Srivastava, *Appl. Organomet. Chem.* 2006, **20**, 851–854.

137 J. C. Stephens, M. A. Khan, K. M. Nicholas, *J. Organomet. Chem.* 2005, **690**, 4727–4733.

138 R. S. Srivastava, M. Kolel-Veetil, K. M. Nicholas, *Tetrahedron Lett.* 2002, **43**, 931–934.

139 S. Singh, K. M. Nicholas, *Synth. Commun.* 2001, **31**, 3087–3097.

3.3 Oxidation of Heteroatoms (N and S)

Olga García Mancheño and Carsten Bolm

3.3.1 Oxidation of Nitrogen Compounds

3.3.1.1 Oxidation of Hydroxylamines to Nitroso Compounds

Hydroxylamines play a significant role in modern industrial chemistry. Their most important chemical properties include differential reactivity of the N and O termini, changes in reactivity with pH and solubility in both aqueous and organic solvents. Hence the study of reactions of hydroxylamine derivatives, especially their oxidation to nitroso compounds, constitutes an important area of investigation.

The oxidation of aromatic hydroxylamines has been widely used in the preparation of nitrosobenzenes. Among the methods described in the literature for this transformation, the most common procedure involves heterogeneous oxidation using iron(III) chloride [140]. This oxidation is normally slow, which can lead to the formation of the corresponding azoxy derivatives through coupling of the formed nitroso compound with the unreacted hydroxylamine. In addition, low yields are sometimes obtained due to the partial instability of the starting hydroxylamine and/or nitroso product. Illustrative examples of this transformation are shown in Table 3.2 [141].

A related reaction is the oxidation of 2-hydroxylaminofluorene with ferric ammonium sulfate (Scheme 3.41) [142]. In this reaction, 2-nitrosofluorene, which

Table 3.2 Oxidation of aryl hydroxylamines with $FeCl_3$.

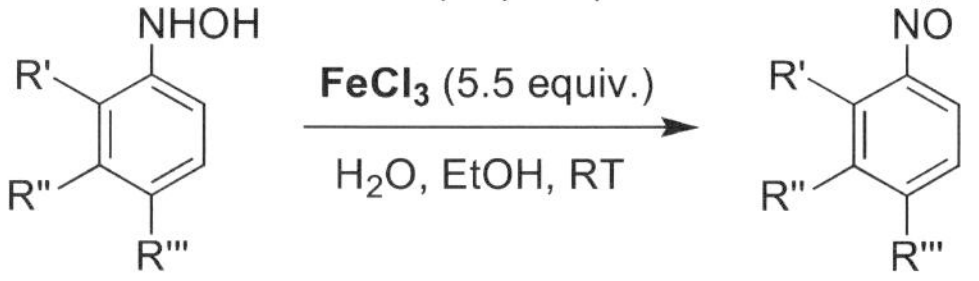

NHOH, R', R'', R''' — **$FeCl_3$ (5.5 equiv.)**, H_2O, EtOH, RT → NO, R', R'', R'''

Entry	R′	R″	R‴	Yield (%)
1	H	OCONHPh	H	93
2	H	OH	H	60
3	H	H	Cl	100
4	CN	Cl	H	36
5	H	Cl	Me	91

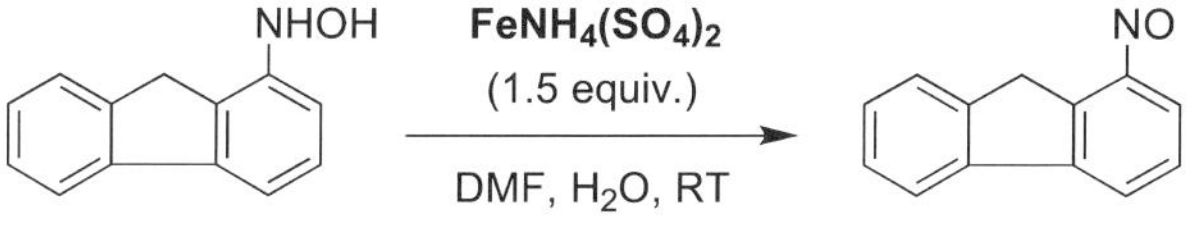

NHOH — **$FeNH_4(SO_4)_2$** (1.5 equiv.), DMF, H_2O, RT → NO

Scheme 3.41 Synthesis of 2-nitrosofluorene.

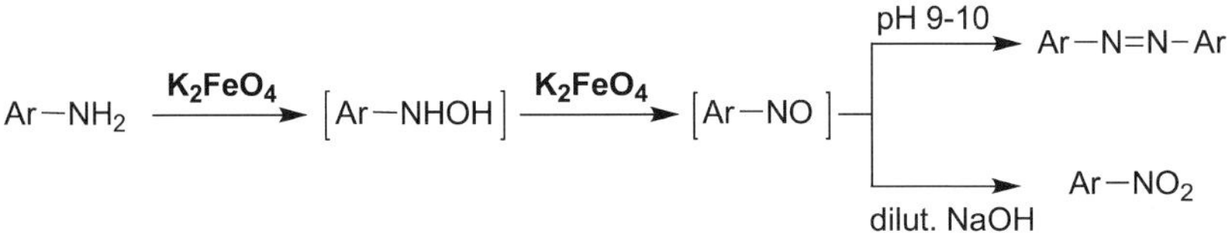

$Ar{-}NH_2 \xrightarrow{K_2FeO_4} [Ar{-}NHOH] \xrightarrow{K_2FeO_4} [Ar{-}NO]$ → (pH 9-10) $Ar{-}N{=}N{-}Ar$; → (dilut. NaOH) $Ar{-}NO_2$

Scheme 3.42 Oxidation of aniline derivatives.

belongs to a family of nitroso compounds with anti-carcinogenic activity, is generated.

3.3.1.2 Oxidation of Arylamines

The classical oxidation of the nitrogen atom of arylamines leads to nitro compounds; however, hydroxylamines and/or the corresponding azobenzenes can also be obtained.

The oxidation of aniline and its *para*-substituted analogues has been described using potassium ferrate (K_2FeO_4) in alkaline aqueous solutions [143]. The product obtained in this oxidation is highly dependent on the pH of the reaction (Scheme 3.42). Thus, when a 1 mol dm^{-3} solution of NaOH was employed, the corresponding nitro compound was isolated. On the other hand, diazobenzene was obtained at pH 9 (0.05 mol dm^{-3} sodium phosphate buffer). In both transformations, ferrate first oxidizes the amine nitrogen to form arylhydroxylamine, which is readily further oxidized to the corresponding nitrosobenzene. At high pH, nitrosobenzenes react with the excess of aniline to give azobenzenes, whereas the use of dilute aqueous NaOH solution produces exclusively nitrobenzenes.

The oxidation of aniline with potassium ferrate/K10 clay has also been described [144]. The reaction at room temperature gave diazobenzene as sole product (63%, Scheme 3.43).

3.3.1.3 Other *N*-Oxidations

Oxidation of Pyridines Fe_2O_3-catalyzed direct oxidation of pyridine by combined use of molecular oxygen (1 atm) and isovaleraldehyde in 1,2-dichloroethane gives pyridine *N*-oxide in 89% isolated yield [145]. Isovaleraldehyde was converted into the corresponding acid under the reaction conditions (Scheme 3.44).

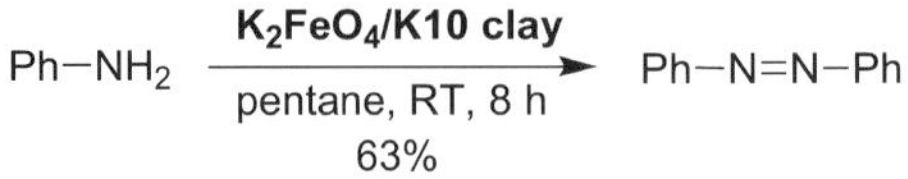

$Ph{-}NH_2 \xrightarrow[\text{pentane, RT, 8 h}]{K_2FeO_4/K10\ clay} Ph{-}N{=}N{-}Ph$ (63%)

Scheme 3.43 Potassium ferrate/K10 clay-catalyzed oxidation of aniline.

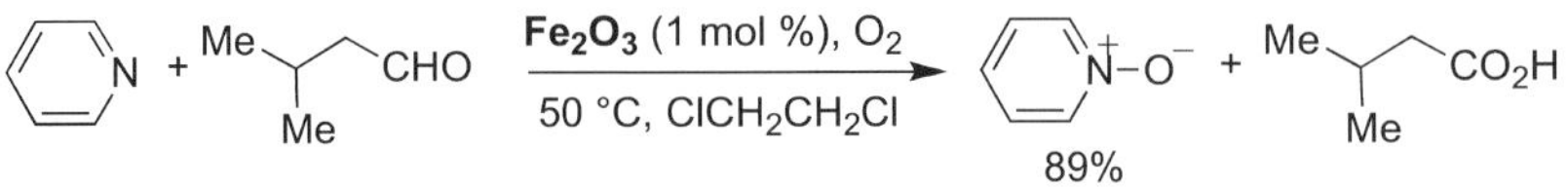

Pyridine + Me_2CHCH_2CHO $\xrightarrow[50\ °C,\ ClCH_2CH_2Cl]{Fe_2O_3\ (1\ mol\ \%),\ O_2}$ pyridine $N^+{-}O^-$ (89%) + $Me_2CHCH_2CO_2H$

Scheme 3.44 Fe_2O_3-catalyzed oxidation of pyridine to its *N*-oxide.

Clayfen

CH_2Cl_2, RT

32-35%

R = H, Me

Scheme 3.45 Example of oxime group oxidation with Clayfen.

Recently, an application of this type of oxidation reaction towards the synthesis of 2-halopyridine *N*-oxide derivatives, which are intermediates of pyrithione antimicrobial agents, has been patented by a Korean company [146]. It is reported that iron (III) oxide (10–100 mol%) provides the best results among all iron catalysts tested, such as $FeCl_3$, $Fe_2(SO_4)_3$ and $Fe(ClO_4)_3$.

Oxidation of Oximes A different type of *N*-oxidation reaction involves the direct oxidation of oximes to nitro compounds. Although a variety of oxidizing agents have been described for this reaction, the use of non-heme iron-based systems is rather limited. In this context, the oxidation of the oxime group in tetrahydro-4*H*-pyrido[1,2-α]pyrimidines was carried out at room temperature in the presence of 50 mol% of Clayfen [K10 montmorillonite-supported iron(III) nitrate] (Scheme 3.45) [147]. Under these conditions the corresponding nitro derivatives were obtained in 32–35% yield. ^{15}N mass studies revealed that the reaction involved the direct oxidation of the hydroxyimino group.

Oxidation of Thiosemicarbazoles Oxidative cyclization to thiadiazoles in good yields from thiosemicarbazones can be performed using $FeCl_3$ in benzene–water at room temperature (Scheme 3.46) [148].

3.3.2 Oxidation of Sulfur Compounds

3.3.2.1 Oxidation of Thiols to Disulfides

The conversion of thiols into disulfides (oxidative S–S coupling) is an important reaction in synthetic organic chemistry and many stoichiometric and catalytic metal reagents have been reported as oxidants for this transformation. However, there are only few reports on the oxidation of thiols using iron-based systems.

In 1966, it was found that stoichiometric amounts of Fe_2O_3 were capable of oxidizing acyclic thiols to disulfides in hydrocarbon media at 55 °C (Scheme 3.47) [149].

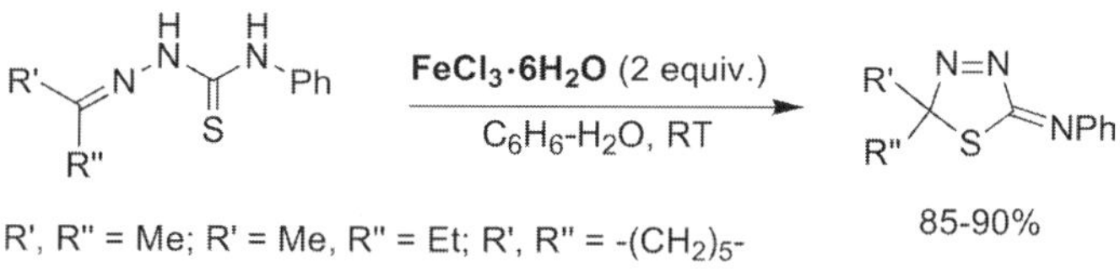

$FeCl_3 \cdot 6H_2O$ (2 equiv.)

C_6H_6-H_2O, RT

85-90%

R', R'' = Me; R' = Me, R'' = Et; R', R'' = $-(CH_2)_5-$

Scheme 3.46 Synthesis of thiadiazoles.

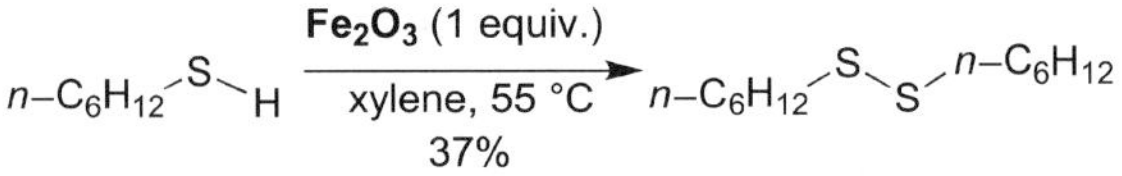

$n\text{-}C_6H_{12}\text{-}S\text{-}H \xrightarrow[\text{xylene, 55 °C}]{Fe_2O_3\text{ (1 equiv.)}} n\text{-}C_6H_{12}\text{-}S\text{-}S\text{-}n\text{-}C_6H_{12}$ 37%

Scheme 3.47 Fe_2O_3-mediated oxidation of thiols to disulfides.

Subsequently, the oxidative coupling of thiols to give disulfides using Clayfen (i. e. K10 clay-supported ferric nitrate) was described by Laszlo's group [150]. It was proposed that the reaction occurred via a thionitrite intermediate (RSNO). This method does not require the use of gaseous nitrogen oxides and uses mild reaction conditions (hydrocarbon solutions at room temperature) to obtain the desired symmetrical disulfides in good yields ($\leq$97%). The same group also studied the performance of potassium ferrate (K_2FeO_4) as an oxidant toward thiols in the presence of K10 clay [144]. Whereas the oxidation of thiophenol was inferior to that with the use of Clayfen, with deactivated substrates such as *tert*-butyl or isopropyl thiols, K_2FeO_4 improved the outcome of the reaction significantly (Table 3.3).

The oxidative coupling of thiols catalyzed by Fe^{III}-exchanged montmorillonite in phosphate buffer (pH 7.2) has also been demonstrated (Scheme 3.48) [151]. This system gives the corresponding disulfide as the sole product, which is in contrast to a previous report [152], where oxidation of thiols catalyzed by ion-exchanged clay gave the sulfide as the major product.

Joshi and coworkers reported the first catalytic oxidation of thiols using an iron(III) complex (Table 3.4) [153]. Fe^{III}-EDTA proved to be an excellent catalyst for the oxidation of a variety of aromatic, aliphatic and heterocyclic thiols under alkaline conditions using molecular oxygen as the primary oxidant. Aromatic thiols were found to be more reactive than aliphatic thiols, for which the reactivity was strongly dependent upon the pH of the medium.

Table 3.3 Synthesis of disulfides using Clayfen-type oxidants.

$R\text{-}S\text{-}H \xrightarrow[\text{RT}]{\text{Fe-clay}} R\text{-}S\text{-}S\text{-}R$

Entry	Thiol	Yield (%)	
		Clayfen	K_2FeO_4/K10 clay
1	PhSH	97	85
2	*t*-BuSH	39	70
3	*i*-PrSH	–	75

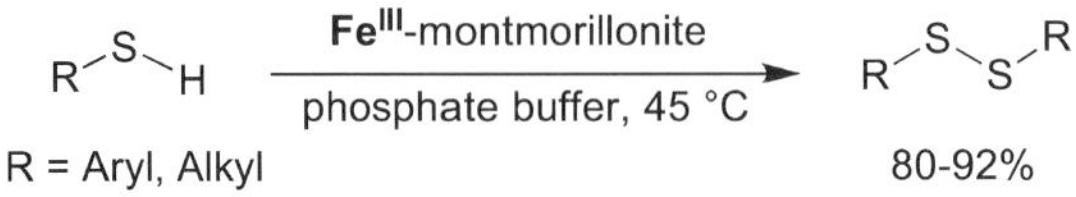

$R\text{-}S\text{-}H \xrightarrow[\text{phosphate buffer, 45 °C}]{Fe^{III}\text{-montmorillonite}} R\text{-}S\text{-}S\text{-}R$

R = Aryl, Alkyl; 80-92%

Scheme 3.48 Oxidative coupling of thiols catalyzed by Fe^{III}-montmorillonite.

Table 3.4 Fe^{III}-EDTA-catalyzed oxidation of thiols with molecular oxygen.

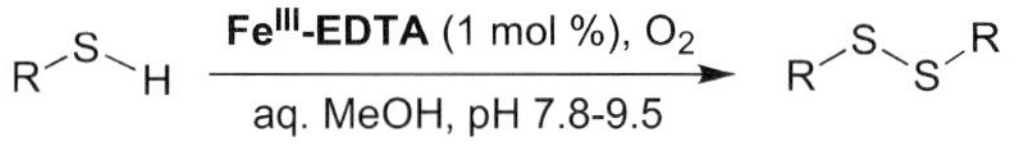

R–S–H → (Fe^{III}-EDTA (1 mol %), O_2; aq. MeOH, pH 7.8-9.5) → R–S–S–R

Entry	R	pH	Yield (%)
1	Ph	7.8	96
2	p-MeO(C_6H_4)	7.8	87
3	2-Py	8.0	84
4	Bn	8.5	98
5	n-Bu	9.5	93
6	n-C_7H_{15}	9.5	87

The oxidative dimerization of thiols was catalyzed by anhydrous $FeCl_3$ and sodium iodide under air at room temperature in acetonitrile (Table 3.5) [154]. The corresponding disulfides were formed in excellent yields (96–99%).

Moreover, the formation of 1,2-dithiacylopentane from 1,3-propanedithiol was achieved in 55% yield under high dilution conditions (Scheme 3.49).

Remarkably, although oxidation of thiols has been more intensively studied, reports on the selective coupling of dithiols are rare due to facile competitive polymerization reactions [155].

3.3.2.2 **Oxidation of Sulfides**

Sulfoxides can be considered as oxidized sulfides and are interesting molecules in organic chemistry. The sulfur atom in sulfoxides presents a lone pair of electrons,

Table 3.5 $FeCl_3$/NaI-catalyzed dimerization of thiols.

R–S–H → ($FeCl_3$ (10 mol %), air, NaI (20 mol %); MeCN, RT) → R–S–S–R

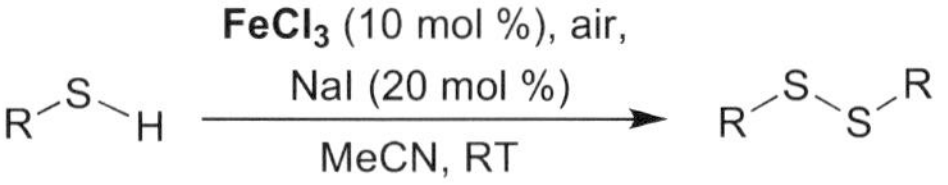

Entry	R	Yield (%)
1	Ph	97
2	Bn	99
3	Cy	98
4	n-Bu	97
5	2-Fur	96

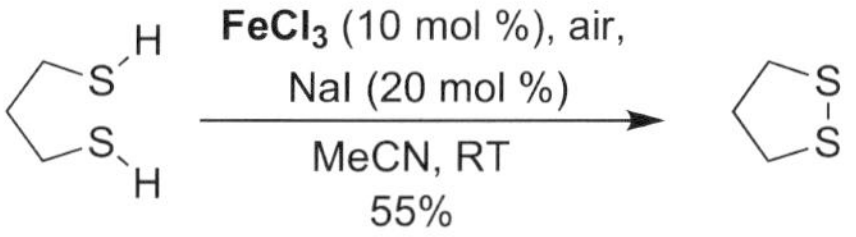

HS–(CH₂)₃–SH → ($FeCl_3$ (10 mol %), air, NaI (20 mol %); MeCN, RT; 55%) → 1,2-dithiolane

Scheme 3.49 Oxidation of dithiols.

Table 3.6 $FeBr_3$-based catalytic oxidation of sulfides with nitric acid.

$R'-S-R''$ —[**Fe cat.** (10 mol %); 1.5 M HNO_3, CH_2Cl_2, RT]→ $R'-S(=O)-R''$

Entry	Sulfide	Yield (%)	
		$FeBr_3$	$(FeBr_3)_2(DMSO)_3$
1	*n*-Bu–S–*n*-Bu	96	99
2	Ph–S–Me	65	81
3	Bn–S–Ph	93	91

providing it with a tetrahedral molecular geometry. Thus, when its two organic residues are dissimilar, the sulfur is a chiral center. Chiral sulfoxides have found application in certain drugs [156] and they are also employed as chiral auxiliaries or ligands. Selective oxidations of sulfides to sulfoxides with iron catalysts have been reported, offering alternatives to existing methods.

Non-asymmetric Oxidations Suárez and coworkers described the oxidation of sulfides using $FeBr_3$ or $(FeBr_3)_2(DMSO)_3$ (10 mol%) and nitric acid as oxidant (Table 3.6) [157]. Both iron catalysts are able to provide sulfoxides in good yields (65–99%).

Oxidation of sulfides to sulfoxides and sulfones was achieved in moderate to high yields with good selectivity by using 1 mol% of Fe_2O_3 as catalyst with molecular oxygen in the presence of isovaleraldehyde (Table 3.7) [145].

Firouzabadi *et al.* described a solvent-free oxidation of sulfides using stoichiometric amounts of $Fe(NO_3)_3{\cdot}9H_2O$. This hydrated iron(III) salt is able to oxidized aryl and alkyl sulfides efficiently at room temperature leading to their corresponding sulfoxides in good yields (92–97%) (Scheme 3.50) [158]. The reaction also proceeds in refluxing ethyl acetate; however, lower yields of the desired sulfoxides are usually obtained.

Table 3.7 Fe_2O_3–O_2–isovaleraldehyde catalytic system.

$R'-S-R''$ —[**Fe_2O_3** (1 mol %), O_2, RCHO (3 equiv.); 40 °C, $ClCH_2CH_2Cl$]→ $R'-S(=O)-R''$ + $R'-S(=O)_2-R''$

Entry	Sulfide	Yield (%)	
		Sulfoxide	Sulfone
1	Et–S–Et	75	10
2	Ph–S–Me	81	8
3	Ph–S–Ph	65	Trace

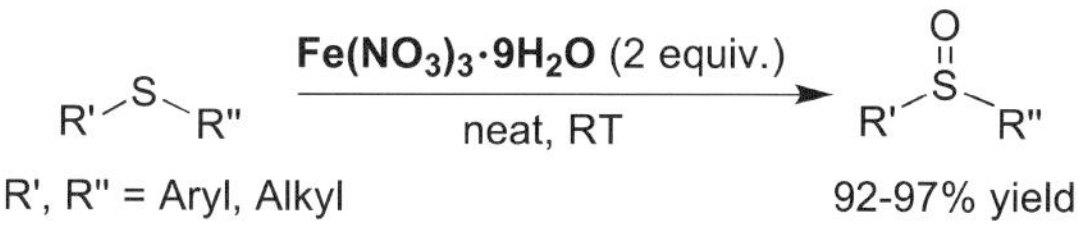

R'–S–R'' → (**$Fe(NO_3)_3 \cdot 9H_2O$** (2 equiv.), neat, RT) → R'–S(=O)–R''

R', R'' = Aryl, Alkyl — 92-97% yield

Scheme 3.50 Solvent-free oxidations in the presence of $Fe(NO_3)_3 \cdot 9H_2O$.

Table 3.8 $Fe(NO_3)_3 \cdot 9H_2O$–$FeBr_3$ catalytic system.

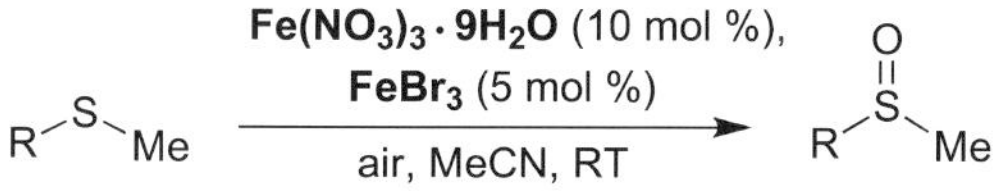

R–S–Me → (**$Fe(NO_3)_3 \cdot 9H_2O$** (10 mol %), **$FeBr_3$** (5 mol %), air, MeCN, RT) → R–S(=O)–Me

Entry	R	Yield (%)
1	Ph	98
2	*p*-CN(C_6H_4)	99
3	*p*-NO_2(C_6H_4)	98
4	*t*-Bu	89

Another successful iron-catalyzed reaction is sulfoxidation, consisting of the use of the binary catalyst $Fe(NO_3)_3 \cdot 9H_2O$–$FeBr_3$. This system was able to catalyze efficiently the oxidation of sulfides at room temperature in MeCN under air (Table 3.8) [159].

In recent years, hypervalent iodines have been employed to oxidize various organic substrates. H_5IO_6 is also capable of oxidizing sulfides to sulfoxides in pyridine [160]. However, it has been established that the oxidation of sulfides in the presence of catalytic amounts of $FeCl_3$ (3 mol%) takes place in shorter reaction times than that without the catalyst, indicating the catalytic effect of this iron salt (Scheme 3.51) [161].

Finally, iron catalysts based on salen-type ligands have been used. These iron(III)–salen complexes were regarded as enzyme models, using PhIO as oxidant (Scheme 3.52) [162]. Initially, the corresponding active iron–oxo complexes were formed by reaction with PhIO and isolated before use. A stoichiometric amount of the iron–oxo complex allowed the efficient oxidation of a variety of aryl methyl sulfides in moderate to good yields.

Asymmetric Oxidations Catalytic asymmetric oxidation of sulfides has attracted great interest in recent decades. The field is dominated by use of titanium, manganese and vanadium complexes, and examples of the use of iron catalysts are less common. The challenging asymmetric oxidation of sulfides with non-heme iron catalysts has been achieved with success in a few cases.

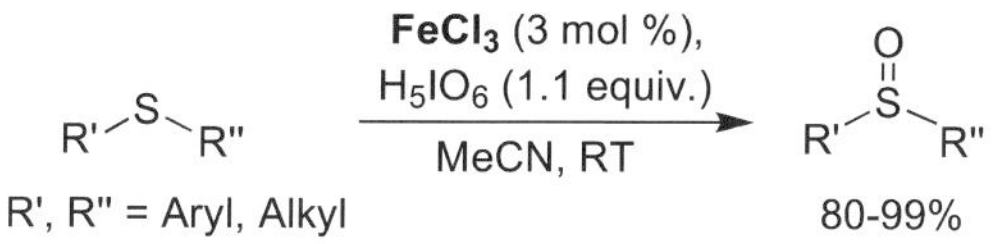

R'–S–R'' → (**$FeCl_3$** (3 mol %), H_5IO_6 (1.1 equiv.), MeCN, RT) → R'–S(=O)–R''

R', R'' = Aryl, Alkyl — 80-99%

Scheme 3.51 $FeCl_3/H_5IO_6$ oxidations.

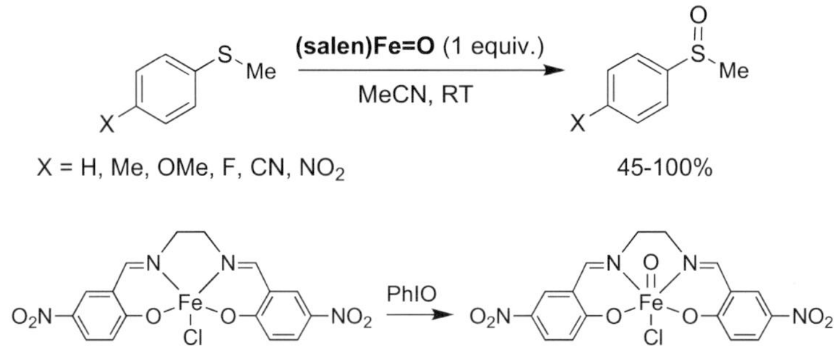

X = H, Me, OMe, F, CN, NO_2 — (salen)Fe=O (1 equiv.), MeCN, RT — 45-100%

PhIO

Scheme 3.52 Use of oxo-iron(salen) oxidants.

The first example of iron-catalyzed asymmetric oxidation of sulfides was described by Fontecave and coworkers in 1997 [163]. An oxo-bridged diiron complex, which contained (−)-4,5-pinenebipyridine as chiral ligand, was reported to catalyze sulfide oxidations with H_2O_2 in acetonitrile, having the potential to transfer an oxygen atom directly to the substrates. However, the enantioselectivity of this process remained rather low (≤40% *ee*, Scheme 3.53).

The major breakthrough in this field was achieved in 2003 by Legros and Bolm [164], who reported a highly enantioselective iron-catalyzed asymmetric sulfide oxidation. Optically active sulfoxides were obtained with up to 96% *ee* in good yields under very simple reaction conditions using $Fe(acac)_3$ as precatalyst in combination with a Schiff base-type ligand (Table 3.9). Furthermore, inexpensive and safe 35% aqueous hydrogen peroxide served as terminal oxidant.

The best results were obtained in the oxidation of aryl methyl sulfides (*ee* >86%, entries 1–7). Moreover, good enantioselectivities have also been achieved with more challenging substrates possessing alkyl groups such as ethyl, benzyl and allyl, which gave the corresponding sulfoxides with 82, 79 and 71% *ee*, respectively (entries 8–10). The presence of 1 mol% of lithium 4-methoxybenzoate (AX) drastically improved the effectiveness of this reaction [164b]. Without this additive, sulfoxides were obtained in low yields (15–44%) and inferior enantioselectivities (26–78% *ee*) [164a]. Based on the observed asymmetric amplification, the intervention of a diiron species with a bridging monocarboxylate has been suggested [164c]. However, to date no evidence on the nature of the catalytic species has been obtained.

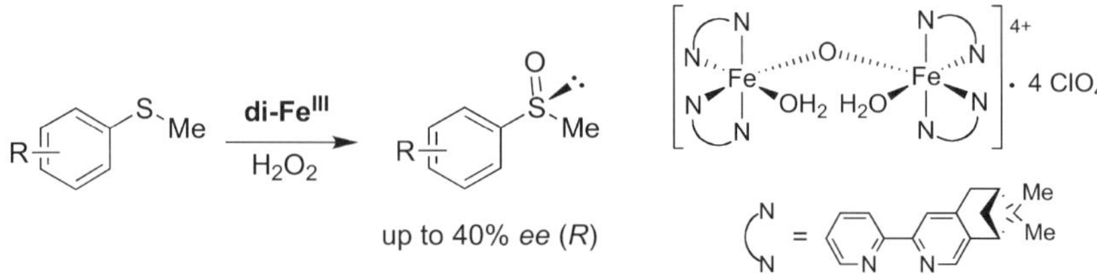

di-Fe^{III}, H_2O_2 — up to 40% *ee* (*R*) — $\cdot\ 4\ ClO_4^-$

Scheme 3.53 Oxo-bridged diiron complex for sulfide oxidations.

Table 3.9 Fe(acac)$_3$–Schiff base ligand-catalyzed asymmetric oxidations.

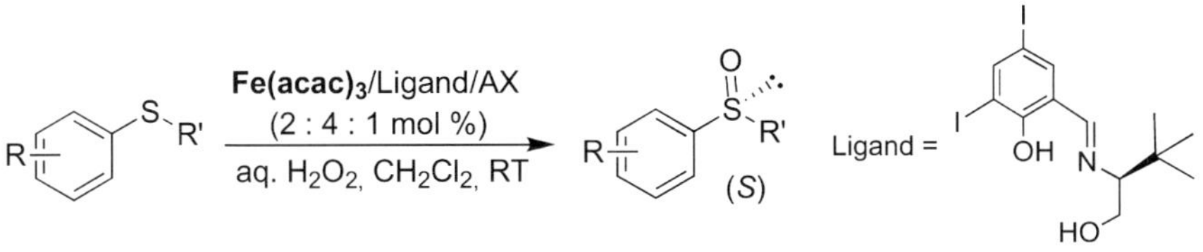

Entry	Sulfide	Yield (%)	*ee* (%)
1	Ph–S–Me	63	90
2	*p*-Tol–S–Me	78	92
3	*p*-MeO(C_6H_4)–S–Me	66	86
4	*p*-Br(C_6H_4)–S–Me	59	94
5	*p*-Cl(C_6H_4)–S–Me	60	92
6	*p*-NO_2(C_6H_4)–S–Me	36	96
7	2-Naph–S–Me	67	95
8	Ph–S–Et	56	82
9	Ph–S–Bn	73	79
10	Ph–S–$CH_2CH{=}CH_2$	63	71

This methodology has been successfully applied to the asymmetric synthesis of (*R*)- and (*S*)-sulindac (90–92% *ee*), which is a chiral sulfoxide marketed as an anti-inflammatory drug (Scheme 3.54) [165].

Shortly thereafter, Bryliakov and Talsi described chiral [iron(salen)Cl] catalysts for the asymmetric sulfide oxidation using PhIO as terminal oxidant (Table 3.10) [166]. Whereas good selectivities in the formation of sulfoxides *vs.* sulfones were achieved, only poor to moderate enantiomeric excesses were obtained with this system (22–62% *ee*).

In addition, these iron(salen) catalysts can oxidize sulfides using other oxidants such as H_2O_2, TBHP or *m*-CPBA with notable chemical selectivity (75–98%); however, no enantioselectivity was observed.

Non-heme Fe^{II} complexes of pentadentate ligands with pyrrolidinyl moieties have also been described for the oxidation of aryl methyl sulfides (Table 3.11) [167]. This system showed high selectivity in the formation of sulfoxides, but provided low

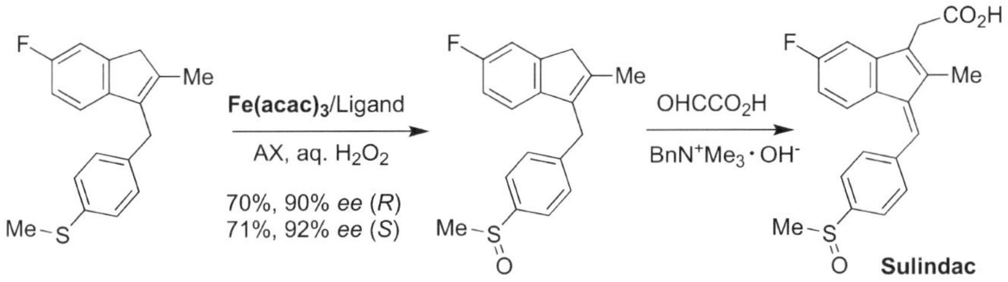

Scheme 3.54 Synthesis of sulindac.

Table 3.10 Chiral Fe(salen)Cl catalysts.

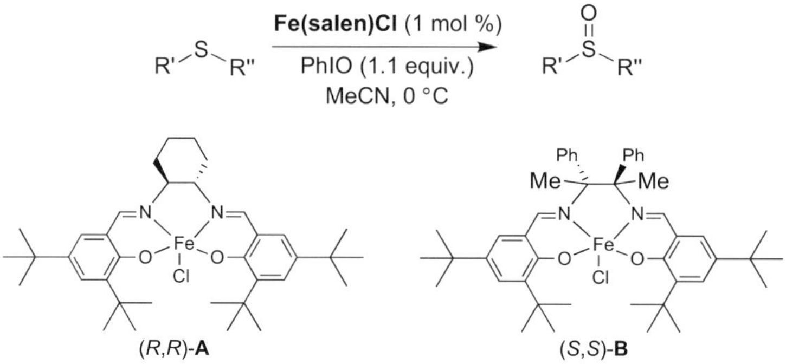

Entry	Sulfide	Fe(salen)Cl	Conversion (%)	*ee* (%)	Configuration
1	Ph–S–Me	(*R*,*R*)-**A**	96	22	*S*
2	Bn–S–Ph	(*R*,*R*)-**A**	95	62	*S*
3	Bn–S–Ph	(*S*,*S*)-**B**	91	62	*R*

enantioselectivities, even when 4-methoxybenzoic acid (AH) was used as an additive ($\leq$27% *ee*).

Recently, Egami and Katsuki developed an effective Fe(salan)-catalyzed oxidation of sulfides with H_2O_2 in water (Table 3.12) [168]. The reaction can be carried out using low catalyst loadings (1 mol%) in air and at room temperature (20 °C). High yields and enantioselectivities (81–96% *ee*) were achieved not only for alkyl aryl sulfoxides but also for alkyl alkyl sulfoxides.

Table 3.11 Pyrrolidinyl-Fe^{II} catalysts for sulfide oxidations.

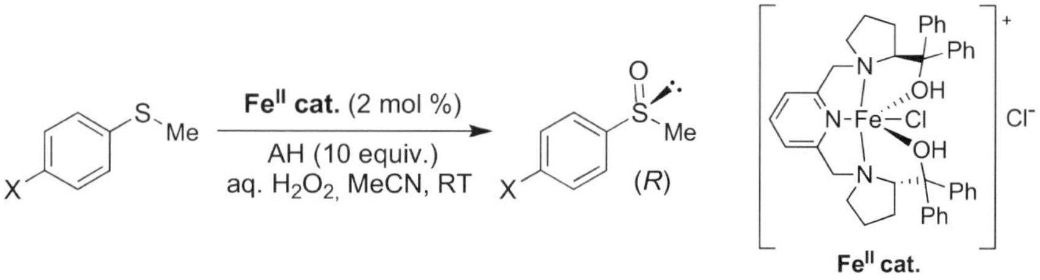

Entry	X	Yield (%)	*ee* (%)
1	H	85	27
2	Me	93	26
3	MeO	91	25
4	NO_2	62	11
5	Br	89	19

Table 3.12 Fe(salan)Cl-catalyzed sulfide oxidations in water.

Fe(salan)Cl (1 mol %), aq. H_2O_2 (1.5 equiv.), H_2O, RT; R'–S–R'' → R'–S(O)–R'' (S)

Fe(salan)Cl

Entry	R′	R″	Yield (%)	*ee* (%)
1	*p*-Tol	Me	88	96
2	*p*-MeO(C_6H_4)	Me	88	95
3	*p*-Cl(C_6H_4)	Me	72	94
4	*o*-Cl(C_6H_4)	Me	86	96
5	*o*-MeO(C_6H_4)	Me	77	93
6	Ph	Et	73	81
7	Bn	Me	85	87
8	*c*-C_6H_{11}	Me	73	88
9	*n*-C_8H_{17}	Me	73	89
10	*n*-$C_{12}H_{25}$	Me	79	94

3.3.2.3 **Oxidative Imination of Sulfur Compounds**

The oxidative imination of sulfides and sulfoxides via nitrene transfer processes leads to *N*-substituted sulfilimines and sulfoximines. This reaction is interesting as chiral sulfoximines are efficient chiral auxiliaries in asymmetric synthesis, a promising class of chiral ligands for asymmetric catalysis and key intermediates in the synthesis of pseudopeptides [169]. However, very few examples of such iron-catalyzed transformations have been described.

Bach and Korber reported in 1998 that iron(II) chloride could be used in combination with *tert*-butyloxycarbonyl azide ($BocN_3$) for nitrene transfer to sulfides and sulfoxides [170]. Whereas usually 1 equiv. of $FeCl_2$ was employed in the imination of sulfoxides, catalytic amounts of this iron salt (25 mol%) could also be applied (Table 3.13).

The reaction proceeds stereospecifically, giving sulfoximines from the corresponding enantiomerically enriched sulfoxides without racemization. The NH-sulfoximines can then be obtained after Boc cleavage with trifluoroacetic acid. Bolm *et al.* applied this reaction in the synthesis of sulfoximines having a benchrotene skeleton (Scheme 3.55) [171].

The imination of sulfides with this catalytic system has been proved to proceed more readily than for the corresponding sulfoxides [169]. Thus, sulfilimines can be obtained in moderate to good yields, especially when acetylacetone or DMF is added (Table 3.14).

Table 3.13 $FeCl_2$-catalyzed sulfoximidations with $BocN_3$.

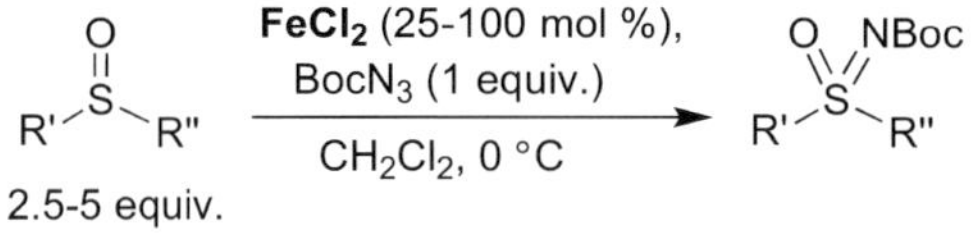

R'–S(=O)–R'' (2.5-5 equiv.) → [**FeCl₂** (25-100 mol %), BocN₃ (1 equiv.), CH₂Cl₂, 0 °C] → R'–S(=O)(=NBoc)–R''

Entry	R′	R″	$FeCl_2$ (mol%)	Yield (%)
1	Ph	Me	100	74
2	Bn	Me	100	70
3	Bn	Me	25	56
4	Bn	Et	100	95
5	*i*-Pr	Me	100	54
6	Me	Me	100	58

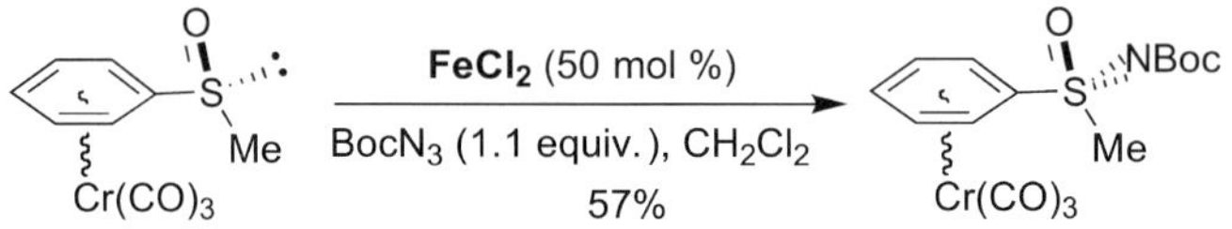

FeCl₂ (50 mol %), BocN₃ (1.1 equiv.), CH₂Cl₂, 57%; Cr(CO)₃, Me, O, NBoc

Scheme 3.55 Synthesis of benchrotene sulfoximines.

As a direct application of this transformation, the combination of $FeCl_2$ and $BocN_3$ was subsequently used for the imination of allyl sulfides, which undergo a [2,3]-sigmatropic rearrangement in the reaction media (Scheme 3.56) [172].

Recently, Van Vranken's group demonstrated that propargyl sulfides are also able to undergo this transformation upon treatment with (dppe)$FeCl_2$, affording *N*-allenylsulfenimides (Scheme 3.57) [173].

Bolm and coworkers discovered that inexpensive and easy to handle iron(III) acetylacetonate [Fe(acac)$_3$] was also capable of catalyzing nitrene transfer to sulfides and sulfoxides using various sulfonyl amides in combination with iodosylbenzene (Scheme 3.58) [174].

Table 3.14 $FeCl_2$-catalyzed iminations of sulfides.

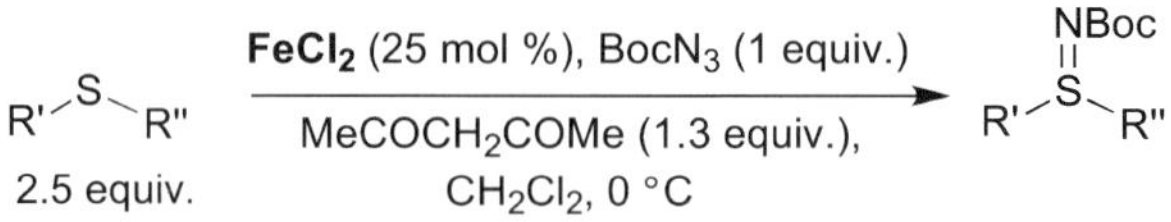

R'–S–R'' (2.5 equiv.) → [**FeCl₂** (25 mol %), BocN₃ (1 equiv.), MeCOCH₂COMe (1.3 equiv.), CH₂Cl₂, 0 °C] → R'–S(=NBoc)–R''

Entry	Sulfide	Yield (%)
1	Ph–S–Me	90
2	Bn–S–Me	61
3	Bn–S–Ph	57
4	Bn–S–Et	67
5	*i*-Pr–S–Me	36

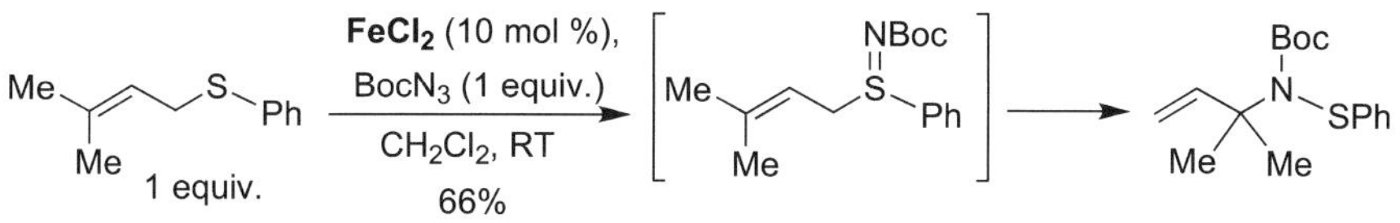

Scheme 3.56 Allyl sulfides imination/sigmatropic rearrangement sequence.

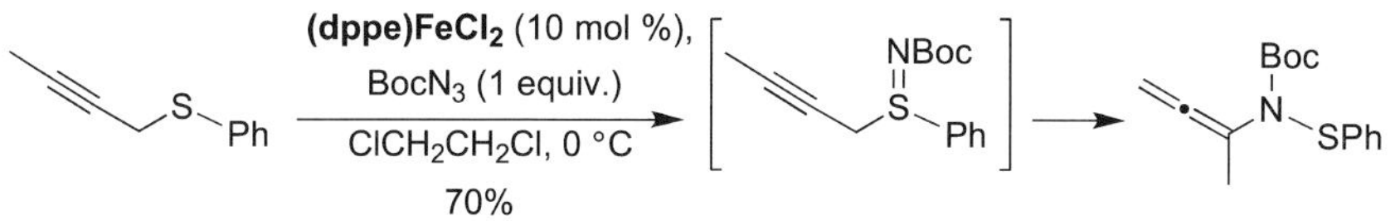

Scheme 3.57 Allenyl sulfides imination/sigmatropic rearrangement sequence.

This catalytic system proved to be highly efficient. Thus, a low catalytic amount of $Fe(acac)_3$ (5–10 mol%) permits the successful imination of a broad variety of substituted sulfides and sulfoxides. The best results were obtained with nosyl sulfonamide as nitrogen source (Figure 3.11). Moreover, as previously observed with $FeCl_2$, this catalyst also iminates sulfides more readily than sulfoxides.

The reaction proceeds in a stereospecific manner, with retention of configuration at sulfur. It thereby constitutes an alternative access to enantiopure sulfoximines from the corresponding optically active sulfoxides. The deprotection of the *N*-nosyl

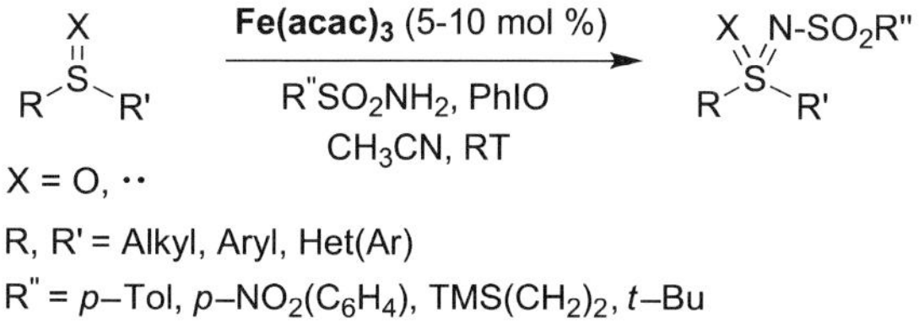

Scheme 3.58 $Fe(acac)_3$-catalyzed iminations of sulfides and sulfoxides.

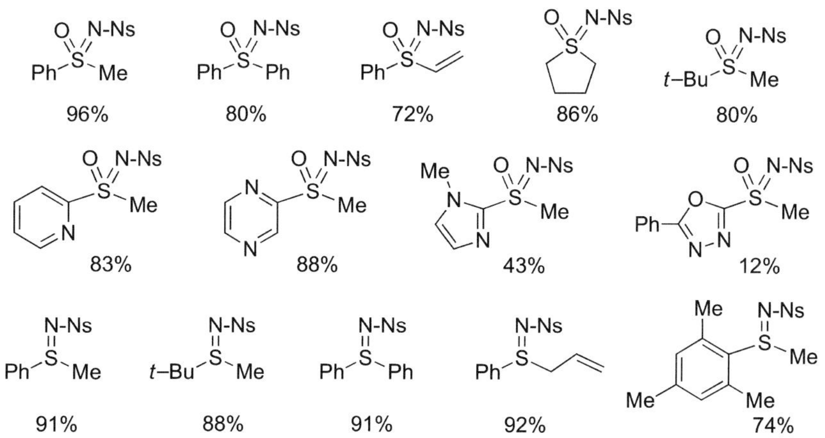

Figure 3.11 Representative examples of $Fe(acac)_3$-catalyzed imination products.

products under standard reaction conditions gives, without epimerization, synthetically valuable NH-sulfoximines in good yields [175].

In analogy with the corresponding known iron–oxo complexes, an iron nitrene complex ($L_nFe{=}NR$) has been proposed as a reactive intermediate in these processes for both catalytic systems described above [176].

References

140 For a recent review on the synthesis of nitroso compounds, see B. G. Gowenlock, G. B. Richter-Addo, *Chem. Rev.* 2004, **104**, 3315–3340, and references therein.

141 I. D. Entwistle, T. Gilkerson, R. A. W. Johnstone, R. P. Telford, *Tetrahedron* 1978, **34**, 213–215.

142 Y. Yost, *J. Med. Chem.* 1969, **12**, 961–961.

143 M. D. Johnson, B. J. Hornstein, *J. Chem. Soc., Chem. Commun.* 1996, 965–966.

144 L. Delaude, P. Laszlo, *J. Org. Chem.* 1996, **61**, 6360–6370.

145 G. Song, F. Wang, H. Zhang, X. Lu, C. Wang, *Synth. Commun.* 1998, **28**, 2783–2787.

146 H. S. Chae, C. B. Jung, G. S. Kim, *Korean Patent* 2002 *KR 2002051353 A.*

147 M. Balogh, P. Pennetreau, I. Hermecz, A. Gerstmans, *J. Org. Chem.* 1990, **55**, 6198–6202.

148 I. Yamamoto, I. Abe, M. Nozawa, M. Konati, J. Motoyoshiya, H. Gotoh, K. Matsuzaki, *J. Chem. Soc., Perkin Trans. I.* 1983, 2297–2301.

149 T. J. Wallece, *J. Org. Chem.* 1966, **31**, 1217–1221.

150 A. Cornélis, N. Depaye, A. Gerstmans, P. Laszlo, *Tetrahedron Lett.* 1983, **24**, 3103–3106.

151 H. M. Meshram, R. Kache, *Synth. Commun.* 1997, **27**, 2403–2406.

152 J. A. Ballantine, R. P. Galvin, R. M. O'Neil, H. Purnell, M. Rayanakorn, J. M. Thomas, *J. Chem. Soc., Chem. Commun.* 1981, 695–696.

153 T. V. Rao, B. Sain, P. S. Murthy, T. S. R. P. Rao, A. K. Jain, G. C. Joshi, *J. Chem. Res. (S)* 1997, 300–301.

154 N. Iranpoor, B. Zeynizadeh, *Synthesis* 1999, 49–50.

155 M. A. Walters, J. Chaparro, T. Siddiqui, F. Williams, C. Ulku, A. L. Rheingold, *Inorg. Chim. Acta* 2006, **359**, 3996–4000.

156 J. Legros, J. R. Dehli, C. Bolm, *Adv. Synth. Catal.* 2005, **347**, 19–31.

157 (a) A. R. Suárez, L. I. Rossi, S. E. Martín, *Tetrahedron Lett.* 1995, **36**, 1201–1204; (b) A. R. Suárez, A. M. Baruzzi, L. I. Rossi, *J. Org. Chem.* 1998, **63**, 5689–5691.

158 H. Firouzabadi, N. Iranpoor, M. A. Zolfigol, *Synth. Commun.* 1998, **28**, 1179–1187.

159 S. E. Martín, L. I. Rossi, *Tetrahedron Lett.* 2001, **42**, 7147–7151.

160 D. H. R. Barton, W. Li, J. A. Smith, *Tetrahedron Lett.* 1998, **39**, 7055–7058.

161 S. S. Kim, K. Nehru, S. S. Kim, D. W. Kim, H. C. Jung, *Synthesis* 2002, 2484–2486.

162 V. K. Sivasubramanian, M. Ganesan, S. Rajagopal, R. Ramaraj, *J. Org. Chem.* 2002, **67**, 1506–1514.

163 (a) C. Duboc-Toia, S. Ménage, C. Lambeaux, M. Fontecave, *Tetrahedron Lett.* 1997, **38**, 3727–3730; (b) C. Duboc-Toia, S. Ménage, R. Y. N. Ho, L. Que, Jr., C. Lambeaux, M. Fontecave, *Inorg. Chem.* 1999, **38**, 1261–1268; (c) Y. Mekmouche, H. Hummel, R. Y. N. Ho, L. Que, Jr., V. Schünemann, F. Thomas, A. X. Trautwein, C. Lebrun, K. Gorgy, J.-C. Leprêtre, M.-N. Collomb, A. Deronzier, M. Fontecave, S. Ménage, *Chem. Eur. J.* 2002, **8**, 1196–1204.

164 (a) J. Legros, C. Bolm, *Angew. Chem.* 2003, **115**, 5645–5647; *Angew. Chem. Int. Ed.* **42**, 2003, 5487–5489; (b) J. Legros, C. Bolm, *Angew. Chem.* 2004, **116**,

4321–4324; *Angew. Chem. Int. Ed.* 2004, **43**, 4225–4228; (c) J. Legros, C. Bolm, *Chem. Eur. J.* 2005, **11**, 1086–1092.

165 A. Korte, J. Legros, C. Bolm, *Synlett* 2004, 2397–2399.

166 K. P. Bryliakov, E. P. Talsi, *Angew. Chem.* 2004, **116**, 5340–5342; *Angew. Chem. Int. Ed.* 2004, **43**, 5228–5230;

167 S. Gosiewska, M. Lutz, A. L. Spek, R. J. M. K. Gebbink, *Inorg. Chim. Acta* 2007, **360**, 405–417.

168 H. Egami, T. Katsuki, *J. Am. Chem. Soc.* 2007, **129**, 8940–8941.

169 For recent reviews concerning sulfoximines, see (a) M. Reggelin, C. Zur, *Synthesis* 2000, 1–64; (b) M. Harmata, *Chemtracts* 2003, **16**, 660–666; (c) H. Okamura, C. Bolm, *Chem. Lett.* 2004, **33**, 482–487; (d) R. Bentley, *Chem. Soc. Rev.* 2005, **34**, 609–624; (e) C. Bolm,in *Asymmetric Synthesis with Chemical and Biological Methods*, ed. D. Enders, K.-E. Jäger, Wiley-VCH, Weinheim, 2007, pp. 149–176.

170 (a) T. Bach, C. Körber, *Tetrahedron Lett.* 1998, **39**, 5015–5016; (b) T. Bach, C. Körber, *Eur. J. Org. Chem.* 1999, 1033–1039.

171 C. Bolm, K. Muñiz, N. Aguilar, M. Kesselgruber, G. Raabe, *Synthesis* 1999, 1251–1260.

172 T. Bach, C. Körber, *J. Org. Chem.* 2000, **65**, 2358–2367.

173 J. P. Bacci, K. L. Greenman, D. L. Van Vranken, *J. Org. Chem.* 2003, **68**, 4955–4958.

174 (a) O. García Mancheño, C. Bolm, *Org. Lett.* 2006, **8**, 2349–2352; (b) O. García Mancheño, C. Bolm, *Chem. Eur. J.* 2007, **13**, 6674–6681.

175 (a) T. Fukuyama, C.-K. Jow, M. Cheung, *Tetrahedron Lett.* 1995, **36**, 6373–6374; (b) G. Y. Cho, C. Bolm, *Org. Lett.* 2005, **7**, 4983–4985.

176 (a) For a recent characterization of a non-heme iron-nitrene complex, see: E. B. Winkler, T. A. Jackson, M. P. Jensen, A. Stubna, G. Juhász, E. L. Bominaar, E. Münck, L. Que, Jr., *Angew. Chem.* 2006, **118**, 7554–7557; *Angew. Chem. Int. Ed.* 2006, **45**, 7394–7397; (b) for a related theoretical study, see: Y. Moreau, H. Chen, E. Derat, H. Hirao, C. Bolm, S. Shaik, *J. Phys. Chem. B*, 2007, **111**, 10288–10299.

4
Reduction of Unsaturated Compounds with Homogeneous Iron Catalysts

Stephan Enthaler, Kathrin Junge, and Matthias Beller

4.1
Introduction

Within the different molecular transformations, reduction processes play one of the major roles in organic chemistry. Apart from stoichiometric reactions, transition metal-catalyzed reactions offer an efficient and versatile strategy and present a key technology for the advancement of "green chemistry", specifically for waste prevention, reducing energy consumption, achieving high atom efficiency and creating advantageous economics [1]. Among the most popular and extensively studied catalytic reactions with respect to industrial applications is the hydrogenation of unsaturated compounds containing C=C, C=O and C=N bonds, since addition of molecular hydrogen provides high atom efficiency [2, 3]. In this regard, for activation of the molecular hydrogen or hydrogen donors, transition metal catalysts are essential. Commonly expensive and rare late transition metals (Ir, Rh or Ru) are applied as the catalyst core. Therefore, substitution by ubiquitously available, inexpensive and less toxic metals is one of the major challenges for future research. Consequently, the use of iron catalysts is especially desirable [The price for iron is relatively low (the current price for 1 t of iron is ~US$300) in comparison with rhodium (US$5975 per troy ounce), ruthenium (US$870 per troy ounce) or iridium (US$460 per troy ounce), due to the widespread abundance of iron in the Earth's crust and its easy accessibility. Furthermore, iron is an essential trace element (daily dose for humans 5–28 mg) and plays an important role in a wide range of biological processes]. This chapter summarizes the impact of iron catalysts on hydrogenation and hydrosilylation chemistry and is mainly directed to the reduction of C=C, C=O and C=N bonds. Further catalytic applications have been summarized elsewhere, e.g. reduction of nitro compounds and dehalogenations [4].

4.2
Hydrogenation of Carbonyl Compounds

The relevance of carbonyl hydrogenation processes is impressively emphasized by the broad scope of applications, for example, as building blocks and synthons for

Iron Catalysis in Organic Chemistry. Edited by Bernd Plietker

ISBN: 978-3-527-31927-5

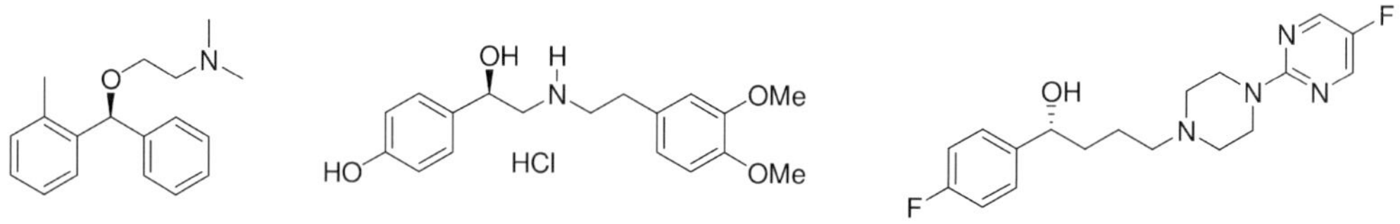

Orphenadrin (1) anticholinergic drug

(*R*)-Denopamine-Hydrochloride (2) β_1-receptor antagonist

BMS 181100 (3) antipsychotica

Figure 4.1 Selected pharmaceuticals based on chiral alcohols [6].

pharmaceuticals, agrochemicals, polymers, synthesis of natural compounds, auxiliaries, ligands and key intermediates in organic syntheses (Figure 4.1) [5].

In the past, this field has been dominated by ruthenium, rhodium and iridium catalysts with extraordinary activities and furthermore superior enantioselectivities; however, some investigations were carried out with iron catalysts. Early efforts were reported on the successful use of hydridocarbonyliron complexes $[HFe_m(CO)_n^-]$ as reducing reagent for α, β-unsaturated carbonyl compounds, dienes and C=N double bonds, albeit complexes were used in stoichiometric amounts [7]. The first catalytic approach was presented by Markó *et al.* on the reduction of acetone in the presence of $Fe_3(CO)_{12}$ or $Fe(CO)_5$ [8]. In this reaction, the hydrogen is delivered by water under more drastic reaction conditions (100 bar, 100 °C). Addition of NEt_3 as co-catalyst was necessary to obtain reasonable yields. The authors assumed a reaction of $Fe(CO)_5$ with hydroxide ions to yield $HFe(CO)_4^-$ with liberation of carbon dioxide since basic conditions are present and exclude the formation of molecular hydrogen via the water gas shift reaction. $HFe(CO)_4^-$ is believed to be the active catalyst, which transfers the hydride to the acceptor. The catalyst presented displayed activity in the reduction of several ketones and aldehydes (Scheme 4.1) [9].

Later on, Vancheesan's group reported the transfer hydrogenation of ketones utilizing 2-propanol or 1-phenylethanol as hydrogen source (Scheme 4.2) [10]. A phase transfer catalyst was essential to support the hydrogenation catalyst $Fe_3(CO)_{12}$ or $Fe(CO)_5$ since a liquid–liquid biphasic system was used as solvent. Under mild reaction conditions, several ketones were hydrogenated to the corresponding alcohols with turnover frequencies of up to 13 h^{-1}. Mechanistic investigations indicated a similar process as reported by Markó *et al.* [8].

More recently, Chen *et al.* reported an asymmetric transfer hydrogenation (Scheme 4.3) based on $[Et_3NH][HFe_3(CO)_{11}]$ and chelating chiral ligands [11]. In the presence of enantiopure diaminodiphosphine iron catalysts they claimed good conversion and enantioselectivity up to 98% *ee* (substrate 5j).

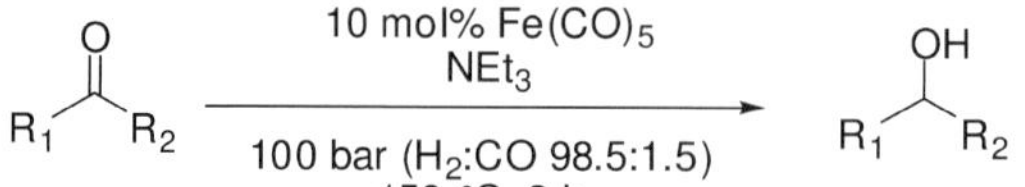

4 → **5**: 10 mol% $Fe(CO)_5$, NEt_3; 100 bar (H_2:CO 98.5:1.5), 150 °C, 3 h

	conversion
5a: $R_1 = CH_3$; $R_2 = CH_3$:	>99%
5b: $R_1 = C_6H_5$; $R_2 = CH_3$:	62%
5c: R_1 = *i*-Bu; $R_2 = CH_3$:	30%
5d: $R_1 = C_6H_5$; R_2 = H:	95%
5e: R_1 = *n*-Pr; R_2 = H:	98%
6: cyclohexanone:	>99%

Scheme 4.1 Reduction of ketones and aldehydes by using the $Fe(CO)_5$–NEt_3 system according to Markó *et al.*

4 mol% $Fe_3(CO)_{12}$
PTC: $BzEt_3NCl$ or Aliquat 336
NaOH
water/2-PrOH
or water/1-phenylethanol
28 °C, 2.5-3 h

$R_1C(O)R_2$ (**4**) → $R_1CH(OH)R_2$ (**5**)

	yield
5b: $R_1 = C_6H_5$; $R_2 = CH_3$:	36%
5f: R_1 = 4-Cl-C_6H_4; $R_2 = CH_3$:	42%
5g: R_1 = 4-Me-C_6H_4; $R_2 = CH_3$:	28%
5h: R_1 = Et; $R_2 = CH_3$:	20%
6: cyclohexanone:	78%

Scheme 4.2 Phase transfer catalyzed transfer hydrogenation of ketones by Vancheesan *et al.*

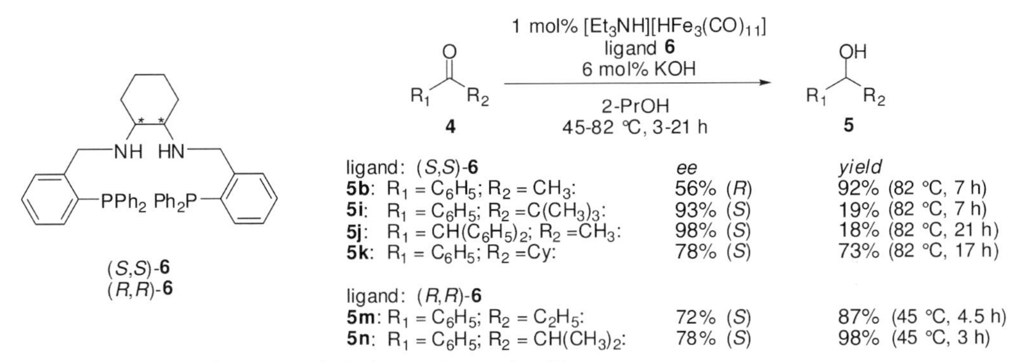

Scheme 4.3 Enantioselective transfer hydrogenation catalyzed by iron complexes according to Chen *et al.*

In 2006, we reported the application of an easy to adopt *in situ* concept composed of an iron source, e.g. $FeCl_2$ or $Fe_3(CO)_{12}$, monodentate phosphines and tridentate nitrogen ligands in the transfer hydrogenation of aliphatic and aromatic ketones using 2-propanol as hydrogen source (Scheme 4.4) [12]. The influence of different reaction parameters on the reduction of the model substrate acetophenone **5b** were studied in detail. A crucial influence of base and base concentration was displayed since only sodium isopropoxide gave a reasonable amount of product. In comparison with ruthenium-based transfer hydrogenations, a higher temperature is necessary. Notably, the advantage of this *in situ* concept is underlined by easy tuneability of the iron catalyst, due to the broad availability of simple phosphines and amines. Numerous phosphorus and amine ligands were subjected to the model reaction. However, the best results were obtained with a catalyst composed of $Fe_3(CO)_{12}$, triphenylphosphine and 2,2′ : 6′,2″-terpyridine or, surprisingly, 3 equiv. of pyridine with respect to iron. The latter resembles a straightforward catalyst with regard to industrial applications. After several optimization steps, an active catalyst was

0.5 mol% $FeCl_2$
0.5 mol% PPh_3/0.5 mol% TerPy
25 mol% NaO-*i*-Pr
2-PrOH, 100 °C, 7 h

$R_1C(O)R_2$ (**4**) → $R_1CH(OH)R_2$ (**5**)

5b: $R_1 = C_6H_5$; $R_2 = CH_3$:	yield: 95%
5f: R_1 = 4-Cl-C_6H_4; $R_2 = CH_3$:	yield: 97%
5g: R_1 = 4-Me-C_6H_4; $R_2 = CH_3$:	yield: 83%
5m: $R_1 = C_6H_5$; $R_2 = C_2H_5$:	yield: 92%
5o: R_1 = 4-MeO-C_6H_4; $R_2 = CH_3$:	yield: 75%
5p: R_1 = 2-MeO-C_6H_4; $R_2 = CH_3$:	yield: >99%
5q: $R_1 = C_6H_5$; $R_2 = CH_2Cl$:	yield: 8%
5r: R_1 = Cy; $R_2 = CH_3$:	yield: 93%
5s: R_1 = *t*-Bu; $R_2 = CH_3$:	yield: >99%

Scheme 4.4 *In situ* catalyst based on $FeCl_2$–terpy–PPh_3 in the transfer hydrogenation of ketones.

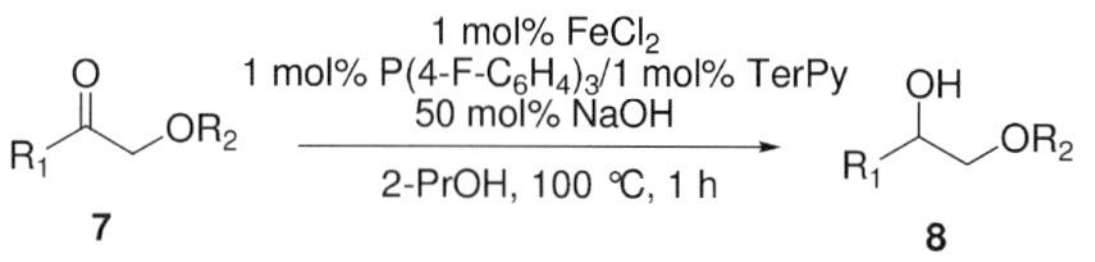

1 mol% $FeCl_2$, 1 mol% $P(4\text{-}F\text{-}C_6H_4)_3$/1 mol% TerPy, 50 mol% NaOH, 2-PrOH, 100 °C, 1 h; **7** → **8**

8a: $R_1 = C_6H_5$; $R_2 = CH_3$: yield: 92%
8b: $R_1 = C_6H_5$; $R_2 = C_6H_5$: yield: 94%
8c: $R_1 = C_6H_5$; $R_2 = 4\text{-Cl-}C_6H_4$: yield: >99%
8d: $R_1 = C_6H_5$; $R_2 = 2,6\text{-}(i\text{-Pr})_2C_6H_4$: yield: 45%
8e: $R_1 = 4\text{-Me-}C_6H_5$; $R_2 = C_6H_5$: yield: 99%
8f: $R_1 = 4\text{-MeO-}C_6H_5$; $R_2 = C_6H_5$: yield: 85%
8g: $R_1 = t\text{-Bu}$; $R_2 = C_6H_5$: yield: 22%

Scheme 4.5 Reduction of α-substituted ketones in the presence of iron catalyst.

developed which exhibited activity comparable to that of $Ru_3(CO)_{12}$-based catalysts and was successful in the reduction of several ketones with good to excellent conversions [13].

Very recently, the effectiveness of $FeCl_2$–terpy–PAr_3 catalysts was demonstrated in the hydrogenation of α-substituted ketones, which are interesting 1,2-diol precursors (Scheme 4.5) [14]. Excellent yields, chemoselectivities and activities [turnover frequencies (TOFs) up to 2000 h^{-1}) were achieved under optimized conditions.

Apart from synthetic aspects, we obtained some useful information concerning the mechanism, even if the "real" catalyst structure so far remains unclear. Various experiments proved the presence of a homogeneous catalyst. Following the original work on Ru, Rh and Ir catalysts, deuterated hydrogen donors were applied since for transfer hydrogenation two common mechanisms are established, nominated as direct hydrogen transfer via formation of a six-membered cyclic transition state constituted of metal, hydrogen donor and acceptor, and second the hydridic route, which is subdivided into two pathways, the monohydride and dihydride mechanisms (Scheme 4.6). More specifically, the formation of monohydride–metal complexes promotes an exclusive hydride transfer from carbon (donor) to carbonyl carbon (acceptor), whereas a hydride transfer via dihydride-metal complexes leads to no accurate prediction of hydride resting state, for the reason that the former hydride was transferred to carbonyl carbon (acceptor) as well as to the carbonyl oxygen (acceptor) [15].

To specify the position and the nature of the transferred hydride, the reaction was performed with 2-propanol-d_1 as solvent/donor, sodium 2-propylate as base and $Fe_3(CO)_{12}/PPh_3$/TerPy as catalyst under optimized conditions. In the transfer hydrogenation of acetophenone a mixture of two deuterated 1-phenylethanols was obtained (Scheme 4.7, **9a** and **9b**). The ratio between **9a** and **9b** (85 : 15) indicated a specific migration of the hydride, albeit some scrambling was detected. However, the incorporation is in agreement with the monohydride mechanism, implying the formation of metal monohydride species in the catalytic cycle.

Parallel to the work on *in situ* three-component catalysts, a nature-inspired *in situ* catalyst based on iron porphyrin complexes, was developed, since high stability of the

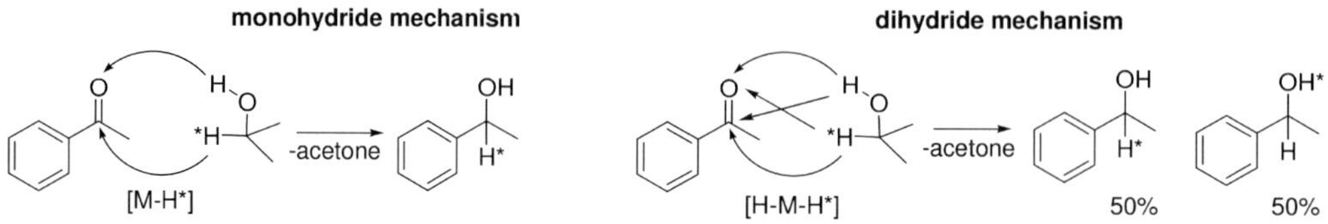

monohydride mechanism: [M-H*], -acetone
dihydride mechanism: [H-M-H*], -acetone, 50%, 50%

Scheme 4.6 Hydridic route established for transfer hydrogenation with Rh-, Ru- and Ir-catalysts.

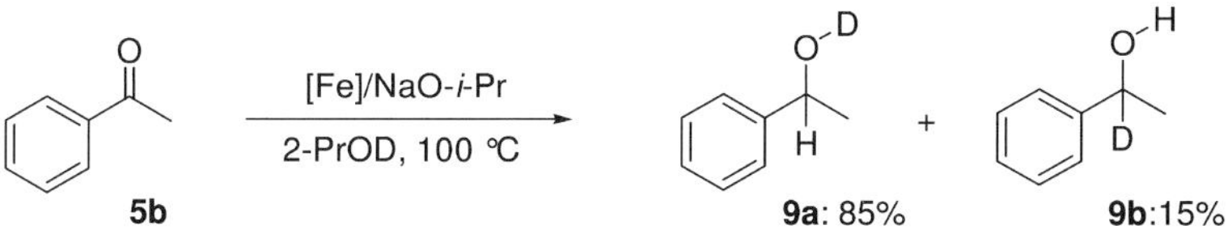

[Fe]/NaO-*i*-Pr

2-PrOD, 100 °C

5b → **9a**: 85% + **9b**:15%

Scheme 4.7 Deuterium incorporation catalyzed by the $Fe_3(CO)_{12}$–terpy–PPh_3 system under transfer hydrogenation conditions.

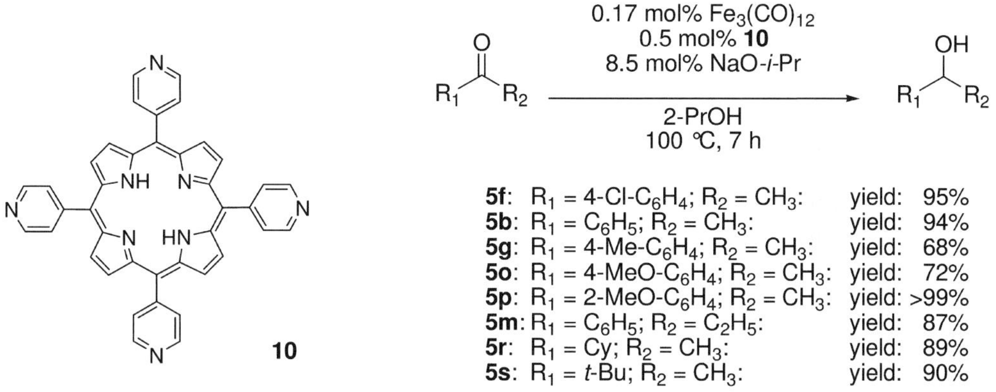

0.17 mol% $Fe_3(CO)_{12}$
0.5 mol% **10**
8.5 mol% NaO-*i*-Pr
2-PrOH
100 °C, 7 h

10

5f: R_1 = 4-Cl-C_6H_4; R_2 = CH_3: yield: 95%
5b: R_1 = C_6H_5; R_2 = CH_3: yield: 94%
5g: R_1 = 4-Me-C_6H_4; R_2 = CH_3: yield: 68%
5o: R_1 = 4-MeO-C_6H_4; R_2 = CH_3: yield: 72%
5p: R_1 = 2-MeO-C_6H_4; R_2 = CH_3: yield: >99%
5m: R_1 = C_6H_5; R_2 = C_2H_5: yield: 87%
5r: R_1 = Cy; R_2 = CH_3: yield: 89%
5s: R_1 = *t*-Bu; R_2 = CH_3: yield: 90%

Scheme 4.8 Biomimetic transfer hydrogenation of ketones with iron porphyrin catalysts.

complexes is known and inertness against oxygen and moisture is feasible [16]. The porphyrin catalysts achieved even higher activities than three-component system in the transfer hydrogenation of ketones (Scheme 4.8). After optimization of reaction parameters, TOFs up to 642 h^{-1} at low catalyst loadings (0.01 mol%) were attained. Ligand screening emphasized porphyrin **10** as a module of a highly active catalyst, suitable for further investigations on substrate variations. Various ketones were reduced in good to excellent yields. Notably, naturally occurring hemin was also used in this study. Even if lower activity was attained, it is a worthwhile system, due to its abundance and easy handling. Advantageously, no pre-catalyst formation is necessary.

Applying Fe porphyrin catalysts and 2-propanol as hydrogen donor, various α-hydroxyl-protected ketones are reduced to the corresponding monoprotected 1,2-diols in good to excellent yields after optimization (Scheme 4.9) [17]. Remarkably, addition of small amounts of water (5–10 mol%) boosted the catalyst activity up to 2500 h^{-1} at low catalyst loadings (0.01 mol%).

4.3
Hydrogenation of Carbon–Carbon Double Bonds

Since the 1960s, tremendous efforts have been made in the field of homogeneous hydrogenation of C–C double bonds, as emphasized by innumerable reports, and

1 mol% $FeCl_2$
1 mol% **11**
50 mol% NaOH
2-PrOH, 100 °C, 2 h

11

8a: $R_1 = C_6H_5$; $R_2 = CH_3$: yield: >99%
8b: $R_1 = C_6H_5$; $R_2 = C_6H_5$: yield: 92%
8c: $R_1 = C_6H_5$; $R_2 = 4\text{-Cl-}C_6H_4$: yield: >99%
8d: $R_1 = C_6H_5$; $R_2 = 2,6\text{-}(i\text{-Pr})_2C_6H_4$: yield: 74%
8e: $R_1 = 4\text{-Me-}C_6H_5$; $R_2 = C_6H_5$: yield: >99%
8f: $R_1 = 4\text{-MeO-}C_6H_5$; $R_2 = C_6H_5$: yield: 83%
8g: $R_1 = t\text{-Bu}$; $R_2 = C_6H_5$: yield: 48%

Scheme 4.9 Reduction of α-substituted ketones in the presence of iron porphyrin catalysts.

later rewarded by integration into several industrial processes (Figure 4.2). However, an even broader use in industry is expected owing to the availability of similar active and selective hydrogenation catalysts containing inexpensive metals (e.g. Fe and Cu).

Indeed iron catalysts are known to transfer hydrogen to C–C double bonds. Early examples were carried out with metal salts activated by aluminum compounds in the hydrogenation of non-funtionalized alkenes under comparatively mild reaction conditions (0–35 °C and low hydrogen pressure) [19]. Some research groups focused on the application of iron carbonyls due to the easier removal of the carbonyl ligands, for generating an active site, compared with halogens and avoidance of activating reagents [20]. Unsaturated carboxylic acid derivatives, e.g. methyl linoleate and methyl linolenate, which contain two and three non-conjugated double bonds, respectively, were described in the mid-1960s (Scheme 4.10) [21, 22]. As catalyst precursor $Fe(CO)_5$ was utilized at high temperature. The authors proposed as the first step an isomerization of the double bonds catalyzed by $Fe(CO)_5$ to obtain various dienes or trienes. A hydrogenation process was observed in the case of conjugated dienes to yield monoenes via an iron–carbonyl–diene complex (**18**), whereas dienes were formed in the hydrogenation of methyl linolenate (**16**). The activity of the system was improved when the iron–carbonyl–diene complex **18** was used instead of $Fe(CO)_5$ as precursor. Analysis of the reaction mixture revealed monoenes as major products with unselective double bond distribution accompanied by small amounts of fully saturated compounds. Later, Cais and Maoz studied the reaction of methyl hexa-2,4-dienoate as a model for fatty acids in more detail and reported several mechanistic considerations [23].

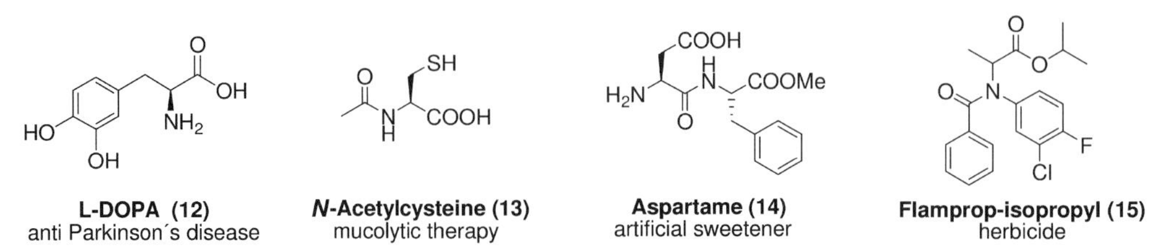

Figure 4.2 Selected pharmaceuticals that are accessible by asymmetric C–C double bond hydrogenation [5, 18].

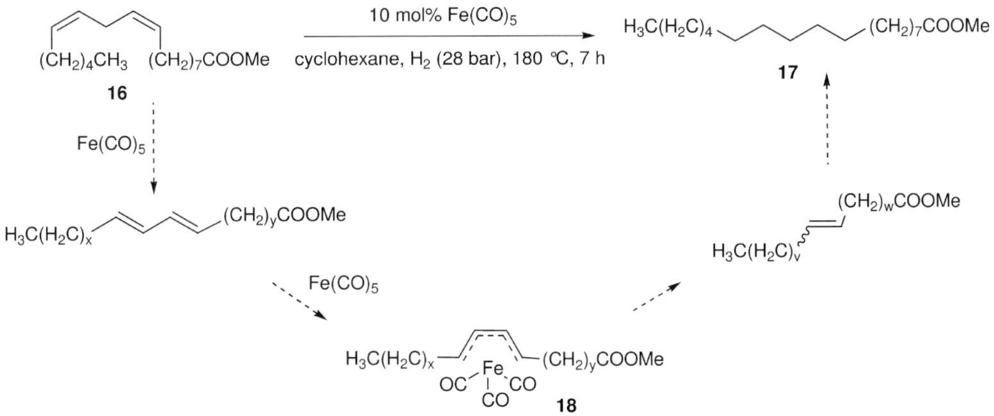

Scheme 4.10 Fe-catalyzed hydrogenation of methyl linoleate.

Tajima and Kunioka described the hydrogenation of conjugated dialkenes, in particular 1,3-butadiene, in the presence of activated [$CpFe(CO)_2Cl$] (Scheme 4.11) [24]. $AlEt_3$ and PhMgBr were tested as activation reagents; the $AlEt_3$-activated complex showed the best performance. The authors proposed the formation of an iron–alkyl species that undergoes hydrogenolysis to create an active metal hydride. The composition of the final mixture was dominated by *cis*- and *trans*-2-butene (ratio ~1 : 1). Furthermore, traces of 1-butene were observed, whereas the catalyst was completely inactive for the hydrogenation of the monoalkene to yield *n*-butane.

Nishiguchi and Fukuzumi reported on the transfer hydrogenation of 1,5-cyclooctadiene (cod) in the presence of catalytic amounts of $FeCl_2(PPh_3)_2$ (10 mol%) at high temperatures (up to 240 °C) [25]. As hydrogen source polyhydroxybenzenes, e.g. pyrogallol, pyrocatechol or hydroquinone, have been proven to be superior for this reaction, whereas primary and secondary alcohols led to lower conversion. The hydrogenation of 1,5-cyclooctadiene led to cyclooctane as the main product, but the unsaturated cyclooctene and isomerization products were also observed.

Later, other groups recognized a reactivity increase with respect to hydrogenation rate when $Fe(CO)_5$ was treated with light [26, 27]. The authors assumed the expulsion of one carbonyl ligand to form an unsaturated species induced by near-ultraviolet irradiation. After complexation of the alkene again activation by light removed another carbonyl ligand to allow hydrogen coordination/activation. A number of

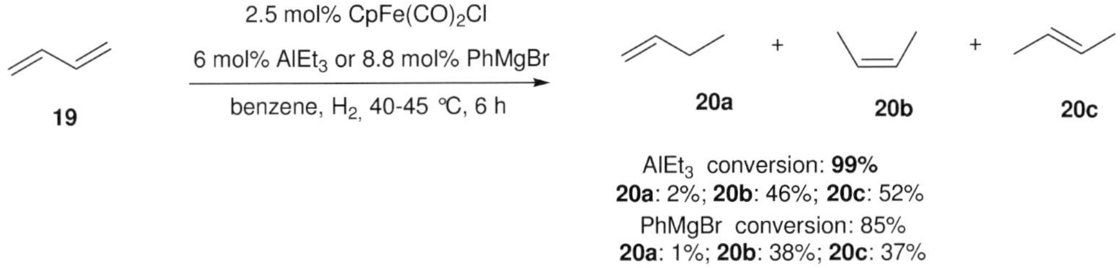

$AlEt_3$ conversion: **99%**
20a: 2%; **20b**: 46%; **20c**: 52%
PhMgBr conversion: 85%
20a: 1%; **20b**: 38%; **20c**: 37%

Scheme 4.11 Hydrogenation of butadiene with $CpFe(CO)_2Cl$ (Cp = cyclopentadienyl).

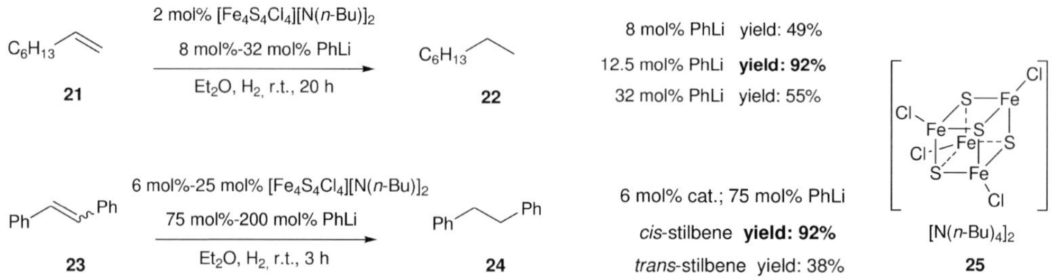

C_6H_{13} **21** → 2 mol% $[Fe_4S_4Cl_4][N(n\text{-}Bu)]_2$, 8 mol%-32 mol% PhLi, Et_2O, H_2, r.t., 20 h → C_6H_{13} **22**

8 mol% PhLi yield: 49%
12.5 mol% PhLi **yield: 92%**
32 mol% PhLi yield: 55%

Ph–CH=CH–Ph **23** → 6 mol%-25 mol% $[Fe_4S_4Cl_4][N(n\text{-}Bu)]_2$, 75 mol%-200 mol% PhLi, Et_2O, H_2, r.t., 3 h → Ph–CH₂CH₂–Ph **24**

6 mol% cat.; 75 mol% PhLi
cis-stilbene **yield: 92%**
trans-stilbene yield: 38%

$[N(n\text{-}Bu)_4]_2$ **25**

Scheme 4.12 Bio-inspired iron catalyst in the hydrogenation of C–C double bonds.

unsubstituted alkenes (e.g. ethylene, propylene, *cis*-3-hexene, cyclopentene, cyclooctene) were hydrogenated to give alkanes in moderate yield (up to 50%), but the catalyst was inactive for sterically demanding alkenes (e.g. 1,2-dimethylcyclohexene), aldehydes and nitriles. Furthermore, alkynes were subjected to this photocatalyzed reduction, but only low yields (<5%) of the corresponding alkenes were obtained.

Inspired by the properties of metalloproteins and metalloenzymes in biological processes, Inoue's group established a hydrogenation protocol based on active sites of enzyme hydrogenases, which contain an iron–sulfur cluster (Scheme 4.12) [28, 29]. For their biomimetic approach, the Fe_4S_4 cluster **24** was used as an active-site model and tested in the hydrogenation of octene (**21**) and stilbene (**23**) with molecular hydrogen. A necessity for activation of the cluster with phenyllithium was reported. The amount of activation reagent played a crucial role in the catalytic activity and selectivity; thereby, an optimum in the hydrogenation of octene was attained when 12.5 mol% phenyllithium and 75 mol% for stilbene hydrogenation, respectively, were applied. Noteworthy, in the case of defined internal octenes, isomerization of the double bond takes place to give different internal octenes, due to β-hydride elimination. 1-Octene is extracted to some extent faster from this mixture by hydrogenation [28]. The final composition of *cis*-stilbene hydrogenation indicated also an isomerization, because significant amounts of *trans*-stilbene were observed, whereas the hydrogenation of *trans*-stilbene was unaffected by isomerization [29].

At the beginning of the 1990s, a switch to iron catalysts stabilized by either phosphines or nitrogen ligands took place. Bianchini and coworkers carried out a detailed comparative study on the transfer hydrogenation of α, β-unsaturated ketones using non-classical trihydride iron, ruthenium and osmium complexes containing tetradentate phosphines (Scheme 4.13) [30, 31]. Complex **26** furthermore exhibited activity in the hydrogenation of alkynes to yield alkenes [30]. In the presence of cyclopentanol as hydrogen donor, several α, β-unsaturated ketones were hydrogenated to the corresponding saturated ketones with good to excellent selectivity under mild reaction conditions (Scheme 4.13). In some cases, e.g. with benzylideneacetone or 2,3-cyclohexenone, the unsaturated or saturated alcohols were formed by the catalyst **26**, and good selectivity was observed. Noteworthily, no co-catalyst, e.g. base, was necessary to activate the catalyst or the hydrogen donor, which is needed for other catalyst systems. However, for carbonyl hydrogenation, e.g. acetophenone and cyclohexanone, the Fe catalyst displayed only low activity. No activity was found for aldehydes and C=C systems without additional carbonyl functionality.

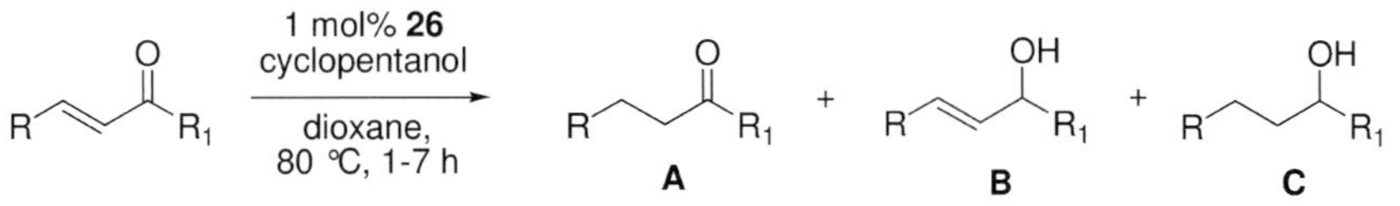

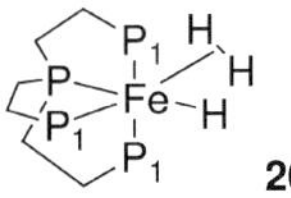

1 mol% **26**
cyclopentanol
dioxane,
80 °C, 1-7 h

27a: R = C_6H_5; R_1 = CH_3: conv.: 95%, **B**: 95%
27b: R = C_6H_5; R_1 = C_6H_5: conv.: 30%, **A**: 30%
27c: R = CH_3; R_1 = C_6H_5: conv.: 7%, **A**: 7%
27d: R = CH_3; R_1 = C_2H_5: conv.: 19%, **A**: 19%
27e: R = CH_3; R_1 = CH_3: conv.: 100%, **A**: 100%
28: 2-cyclohexene: conv.: 72%, **B**: 44%, **C**: 28%
29: 2-cyclopentene: conv.: 100%, **A**: 84%, **B**: 16%

P_1 = PPh_2

Scheme 4.13 Application of non-classical iron trihydride complex in transfer hydrogenation according to Bianchini and coworkers.

Reduction of styrene was reported by Kano *et al.* using $NaBH_4$ as hydrogen source in the presence of iron porphyrin complex **30** (Scheme 4.14) [32]. Good TOFs up to 81 h^{-1} and good chemoselectvities were obtained in protic solvents with a catalyst loading of 1 mol% at room temperature. The reaction has been proven to be a radical process since to some extent 2,3-diphenylbutane was detected. A similar approach was used in the hydrogenation of α, β-unsaturated esters by Sakaki and coworkers, who reported an improvement in catalyst activity (TOF = 4580 h^{-1}) [33]. A detailed study of the reaction mechanism displayed a crucial influence of the protic solvent because the hydride is assigned by $NaBH_4$ and the proton by methanol.

More recently, Chirik's group has shown elegantly the application of low-valent iron complexes in the hydrogenation of various C–C double and triple bonds [34]. Based on the mentioned work of Wrighton *et al.* [26], an approach to stabilized 14-electron $L_3Fe(0)$ fragments was presented. The catalyst precursors were synthesized by reduction of dihalogen complexes **33** containing a tridentate pyridinediimine ligand with sodium amalgam or with sodium triethylborohydride under an atmosphere of nitrogen (Scheme 4.15). The bis-dinitrogen complexes **34** obtained are relative labile compounds and a loss of one equivalent of dinitrogen in solution at room temperature or an easy exchange against hydrogen or alkynes occurred. The catalyst activity was studied in the context of C–C double and triple bond hydrogenation. Applying 0.3 mol% of the iron catalyst, simple alkenes such as 1-hexene and cyclohexene were hydrogenated with TOFs up to 1814 h^{-1} under comparatively

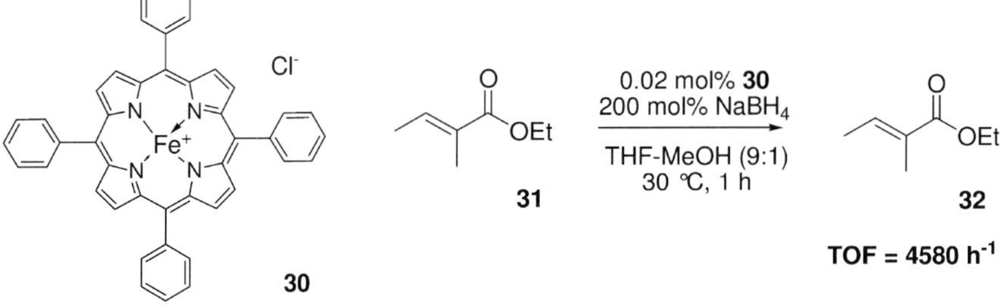

0.02 mol% **30**
200 mol% $NaBH_4$
THF-MeOH (9:1)
30 °C, 1 h

TOF = 4580 h^{-1}

Scheme 4.14 Reduction of α,β-unsaturated ester catalyzed by iron porphyrin **30**.

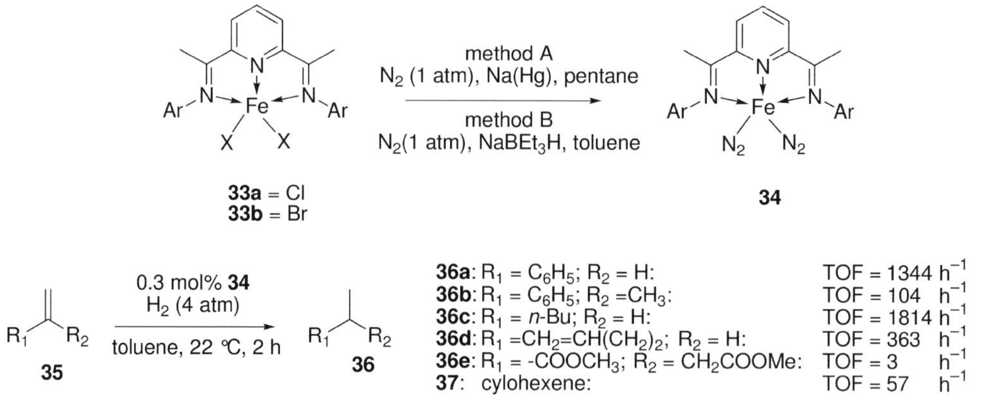

method A
N_2 (1 atm), Na(Hg), pentane
method B
N_2(1 atm), $NaBEt_3H$, toluene

33a = Cl
33b = Br

34

0.3 mol% **34**
H_2 (4 atm)
toluene, 22 °C, 2 h

35 → **36**

36a: $R_1 = C_6H_5$; R_2 = H: TOF = 1344 h^{-1}
36b: $R_1 = C_6H_5$; $R_2 = CH_3$: TOF = 104 h^{-1}
36c: R_1 = *n*-Bu; R_2 = H: TOF = 1814 h^{-1}
36d: $R_1 = CH_2{=}CH(CH_2)_2$; R_2 = H: TOF = 363 h^{-1}
36e: R_1 = -$COOCH_3$; $R_2 = CH_2COOMe$: TOF = 3 h^{-1}
37: cylohexene: TOF = 57 h^{-1}

Scheme 4.15 Preparation of catalyst precursors containing tridentate nitrogen ligands and abilities in catalytic reactions according to Chirik and coworkers.

mild reaction conditions (4 atm hydrogen pressure and room temperature). The reaction was also carried out under preferred solvent-free conditions in neat substrates, leading to comparable activity. The scope and limitation of the catalyst system were demonstrated in the hydrogenation of various alkene units enclosing geminal, internal and trisubstituted alkenes and also dialkenes. The activity attained decreased in the order terminal alkenes > internal alkenes = geminal alkenes > trisubstituted alkenes. In addition, one example of a functionalized alkenes was mentioned. However, lower activity was observed for dimethyl itaconate, albeit using higher catalyst loadings. In the case of cyclohexene, some detailed investigations emphasized deactivation of the catalyst by arene complexation either from the solvent or the aryl groups in the ligand [35]. Noteworthily, on comparing the activity of catalyst **34** (TOF = 1814 h^{-1}) with those of common catalysts, e.g. Pd/C (TOF = 366 h^{-1}), $RhCl(PPh_3)_3$ (10 h^{-1}) or $[Ir(cod)(PCy_3)(py)]PF_6$ (75 h^{-1}) in the hydrogenation of 1-hexene under optimized conditions, a significantly higher value was reached with the iron catalyst.

Subsequently, Chirik's group studied the influence of replacing the imino functionalities in ligand system **34** by phosphino groups (Scheme 4.16) [36]. A different coordination mode was found since only one nitrogen ligand was replaced by hydrogen. The unstable complex 38 was also tested in the hydrogenation of 1-hexene, but no improvement in activity was observed.

Using the same synthetic method as described for catalyst **34**, diimine complexes **39** were also attainable (Scheme 4.17) [37]. Due to the instability of dinitrogen

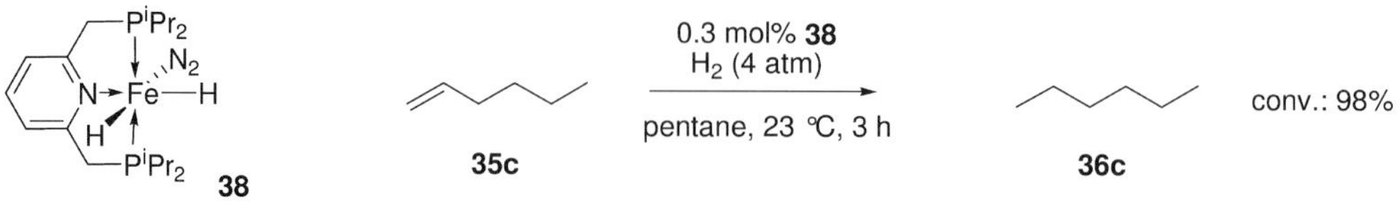

38

0.3 mol% **38**
H_2 (4 atm)
pentane, 23 °C, 3 h

35c → **36c** conv.: 98%

Scheme 4.16 Application of iron–aminodiphosphine complexes in hydrogenation reactions.

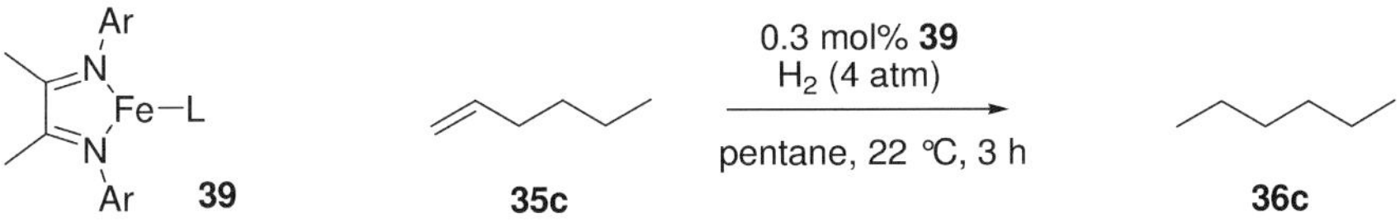

39 + **35c** → **36c** (0.3 mol% **39**, H_2 (4 atm), pentane, 22 °C, 3 h)

39a: Ar = 2,6-(^{i}Pr)$_2$C$_6$H$_3$; L = (Me$_3$Si)$_2$C$_2$ TOF = 4 h^{-1}
39b: Ar = 2,6-(^{i}Pr)$_2$C$_6$H$_3$; L = cod TOF = 90 h^{-1}
39c: Ar = 2,6-(^{i}Pr)$_2$C$_6$H$_3$; L = coe TOF = 90 h^{-1}

Scheme 4.17 Hydrogenation with low-valent iron complexes (cod = 1,5-cyclooctadiene; coe = cyclooctene).

complexes, stabilization was achieved by introducing more suitable ligands, e.g. alkynes, alkenes or dialkenes. However, in the hydrogenation of 1-hexene under comparable conditions, only low activities were observed. Apart from the synthesis and application of low-valent iron complexes, Chirik and co-workers carried out some mechanistic investigations.

The suggestion for the catalytic cycle is presented in Scheme 4.18. Initially, an unsaturated iron complex is formed by expulsion of both dinitrogen molecules. Next, coordination of the alkene takes place, which is preferred since activation of hydrogen is also feasible. After alkene coordination, oxidative addition of hydrogen yields a formally 18-electron complex. Insertion of the alkene gave an alkyl complex, which recreated the starting complex via reductive elimination. Notably, the alkene complex also supports an isomerization of the double bond, hence an extension of possible intermediates is conceivable.

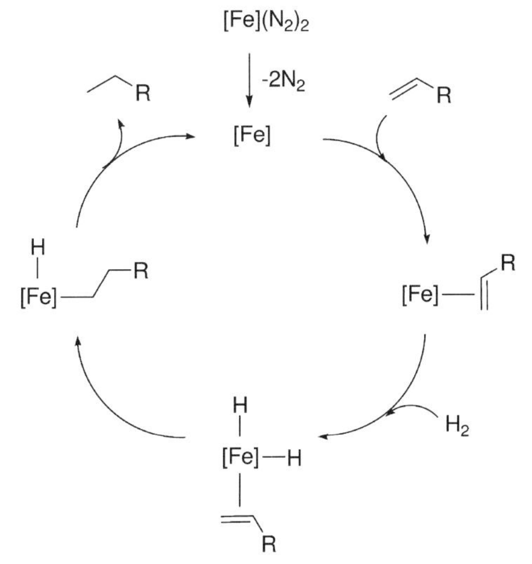

Scheme 4.18 Mechanism for the hydrogenation process proposed by Chirik and coworkers.

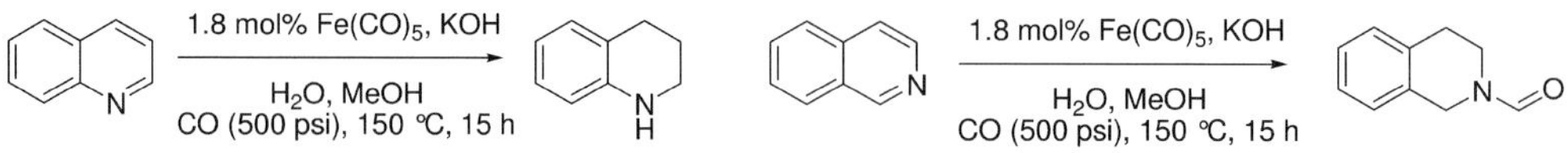

1.8 mol% $Fe(CO)_5$, KOH; H_2O, MeOH; CO (500 psi), 150 °C, 15 h

40 → 41

42 → 43

Scheme 4.19 Hydrogenation of aromatic nitrogen heterocycles.

4.4
Hydrogenation of Imines and Similar Compounds

So far, only a few hydrogenations of unsaturated CN functionalities utilizing iron catalysts have been described. On the one hand, hydrogenation of *N*-benzylideneaniline to yield *N*-benzylaniline was carried out by Radhi and Markó in the presence of catalytic amounts of $[NEt_3H][HFe(CO)_4]$, which was also applied for ketone/aldehyde reduction (see above) [38]. Even though good to excellent activities were obtained in various solvents, for the reaction to proceed harsh conditions are required (150 °C, 100 bar). On the other hand, Kaesz and coworkers reported the reduction of nitrogen heterocycles under water-gas shift conditions [39]. $Fe(CO)_5$ was used as the catalyst at high temperature and pressure. Moderate turnover numbers in the range 18–37 were obtained. In the case of isoquinoline (**42**) as main product, the formylated compound **43** is observed (Scheme 4.19).

4.5
Catalytic Hydrosilylations

In addition to catalytic reductions with molecular hydrogen or hydrogen donors, silanes also represent useful reducing agents [40].

Most examples in the literature on hydrosilylation with iron complexes as catalyst concern $Fe(CO)_5$ or related iron carbonyl compounds [41]. The first use of iron pentacarbonyl was reported for the reaction of silicon hydrides with alkenes at 100–140 °C to form saturated and unsaturated silanes according to Scheme 4.20 [42, 43].

An excess of alkene favored the formation of the unsaturated product **44** whereas the silylated compound **45** dominated at higher silane to alkene ratios. The described reaction was also carried out in the presence of colloidal iron and led to results similar to those obtained with $Fe(CO)_5$. Hydrosilylation of vinyltrimethylsilane **46** with chlorosilanes R_3SiH using the modified iron carbonyl complex $(CH_2{=}CHSiMe_3)Fe(CO)_4$ [44, 45]

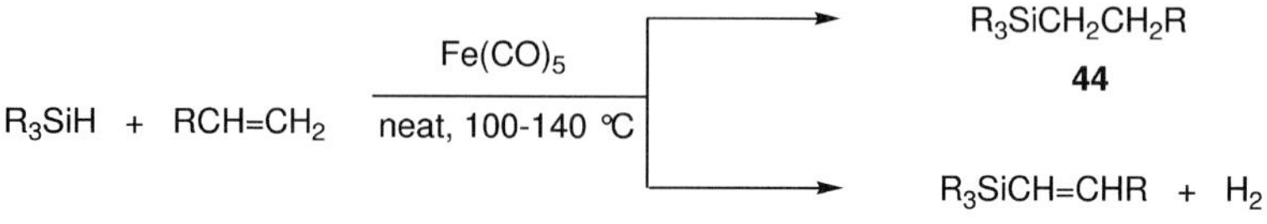

R_3SiH + $RCH{=}CH_2$ —[$Fe(CO)_5$, neat, 100-140 °C]→ $R_3SiCH_2CH_2R$ (**44**) + $R_3SiCH{=}CHR$ + H_2 (**45**)

R = Cl, OC_2H_5, C_2H_5

Scheme 4.20 Fe-catalyzed hydrosilylation of C–C double bonds.

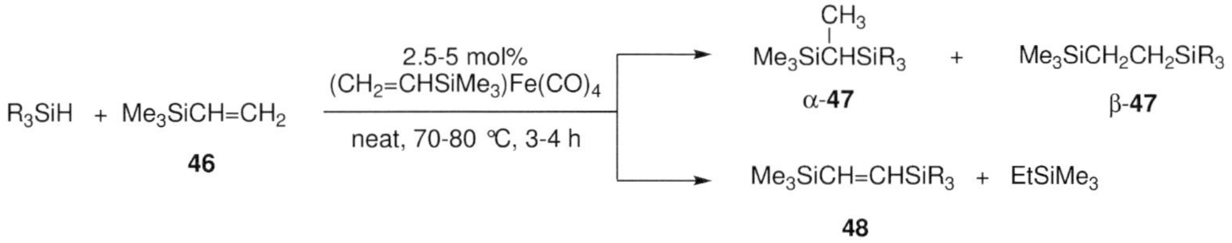

$$R_3SiH + Me_3SiCH{=}CH_2 \ (\mathbf{46}) \xrightarrow[\text{neat, 70-80 °C, 3-4 h}]{\text{2.5-5 mol\% } (CH_2{=}CHSiMe_3)Fe(CO)_4} Me_3SiCH(CH_3)SiR_3 \ (\alpha\text{-}\mathbf{47}) + Me_3SiCH_2CH_2SiR_3 \ (\beta\text{-}\mathbf{47}) \ ; \ Me_3SiCH{=}CHSiR_3 \ (\mathbf{48}) + EtSiMe_3$$

Scheme 4.21 Hydrosilylation of vinyltrimethylsilane **46**.

gave a mixture of α- and β-isomers of the silylated product **47** and the unsaturated product **48** attained by dehydrogenative silylation (Scheme 4.21).

The proportions of the compounds formed, α-**47**, β-**47** and **48**, depend on the nature of the hydrosilane R_3SiH used. There is an indirect relationship between the yield of the unsaturated silane **48** and the catalytic activity of R_3SiH ($Cl_3SiH > Cl_2MeSiH > ClMe_2SiH > Me_3SiH$). With the exception of Me_3SiH, the α-isomer is mainly formed in all reactions studied. Iron pentacarbonyl is known to hydrosilylate functionalized alkenes (Scheme 4.22) [46]. For instance, hydrosilylation of acrolein (**49**) with Et_3SiH gave up to 95% of a mixture of *cis*- and *trans*-$MeCH{=}CHOSiEt_3$ (**50**), whereas acrolein diethyl acetal (**51**) formed the Markovnikov product $Et_3SiCH_2CH_2CH(OEt)_2$ (**52**) and Et_3SiOEt via C–O bond cleavage. Allylic alcohol reacts in the presence of $Fe(CO)_5$ to give $CH_2{=}CHCH_2OSiEt_3$ (**54**) and $PrOSiEt_3$ (**55**).

In 1993, Murai's group examined the effectiveness of the iron-triad carbonyl complexes $Fe(CO)_5$, $Fe_2(CO)_9$ and $Fe_3(CO)_{12}$ as catalysts for the reaction of styrene with triethylsilane [47]. Whereas $Fe(CO)_5$ showed no catalytic activity, $Fe_2(CO)_9$ and $Fe_3(CO)_{12}$ formed selectively β-silylstyrene **57a** and ethylbenzene **58**. Interestingly, $Fe_3(CO)_{12}$ is the catalyst that exhibited the highest selectivity. This trinuclear iron carbonyl catalyst was also successfully applied in the reaction of different *para*-substituted styrenes with Et_3SiH giving only the (*E*)-β-triethylstyrenes in 66–70% yield (Scheme 4.23).

Hydrosilylation reactions catalyzed by iron carbonyl compounds often occur under drastic thermal conditions. Schroeder and Wrighton reported a photocatalyzed reaction of trialkylsilanes with alkenes in the presence of $Fe(CO)_5$ at low temperatures (0–50 °C) [48]. It is well known that irradiation of mononuclear metal carbonyls leads

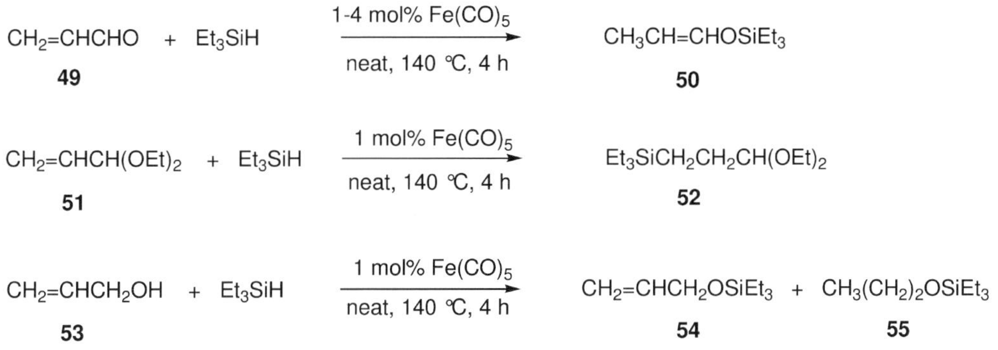

$$CH_2{=}CHCHO \ (\mathbf{49}) + Et_3SiH \xrightarrow[\text{neat, 140 °C, 4 h}]{\text{1-4 mol\% } Fe(CO)_5} CH_3CH{=}CHOSiEt_3 \ (\mathbf{50})$$

$$CH_2{=}CHCH(OEt)_2 \ (\mathbf{51}) + Et_3SiH \xrightarrow[\text{neat, 140 °C, 4 h}]{\text{1 mol\% } Fe(CO)_5} Et_3SiCH_2CH_2CH(OEt)_2 \ (\mathbf{52})$$

$$CH_2{=}CHCH_2OH \ (\mathbf{53}) + Et_3SiH \xrightarrow[\text{neat, 140 °C, 4 h}]{\text{1 mol\% } Fe(CO)_5} CH_2{=}CHCH_2OSiEt_3 \ (\mathbf{54}) + CH_3(CH_2)_2OSiEt_3 \ (\mathbf{55})$$

Scheme 4.22 Hydrosilylation of functionalized alkenes.

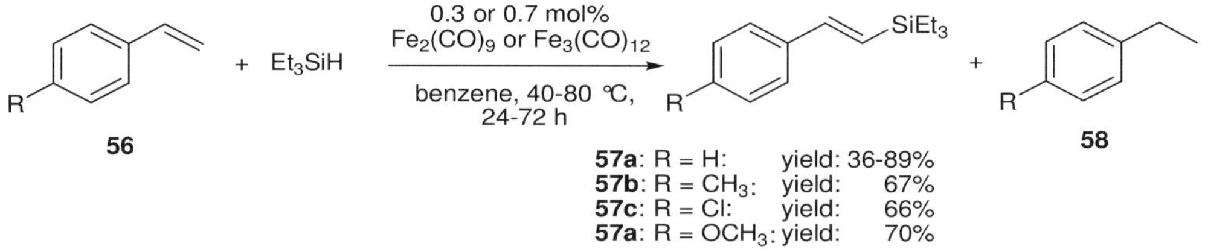

56 + Et_3SiH → (0.3 or 0.7 mol% $Fe_2(CO)_9$ or $Fe_3(CO)_{12}$; benzene, 40-80 °C, 24-72 h) 57 ($SiEt_3$) + 58

57a: R = H: yield: 36-89%
57b: R = CH_3: yield: 67%
57c: R = Cl: yield: 66%
57a: R = OCH_3: yield: 70%

Scheme 4.23 Hydrosilylation of styrene derivatives.

to efficient dissociative loss of CO to yield coordinative unsaturated, 16-electron, intermediates (Scheme 4.24) [26].

Irradiation of $Fe(CO)_5$ in the presence of a trialkylsilane (R_3SiH) and alkene initially yields a mixture of $Fe(CO)_4$(alkene) and (H)(SiR_3)$Fe(CO)_4$, which were determined by infrared spectroscopy. (H)(SiR_3)$Fe(CO)_3$(alkene) is postulated to be the catalytically active species photogenerated from these intermediates according to Scheme 4.24. Continuous irradiation is needed to maintain a steady-state concentration of the repeat unit "$Fe(CO)_3$". Such an iron tricarbonyl complex was also detected by infrared spectroscopy in the photoinduced addition of R_3SiH to 1,3-butadiene with $Fe(CO)_5$ [49].Polymer-anchored iron carbonyl species were reported to catalyze the hydrosilylation of 1-pentene with $HSiEt_3$ to give a mixture of pentyl- and pentenylsilanes [50]. The different pathways of the mechanism of transition metal-catalyzed hydrosilylation are summarized in Scheme 4.25. A commonly proposed mechanism involves the insertion of the alkene into the Fe–H bond of the complex (H)(SiR_3)$Fe(CO)_3$(alkene) followed by reductive elimination of the alkyl and silyl groups to form an alkylsilane (pathway **B**). In an alternative mechanism, insertion of the alkene into the Fe–Si bond of the complex (H)(SiR_3)$Fe(CO)_3$(alkene) is discussed (pathway **A**). This route, which is also suggested for the use of $Fe_3(CO)_{12}$, explains the formation of vinylsilanes as by-products in hydrosilylations [51].

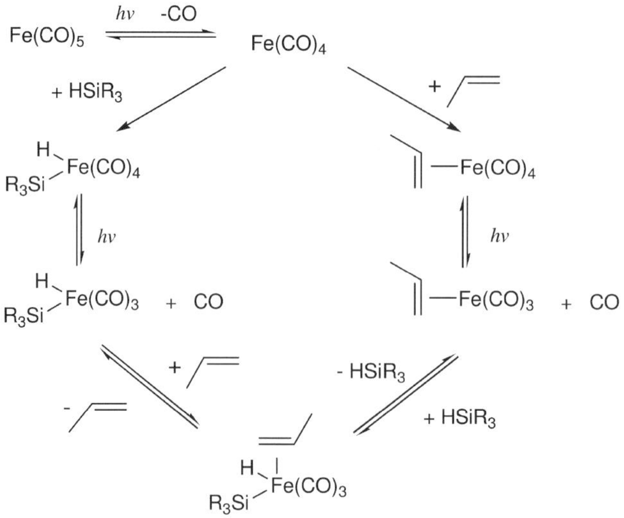

$Fe(CO)_5$ ⇌ (hν, -CO) $Fe(CO)_4$
+ $HSiR_3$ → (H)(R_3Si)$Fe(CO)_4$ ⇌ (hν) (H)(R_3Si)$Fe(CO)_3$ + CO
+ alkene → (alkene)–$Fe(CO)_4$ ⇌ (hν) (alkene)–$Fe(CO)_3$ + CO
+ alkene / - alkene; - $HSiR_3$ / + $HSiR_3$ ⇌ (H)(R_3Si)$Fe(CO)_3$(alkene)

Scheme 4.24 Formation of the catalytic active species by irradiation.

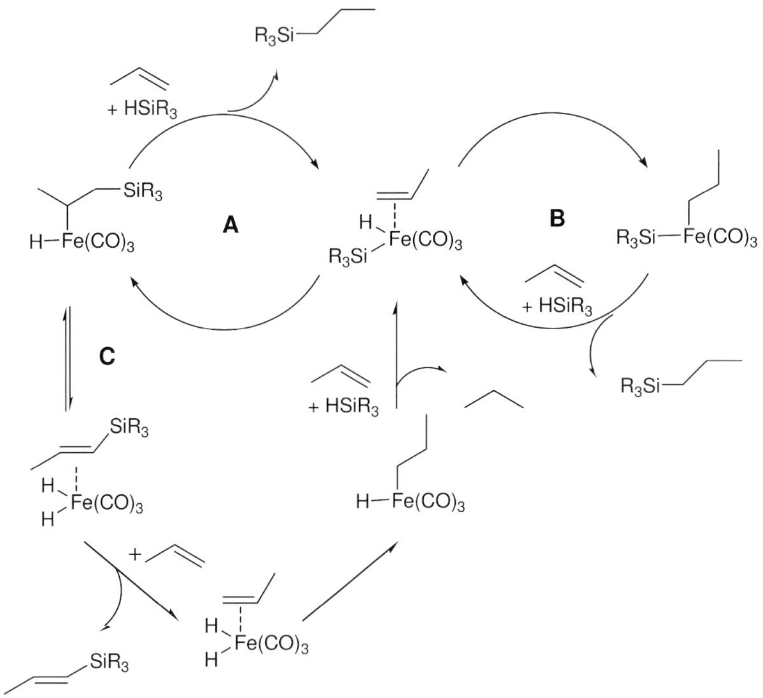

Scheme 4.25 Proposed catalytic cycle for the hydrosilylation of alkenes in the presence of $Fe(CO)_5$.

The photoinduced alkene insertion into the Fe–Si bond of (η^5-C_5Me_5)Fe(CO)R was shown to be reversible (reaction **C**) [52]. A reactivity study of the complex (H)Fe$(CO)_4SiPh_3$ with nucleophiles allowed a better understanding of the underlying reaction processes [53]. Under photochemical conditions, the activation or the substitution of the carbonyl ligands supports the classical mechanism involving insertion of the coordinated alkene into the Fe–Si (**A**) or the Fe–H (**B**) bonds. Under thermal conditions, however, direct addition of the Fe–H group by a radical or ionic process can occur, as observed in the reaction of (H)Fe$(CO)_4SiPh_3$ with isoprene. These mechanistic investigations of photogenerated hydrosilylation have been extended to the reaction of the iron carbonyl complex ($CH_2{=}CHSiMe_3$)Fe$(CO)_4$ with vinylsilane and R_3SiH [45]. Tetrahedral heterometallic clusters containing iron have proven to be suitable catalysts in the photoinitiated hydrosilylation of acetophenone with triethylsilane [54]. Although a chiral FeCoMoS tetrahedron was used, only racemic product and racemic cluster have been isolated. This can be explained by photo-racemization of the chiral catalyst, which proceeds faster than the hydrosilylation reaction. Instead of irradiation, metal vapor generation at $-196\,°C$ was described as the activation method for transition metal catalysts [55]. Iron vapor co-condensed with isoprene catalyzed the hydrosilylation with triethoxysilane. Not surprisingly, exclusive 1,4-addition was observed.

Brunner's group investigated the influence of thermal or photoinduced activation of Fe(Cp)(CO) complexes in the hydrosilylation of acetophenone (**4b**) with diphenylsilane, forming quantitatively the silylated 1-phenylethanol **59** (Scheme 4.26) [56, 57]. Brunner

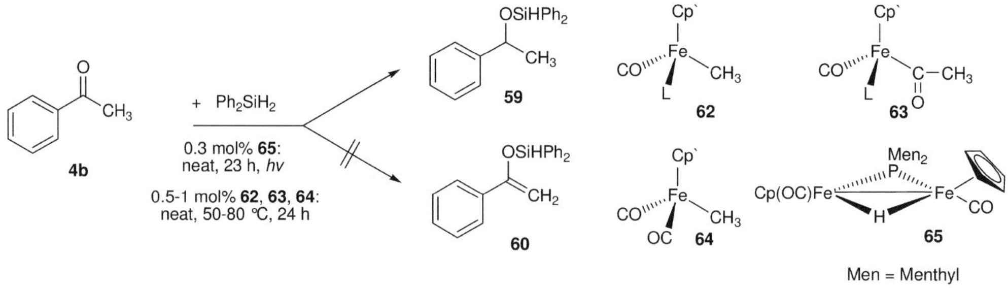

Scheme 4.26 Asymmetric hydrosilylation with chiral iron complexes.

and coworkers also reported the successful application of [(indenyl)Fe$(CO)_2(CH_3)$] complexes in the hydrosilylation of ketones, but extension to the hydrogenation of styrenes with molecular hydrogen was impossible since catalyst deactivation was observed.

The absence of silyl enol ether **60** can be explained by a hydrosilylation mechanism involving an Fe–Si–H three-center bond rather than an Fe–H species. Although optically active ligands of type **62**, **63** and **64** containing chiral iron atoms were used, only poor enantioselectivities below 10% *ee* were achieved [L = DIOP, *O*-isopropyliden-2,3-dihydroxy-1,4-bis(diphenylphosphino)butane] [58]. The rate-determining step in this reaction is the thermal loss of the phosphine ligand L in **62** and **63** and methyl migration in **64**, followed by addition of phenylsilane and subsequent reaction with acetophenone. Under photoirradiation conditions, hydrosilylation occurred very rapidly with complex **65** as catalyst, producing the silylated (*S*)-1-phenylethanol **59** in 33% *ee*. This was the first appreciable example of an enantioselective iron-catalyzed hydrosilylation of ketones. Recently, Nishiyama and Furuta pursued a different elegant strategy to find an efficient catalytic system based on iron [59]. They combined $Fe(OAc)_2$ and a multi-nitrogen-based ligand such as *N,N,N′,N′*-tetramethylethylenediamine (tmeda), bis-*tert*-butylbipyridine (bipy-*tb*) or bis(oxazolinyl)pyridine (pybox) to catalyze the hydrosilylation of ketones to give the corresponding alcohol after acidic cleavage of the silyl ether (Figure 4.3). The reaction was carried out under mild conditions (THF at 65 °C, 24 h), producing yields up to 95%. Using tmeda as nitrogen ligand, hydrosilylation works also with other aromatic ketones.

The application of chiral tridentate nitrogen ligands leads to the enantioselective reduction of methyl 4-phenylphenyl ketone **66** (Scheme 4.27). Whereas pybox-*bn* **68** gave 37% *ee* of **67**, bopa-*ip* **69** and -*tb* **70** increased the enantioselectivity up to 57% and 79% *ee*, respectively. Recent results from Beller's group revealed that it is possible to perform Fe-catalyzed hydrosilylations of acetophenone also in the presence of chiral phosphine ligands like DuPhos. Enantioselectivities up to 99% *ee* have been achieved for electronically rich and sterically hindered aryl ketones [60]. In addition, diaryl and dialkyl ketones were converted into the corresponding alcohols in good to excellent enantioselectivities (up to 79% *ee*). The same research group also developed a general and highly chemoselective method for hydrosilylation of aldehydes using ferrous acetate and tricyclohexylphosphine as catalyst system and PMHS (polymethylhydrosiloxane) as hydride source [61].

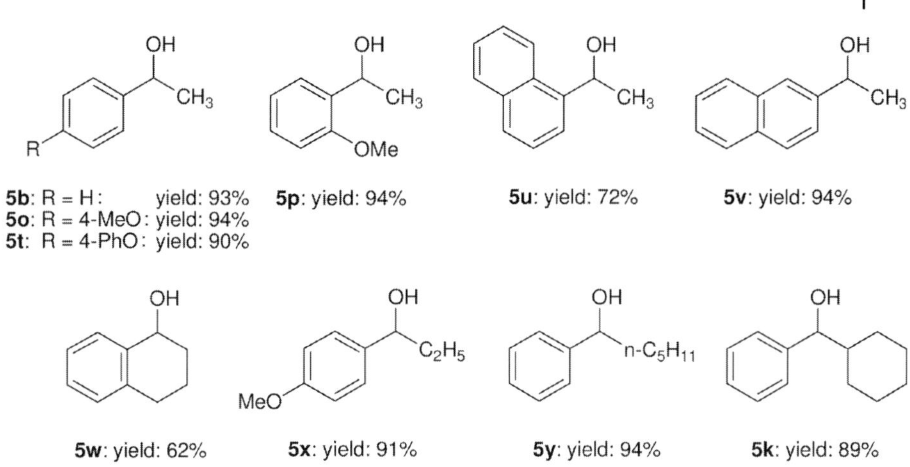

Figure 4.3 Substrate range in the Fe-catalyzed hydrosilylation of ketones in the presence of tmeda.

Iron-catalyzed hydrosilylations of alkenes in the presence of nitrogen ligands were first realized by Chirik and coworkers [34, 35]. They investigated the reaction of 1-hexene with $PhSiH_3$ or Ph_3SiH in the presence of iron catalyst **34** containing a tridentate pyridinediimine ligand (Scheme 4.15), which led to silylation over a course of minutes at ambient temperature. In both cases, the anti-Markovnikov product was formed exclusively. On the basis of these results, a series of alkenes (see Scheme 4.15) were examined. Terminal alkenes such as 1-hexene and styrene reacted most rapidly, followed by *gem*-disubstituted alkenes and internal alkenes. The silylation of alkynes was also examined and proceeded efficiently under mild conditions to give the corresponding silylalkene.

4.6 Conclusion

Since the beginning of transition metal-catalyzed hydrogenations and hydrosilylations more than 40 years ago, a tremendous number of studies have been directed to developing powerful methods for organic synthesis. In the past, Rh, Ir and Ru were clearly the metals of choice for such transformations. Several hundred catalysts are nowadays commercially available to chemists around the world. However, the

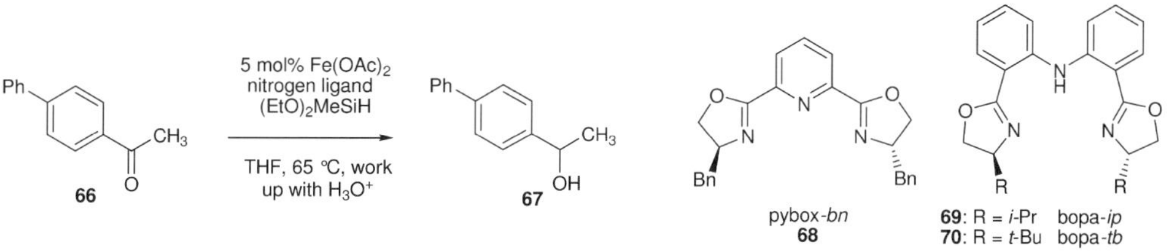

66 → 67: 5 mol% $Fe(OAc)_2$, nitrogen ligand, $(EtO)_2MeSiH$; THF, 65 °C, work up with H_3O^+

pybox-*bn* **68**

69: R = *i*-Pr bopa-*ip*
70: R = *t*-Bu bopa-*tb*

Scheme 4.27 Asymmetric hydrosilylation of ketone **66** with iron catalysts according to Nishiyama and Furuta [59].

accelerated costs and the stronger requirements for pharmaceuticals cast a shadow on these commonly used transition metals.

For the mid-term future, we expect an increased use of more available and "biomimetic" metals. Clearly, iron is an ideal example of this. Unfortunately, simple method transfer from Rh, Ir and Ru to Fe is not readily possible. So far, iron-based hydrogenations and hydrosilylations are far off becoming general methods for organic chemistry. Typically comparatively high catalyst loadings and harsh reaction conditions are required. In order to allow for more synthetic applications, especially more functional group tolerance and also efficient control of stereoselectivity are needed. There are some promising developments such as the recent work of the groups of Chirik and Nishiyama, who demonstrated that these goals can be achieved. It will be interesting to take part in this renaissance of iron chemistry.

References

1 (a) P. T. Anastas, M. M. Kirchhoff, *Acc. Chem. Res.* 2002, **35**, 686–694; (b) P. T. Anastas, M. M. Kirchhoff, T. C. Williamson, *Appl. Catal. A* 2001, **221**, 3–13.

2 (a) H.-U. Blaser, B. Pugin, F. Spindler, *J. Mol. Catal.* 2005, **231**, 1–20; (b) H.-U. Blaser, *Chem. Commun.* 2003, 293–296.

3 H.-U. Blaser,in "Chiral Catalysis – Asymmetric Hydrogenations" Supplement to *Chim. Oggi* 2004, 4–5.

4 For an excellent review, see C. Bolm, J. Legros, J. Le Paih, L. Zani, *Chem. Rev.* 2004, **104**, 6217–6254.

5 (a) I. C. Lennon, J. A. Ramsden, *Org. Process Res. Dev.* 2005, **9**, 110–112; (b) J. M. Hawkins, T. J. N. Watson, *Angew. Chem.* 2004, **116**, 3286–3290; (c) R. Noyori, T. Ohkuma, *Angew. Chem.* 2001, **113**, 40–75; (d) M. Miyagi, J. Takehara, S. Collet, K. Okano, *Org. Process Res. Dev.* 2000, **4**, 346–348; (e) E. N. Jacobsen, A. Pfaltz, H. Yamamoto (eds.), *Comprehensive Asymmetric Catalysis*, Springer, Berlin, 1999. (f) R. Noyori, *Asymmetric Catalysis in Organic Synthesis*, Wiley, New York, 1994.

6 (a) S. Sakuraba, K. Achiwa, *Synlett* 1991, 689–690; (b) S. Sakuraba, N. Nakajima, K. Achiwa, *Synlett* 1992, 829–830; (c) T. Kawaguchi, K. Saito,K. Matsuki, T. Iwakuma, M. Takeda, *Chem. Pharm. Bull.* 1993, **41**, 639–642; (d) T. Ohkuma, M. Koizuma, H. Doucet, T. Pham, M. Kozawa, K. Murata,E. Katayama,T. Yokozawa,T. Ikariya, R. Noyori, *J. Am. Chem. Soc.* 1998, **120**, 13529–13530; (e) T. Ohkuma, H. Ooka, R. Noyori, *J. Am. Chem. Soc.* 1995, **117**, 10417–10418; (f) T. Ohkuma, H. Ikehira, T. Ikariya, R. Noyori, *Synlett* 1997, 467–468.

7 (a) Y. Takegami, Y. Watanabe, I. Kanaya, T. Mitsudo, T. Okajima, Y. Morishita, H. Masada, *Bull. Chem. Soc. Jpn.* 1968, **41**, 2990–2994; (b) R. Noyori, I. Umeda, T. Istigami, *J. Org. Chem.* 1972, **37**, 1542–1545; (c) Y. Watanabe, M. Yashita, T.-a. Mitsudo, M. Tanaka, Y. Takegami, *Tetrahedron Lett.* 1974, **22**, 1879–1880; (d) J. P. Collmann, R. G. Finke, P. L. Matlock, R. Wahren, J. I. Brauman, *J. Am. Chem. Soc.* 1976, **98**, 4685–4687; (e) M. Yamashita, K. Miyoshi, Y. Okada, R. Suemitsu, *Bull. Chem. Soc. Jpn.* 1982, **55**, 1329–1330; (f) Y. Moglie, F. Alonso, C. Vitale, M. Yus, G. Radivoy, *Tetrahedron* 2006, **62**, 2812–2819.

8 L. Markó, M. A. Radhi, I. Ötvös, *J. Organomet. Chem.* 1981, **218**, 369–376.

9 L. Markó, J. Palágyi, *Transition Met. Chem.* 1983, **8**, 207–209.

10 (a) K. Jothimony, S. Vancheesan, *J. Mol. Catal.* 1989, **52**, 301–304; (b) K. Jothimony,

S. Vancheesan, J. C. Kuriacose, *J. Mol. Catal.* 1985, **32**, 11–16.

11 J.-S. Chen, L.-L. Chen, Y. Xing, G. Chen, W.-Y. Shen, Z.-R. Dong, Y.-Y. Li, J.-X. Gao, *Huaxue Xuebao* 2004, **62**, 1745–1750.

12 S. Enthaler, B. Hagemann, G. Erre, K. Junge, M. Beller, *Chem. Asian J.* 2006, **1**, 598–604.

13 H. Zhang, C.-B. Yang, Y.-Y. Li, Z.-R. Donga, J.-X. Gao, H. Nakamura, K. Murata, T. Ikariya, *Chem. Commun.* 2003, 142–143.

14 S. Enthaler, G. Erre, K. Junge, M. Beller, unpublished results.

15 (a) J. S. M. Samec, J.-E. Bäckvall, P. G. Andersson, P. Brandt, *Chem. Soc. Rev.* 2006, **35**, 237–248; (b) T. Ikariya, K. Murata, R. Noyori, *Org. Biomol. Chem.* 2006, **4**, 393–406, and references therein; (c) H. Guan, M. Iimura, M. P. Magee, J. R. Norton, G. Zhu, *J. Am. Chem. Soc.* 2005, **127**, 7805–7814; (d) M. Gómez, S. Janset, G. Muller, G. Aullón, M. A. Maestro, *Eur. J. Inorg. Chem.* 2005, 4341–4351; (e) K. Muñiz, *Angew. Chem.* 2005, **117**, 6780–6785; *Angew. Chem. Int. Ed.* 2005, **44**, 6622–6627; (f) C. Hedberg, K. Källström, P. I. Arvidsson, P. Brandt, P. G. Andersson, *J. Am. Chem. Soc.* 2005, **127**, 15083–15090; (g) A. S. Y. Yim, M. Wills, *Tetrahedron*, 2005, **61**, 7994–8004; (h) S. E. Clapham, A. Hadzovic, R. H. Morris, *Coord. Chem. Rev.* 2004, **248**, 2201–2237; (i) T. Koike, T. Ikariya, *Adv. Synth. Catal.* 2004, **346**, 37–41; (j) P. Brandt, P. Roth, P. G. Andersson, *J. Org. Chem.* 2004, **69**, 4885–4890; (k) J.-W. Handgraaf, J. N. H. Reek, E. J. Meijer, *Organometallics* 2003, **22**, 3150–3157; (l) C. P. Casey, J. B. Johnson, *J. Org. Chem.* 2003, **68**, 1998–2001; (m) C. A. Sandoval, T. Ohkuma, K. Muñiz, R. Noyori, *J. Am. Chem. Soc.* 2003, **125**, 13490–13503; (n) R. Noyori, *Angew. Chem.* 2002, **114**, 2108–2123; *Angew. Chem. Int. Ed.* 2002, **41**, 2008–2022; (o) C. P. Casey, S. W. Singer, D. R. Powell, R. K. Hayashi, M. Kavana, *J. Am. Chem. Soc.* 2001, **123**, 1090–1100; (p) O. Pàmies, J. E. Bäckvall, *Chem. Eur. J.* 2001, **7**, 5052–5058; (q) C. S. Yi, Z. He, *Organometallics* 2001, **20**, 3641–3643; (r) R. Noyori, M. Yamakawa, S. Hashiguchi, *J. Org. Chem.* 2001, **66**, 7931–7944; (s) M. Yamakawa, H. Ito, R. Noyori, *J. Am. Chem. Soc.* 2000, **122**, 1466–1478; (t) D. A. Alonso, P. Brandt, S. J. M. Nordin, P. G. Andersson, *J. Am. Chem. Soc.* 1999, **121**, 9580–9588; (u) D. G. I. Petra, J. N. H. Reek, J.-W. Handgraaf, E. J. Meijer, P. Dierkes, P. C. J. Kamer, J. Brussee, H. E. Schoemaker, P. W. N. M. van Leeuwen *Chem. Eur. J.* 2000, **6**, 2818–2829; (v) A. Aranyos, G. Csjernyik, K. J. Szabó, J. E. Bäckvall, *Chem. Commun.* 1999, 351–352; (w) M. L. S. Almeida, M. Beller, G. Z. Wang, J. E. Bäckvall, *Chem. Eur. J.* 1996, **2**, 1533–1536.

16 S. Enthaler, G. Erre, M. K. Tse, K. Junge, M. Beller, *Tetrahedron Lett.* 2006, **47**, 8095–8099.

17 S. Enthaler, B. Spilker, G. Erre, M. K. Tse, K. Junge, M. Beller, *Tetrahedron* 2008, **64**, 3867–3876.

18 (a) W. S. Knowles, M. J. Sabacky, *J. Chem. Soc., Chem. Commun.* 1968, 1445–1446; (b) W. S. Knowles, *Angew. Chem.* 2002, **114**, 2097–2107; (c) R. Selke, H. Pracejus, *J. Mol. Catal.* 1986, **37**, 213–225; (d) R. Selke, *J. Organomet. Chem.* 1989, **370**, 241–248.

19 (a) H. Pracejus, *Koordinationschemische Katalyse Organischer Reaktionen*, Verlag Theodor Steinkopf, Dresden, 1977; (b) Y. Takegami, T. Ueno, T. Fujii, *Bull. Chem. Soc. Jpn.* 1969, **42**, 1663–1667; (c) J. W. Kalecic, F. K. Schmidt, *Kinet. Katal.* 1966, **7**, 614–618; (d) W. G. Lipovic, F. K. Schmidt, J. W. Kalecic, *Kinet. Katal.* 1967, 1300–1306.

20 Y. Takegami, Y. Watanabe, I. Kanaya, T. Mitsudo, T. Okajima, Y. Morishita, H. Masada, *Bull. Chem. Soc. Jpn.* 1968, **41**, 2990–2994, and references therein.

21 E. N. Frankel, E. A. Emken, H. M. Peters, V. L. Davison, R. O. Butterfield, *J. Org. Chem.* 1964, **29**, 3292–3297.

22 E. N. Frankel, E. A. Emken, V. L. Davison, *J. Org. Chem.* 1965, **30**, 2739–2745.

23 M. Cais, N. Maoz, *J. Chem. Soc. A* 1971, 1811–1820.

24 Y. Tajima, E. Kunioka, *J. Org. Chem.* 1968, **33**, 1689–1690.

25 T. Nishiguchi, K. Fukuzumi, *Bull. Chem. Soc. Jpn.* 1972, **45**, 1656–1660.

26 M. A. Schroeder, M. S. Wrighton, *J. Am. Chem. Soc.* 1976, **98**, 551–558.

27 (a) R. L. Whetten, K.-J. Fu, E. R. Grant, *J. Am. Chem. Soc.* 1982, **104**, 4270–4272; (b) M. E. Miller, E. R. Grant, *J. Am. Chem. Soc.* 1984, **106**, 4635–4636; (c) H. Nagorski, M. J. Mirbach, *J. Organomet. Chem.* 1985, **291**, 199–204.

28 H. Inoue, M. Sato, *J. Chem. Soc., Chem., Commun.* 1983, 983–984.

29 H. Inoue, M. Suzuki, *J. Chem. Soc., Chem. Commun.* 1980, 817–818.

30 C. Bianchini, A. Mell, M. Peruzzini, P. Frediani, C. Bohanna, M. A. Esteruelas, L. A. Oro, *Organometallics* 1992, **11**, 138–145.

31 C. Bianchini, E. Farnetti, M. Graziani, M. Peruzzini, A. Polo, *Organometallics* 1993, **12**, 3753–3761.

32 K. Kano, M. Takeuchi, S. Hashimoto, Z.-i. Yoshida, *J. Chem. Soc., Chem. Commun.* 1991, 1728–1729.

33 (a) S. Sakaki, T. Sagara, T. Arai, T. Kojima, T. Ogata, K. Obkubo, *J. Mol. Catal.* 1992, **75**, L33–L37; (b) S. Sakaki, T.-i. Kojima, T. Arai, *J. Chem. Soc., Dalton Trans.* 1994, 7–11.

34 S. C. Bart, E. Lobkovsky, P. J. Chirik, *J. Am. Chem. Soc.* 2004, **126**, 13794–13807.

35 A. M. Archer, M. W. Bouwkamp, M.-P. Cortez, E. Lobkovsky, P. J. Chirik, *Organometallics* 2006, **25**, 4269–4278.

36 R. J. Trovitch, E. Lobkovsky, P. J. Chirik, *Inorg. Chem.* 2006, **45**, 7252–7260.

37 S. C. Bart, E. J. Hawrelak, E. Lobkovsky, *Organometallics* 2005, **24**, 5518–5527.

38 M. A. Radhi, L. Markó, *J. Organomet. Chem.* 1984, **262**, 359–364.

39 T. J. Lynch, M. Banah, H. D. Kaesz, C. R. Porter, *J. Org. Chem.* 1984, **49**, 1266–1270.

40 K. Yamamoto, T. Hayashi, H. Nishiyama, in M. Beller, C. Bolm (eds.), *Transition Metals for Organic Synthesis*, 2nd edn., Wiley-VCH, Weinheim, 2004, 167–191.

41 B. Marciniec, *Comprehensive Handbook on Hydrosilylation*, Pergamon Press, Oxford, 1992.

42 R. Kh. Freidlina, E. C. Chukovskaya, J. Tsao, A. N. Nesmeyanov, *Dokl. Akad. Nauk. SSSR* 1960, **132**, 374–377.

43 A. N. Nesmeyanov, R. Kh. Freidlina, E. C. Chukovskaya, R. G. Petrova, A. B. Belyavsky, *Tetrahedron* 1962, **17**, 61–68.

44 G. V. Nurtdinova, G. Gailiunas, V. P. Yur'ev, *Izv. Akad. Nauk SSSR, Ser. Khim.* 1981, **11**, 2652.

45 G. Gailiunas, G. V. Nurtdinova, V. P. Yur'ev, G. A. Tolstikov, S. R. Rafikov, *Izv. Akad. Nauk SSSR, Ser. Khim.* 1982, **4**, 914–920.

46 I. V. Savos'kina, N. A. Kuz'min, G. G. Galust'yan, *Izv. Akad. Nauk SSSR, Ser. Khim.* 1990, **2**, 483–485.

47 F. Kakiuchi, Y. Tanaka, N. Chatani, S. Murai, *J. Organomet. Chem.* 1993, **456**, 45–47.

48 M. A. Schroeder, M. S. Wrighton, *J. Organomet. Chem.* 1977, **128**, 345–358.

49 I. Fischler, F.-W. Grevels, *J. Organomet. Chem.* 1980, **204**, 181–190.

50 C. U. Pittman, Jr., W. D. Honnick, M. S. Wrighton, R. D. Sanner, R. G. Austin, *Fundam. Res. Homogeneous Catal.* 1979, **3**, 603–619.

51 R. G. Austin, R. S. Paonessa, P. J. Giordano, M. S. Wrighton, *Adv. Chem. Ser.* 1978, **168**, 189–214.

52 C. L. Randolph, M. S. Wrighton, *J. Am. Chem. Soc.* 1986, **108**, 3366–3374.

53 G. Bellachioma, G. Cardaci, E. Colomer, R. J. P. Corriu, A. Vioux, *Inorg. Chem.* 1989, **28**, 519–525.

54 C. U. Pittman, Jr., M. G. Richmond, M. Absi-Halabi, H. Beurich, F. Richter, H. Vahrenkamp, *Angew. Chem.* 1982, **94**, 805–806; *Angew. Chem. Int. Ed. Engl.* 1982, **21**, 786–787.

55 A. J. Cornish, M. F. Lappert, J. J. MacQuitty, R. K. Maskell, *J. Organomet. Chem.* 1979, **177**, 153–161.

56 H. Brunner, K. Fisch, *Angew. Chem.* 1990, **102**, 1189–1191; *Angew. Chem. Int. Ed. Engl.* 1990, **29**, 1131–1132.

57 H. Brunner, K. Fisch, *J. Organomet. Chem.* 1991, **412**, C11–C13.

58 H. Brunner, R. Eder, B. Hammer, U. Klement, *J. Organomet. Chem.* 1990, **394**, 555–567.

59 H. Nishiyama, A. Furuta, *Chem. Commun.* 2007, 760–762.

60 N. Shaikh, K. Junge, S. Enthaler, M. Beller, *Angew. Chem.* 2008, **18**, 2531–2535; *Angew. Chem. Int. Ed.* 2008, **47**, 2497–2501.

5
Iron-catalyzed Cross-coupling Reactions

Andreas Leitner

5.1
Introduction

In recent decades, transition metal-catalyzed carbon carbon bond-forming reactions have been developed into an indispensable tool for organic synthesis [1]. In particular, nickel- and palladium-catalyzed cross-coupling processes display an impressive application profile and have frequently been used in academia and industry [2]. Several important discoveries based on rational ligand design and diligent optimization of reaction conditions have been accomplished and the substrate scope of these methodologies has been significantly extended even to unreactive aryl chlorides and molecules bearing sensitive functional groups [3]. In view of developing environmental benign chemical processes to reduce waste and production costs, new transition metal cross-coupling processes are highly desirable. Especially rapid increasing prices of precious metals together with high costs of ligands can impede the application of transition metal-catalyzed methodologies on an industrial scale. Therefore, the development of catalysts based on inexpensive, low-toxicity metals and cheap ligands is an appealing challenge in chemical research. The implementation of iron salts as pre-catalysts in cross coupling reactions opens up new avenues in this context [4]. This chapter summarizes the synthetic scope and some mechanistic discussions of these powerful methods.

5.2
Cross-coupling Reactions of Alkenyl Electrophiles

In 1971, a year before the groups of Corriu and Kumada [5] independently reported the groundbreaking work on the topic of nickel-catalyzed cross-coupling reactions of aryl and vinyl halides with Grignard reagents, Tamura and Kochi described an iron-catalyzed vinylation reaction of Grignard reagents with vinyl halides (Scheme 5.1) [6].

Iron Catalysis in Organic Chemistry. Edited by Bernd Plietker

ISBN: 978-3-527-31927-5

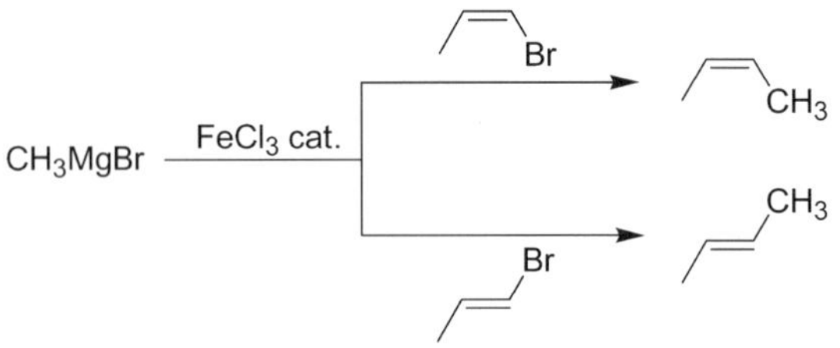

Scheme 5.1 Stereospecific iron-catalyzed cross-coupling reaction reported by Kochi and Tamura in 1971.

This process proceeded highly stereospecifically with retention of the configuration of the vinylic halide. Interestingly, (*E*)-alkenyl bromides reacted an order of magnitude faster than the corresponding (*Z*)-substrates. The catalyst has been described as an iron(I) species formed by reduction of an iron(III) precursor with the Grignard reagent. The reduced iron species appeared to be metastable and tended to deactivate, most probably by aggregation. Although high yields of cross-coupling product were obtained with methylmagnesium bromide, under the same conditions ethylmagnesium bromide afforded ethane and ethylene as side-products in significant amounts [7]. These side-reactions can be attributed to the available β-hydrogens in EtMgBr. Kochi's mechanistic proposal is depicted in Scheme 5.2 [8]. The catalyst can be formed by reduction of the pre-catalyst, naturally an iron salt in the oxidation state (+II) or (+III), by the Grignard reagent. Subsequent oxidative addition of the vinyl halide to the formal iron(I) species produces an iron(III) species. Transmetallation of the alkyl group from the organomagnesium compound to iron and final reductive elimination deliver the cross-coupling product with recovery of the iron(I) catalyst. A mechanistic delineation of this catalytic system is difficult to accomplish, since all organometallic intermediates are extremely unstable and impractical to isolate.

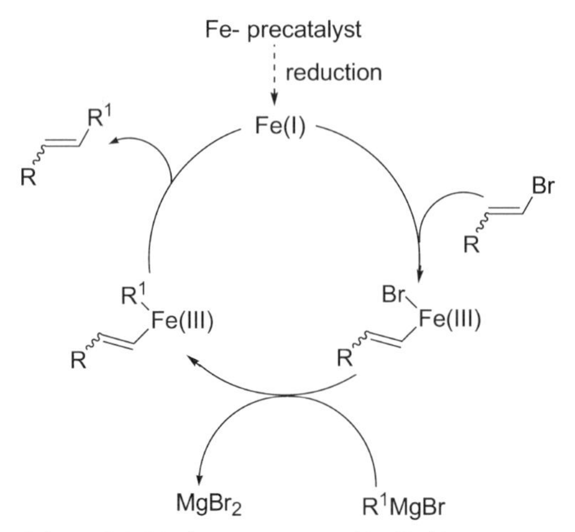

Scheme 5.2 Mechanism proposed by Kochi.

Later, Molander *et al.* extended this reaction to aryl-Grignard reagents by lowering the temperature and optimizing the solvent [9]. Futhermore, use of excess of alkenyl halide, as described in the initial mechanistic work of Kochi's group, proved to be unnecessary.

The rather limited synthetic scope of the iron-catalyzed cross-coupling reaction between alkenyl electrophiles and alkyl-Grignard reagents was significantly improved by Cahiez and Avedissian by introducing NMP (*N*-methylpyrrolidone) as co-solvent [10]. As depicted in Scheme 5.3, only 5% of the desired product could be obtained in pure THF. Addition of 9 equiv. of NMP (with respect to the Grignard reagent) enhanced the yield up to 85%. Most probably, NMP stabilizes organometallic iron intermediates, which are accountable for the turnover of the whole catalytic process.

Bu(Bu)C=CHCl + BuMgCl —[1% Fe(acac)$_3$, -5°C to 0°C, 15 min]→ Bu(Bu)C=CHBu
in THF: 5% yield
in THF/NMP: 85% yield

Scheme 5.3 Effect of NMP as additive.

Under these mild reaction conditions, cross-coupling reactions of vinyl halides (X = I, Br, Cl) proceed chemo- and stereoselectively with high yields (Table 5.1, entries 1–3). Increase of the steric bulk of the Grignard reagent leads to a general trend of decreasing yields in the order Me $\approx 1° > 2° > 3°$ (entries 4–7). The latter are prone to give reduced products over the decomposition pathway of the organometallic intermediate through β-hydrogen elimination [11]. The configuration of the vinylic double bond has no significant influence on the rate of the reaction (entries 10 and 11). Reactions with cyclic Grignard reagents and cyclic alkenyl halides display high conversions (entries 12 and 13). Vinyl–vinyl and aryl–vinyl cross-coupling reactions were conducted at higher temperatures of 15–20 °C with reasonable yields (entries 14 and 15).

Organomanganese reagents display a similar reactivity profile (Table 5.1, entries 16 and 17) [12]. Although the reaction rates are lower, yields and selectivities are generally high.

Furthermore, bromothioethenes were reported as fairly useful electrophiles for the coupling reaction with secondary Grignard reagents (Table 5.1, entries 18 and 19) [13]. These transformations occur highly stereo- and chemoselectively at the vinyl bromide moiety whereas the thiophene functionality remains conserved under the described reaction conditions.

As exemplified in Table 5.2, the iron-catalyzed cross-coupling reaction with Grignard reagents proceeds with high yields even if reactive functional groups such as esters, ketones, alkyl chlorides and protected amines [14] are present (entries 1–4, 6 and 10). Free alcohol groups can be tolerated by protection through deprotonation and generation of the corresponding magnesium salt (entry 5) [15]. In particular, the scope of this methodology has been advanced by introducing functionalized aryl-Grignard reagents developed by Knochel and co-workers (entries 6–9). Most impressive, an aryl nonaflate (Nf) moiety (entry 8) remains intact during the cross-coupling event [16].

Alkenyl triflates were established by Fürstner's group as suitable substrates (Table 5.3) [17]. A variety of alkenyl triflates of ketones, β-keto esters and cyclic

Table 5.1 Iron catalyzed cross-coupling reactions of alkenyl halides.

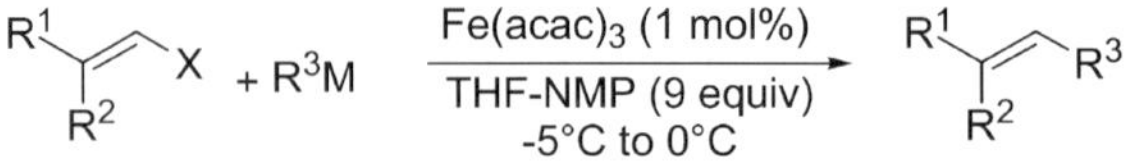

R^1, R^2, X + R^3M → (Fe(acac)$_3$ (1 mol%); THF-NMP (9 equiv); -5°C to 0°C) → R^1, R^2, R^3

Entry	Alkenyl halide	RM	Yield (%)	Entry	Alkenyl halide	RM	Yield (%)
	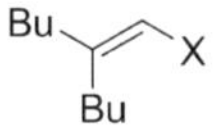 Bu, Bu, X			13	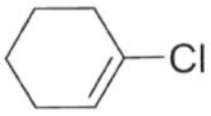 Cl (cyclohexenyl)	*n*-BuMgCl	75
1	X = I	*n*-BuMgCl	83				
2	X = Br		82				
3	X = Cl		85				
4	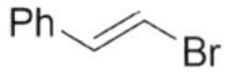 Ph, Br	MeMgCl	88	14[b]	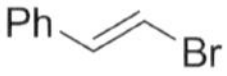 Ph, Br	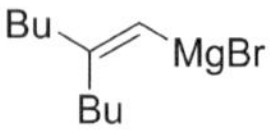 Bu, Bu, MgBr	60
5		*n*-BuMgCl	84				
6		*i*-PrMgCl	73				
7		*t*-BuMgCl	64				
8	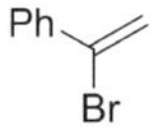 Ph, Br	MeMgCl	60	15[b]	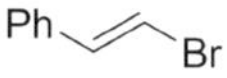 Ph, Br	PhMgCl	90
9	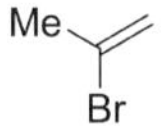 Me, Br	*n*-C_8H_{17}MgCl	86	16[c]	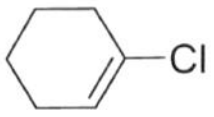 Cl (cyclohexenyl)	*n*-BuMnCl	80
10[a]	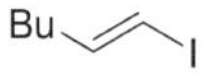 Bu, I	*n*-BuMgCl	75	17[c]	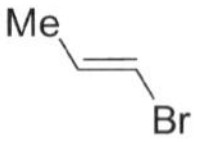 Me, Br	PhMnCl	71
11[a]	 Bu, I	*n*-BuMgCl	84	18[d]	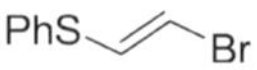 PhS, Br	*i*-PrMgCl	80
12	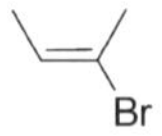 Br	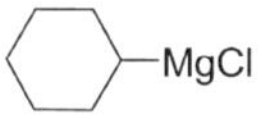 MgCl (cyclohexyl)	87	19[d]	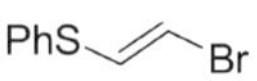 PhS, Br	*s*-BuMgCl	76

Reactions conducted with RMgX (1.1 equiv.) and Fe(acac)$_3$ (1 mol%) in THF-NMP (2–9 equiv.) at –5 to 0 °C (entries 1–13).

[a] Reaction performed with 0.1 mol% Fe(acac)$_3$.

[b] Reaction temperature 15–20 °C.

[c] 1.4 equiv. of RMnX was applied.

[d] Reaction performed with 2 equiv. of RMgX and 0.7 mol% Fe(DBM)$_3$ as pre-catalyst.

Table 5.2 Functional group tolerance.

$R^1(R^2)C=CH-X + R^3M \xrightarrow{Fe(acac)_3\ (cat.)} R^1(R^2)C=CH-R^3$

Entry	Alkenyl halide	RMgX	Yield (%)	Entry	Alkenyl halide	RMgX	Yield (%)
1	Cl, OAc	*n*-BuMgCl	80	6[b]	I, N(SO$_2$CF$_3$), Ph	EtO$_2$C, MgBr	69
2	O, Cl	*n*-BuMgCl	80	7[b]	Bu, I	NC, MgBr	60
3	O, I	MeMgCl	68	8[b]	Bu, I	CO$_2$Et, NfO, MgBr	73
4	Br, Cl	*n*-BuMgCl	79	9[b]	Bu, I	TIPSO, MgBr	62
5[a]	HO, Cl	*n*-$C_{12}H_{25}$MgBr	85	10[c]	Br, N, Boc	*n*-BuMgCl	64

Reactions conducted with RMgX (1.1equiv.) and Fe(acac)$_3$ (1 mol%) in THF–NMP (2–9 equiv.) at –5 to 0 °C (entry 1–4).
[a] 3 equiv. of RMgBr used.
[b] Fe(acac)$_3$ (5 mol%), THF, –20 °C, 15–30 min.
[c] Fe(acac)$_3$ (2 ,mol%), THF-NMP, 0 °C, 30 min.

1,3-diketones can be cross-coupled with Grignard reagents in good to excellent yields by applying Fe(acac)$_3$ as pre-catalyst of choice (Table 5.3, entries 1–13). In accord with reactions of alkenyl halides, this cross-coupling reaction outperforms the uncatalytic nucleophilic attack on other electrophilic sites in the substrates. This enables high functional group tolerance to e.g. esters, enones, ethers, carbamates, acetals and lactones.

Moreover, functionalized arylcopper reagents have been reported as appropriate nucleophiles with alkenyl and dienyl sulfonates (Table 5.3, entries 14–18) [18]. Although aryl sulfonates showed little reactivity towards arylcopper reagents, alkenyl sulfonates proved to be much more reactive under the described reaction conditions.

Table 5.3 Iron catalyzed cross-coupling reactions of alkenyl sulfonates.

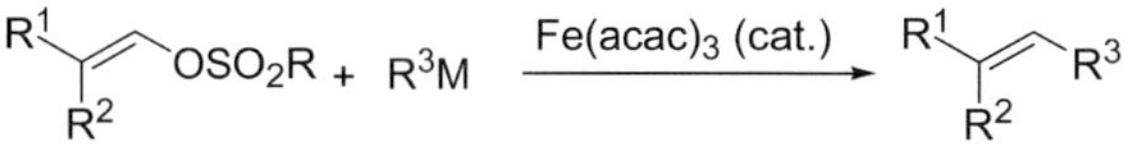

$$R^1(R^2)C{=}CH{-}OSO_2R + R^3M \xrightarrow{Fe(acac)_3\ (cat.)} R^1(R^2)C{=}CH{-}R^3$$

Entry	Alkenyl sulfonate	R–M	Yield (%)
1	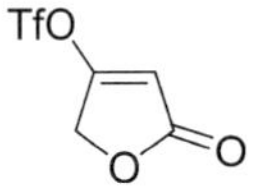 TfO, O, O	CH_3MgBr	70
2		n-$C_6H_{13}MgBr$	68
3		Me_3SiCH_2MgBr	80
4		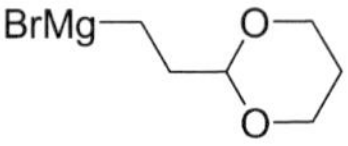 BrMg, O, O	67
5		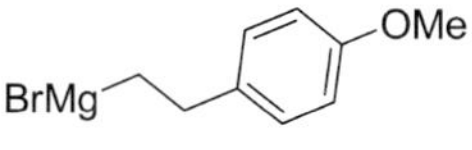 BrMg, OMe	84
6[a]	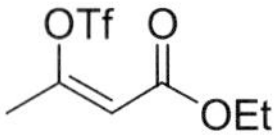 OTf, O, OEt	PhMgBr	83
7[a]		$C_{14}H_{29}MgBr$	98
8[b]	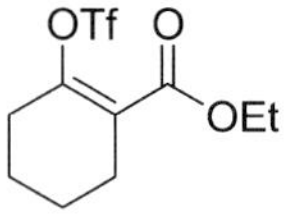 OTf, O, OEt	PhMgBr	53
9[b]		$C_{14}H_{29}MgBr$	79
10	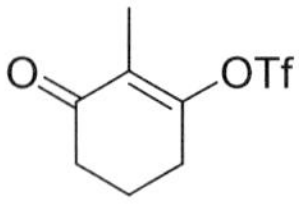 O, OTf	CH_3MgBr	73
11		C_4H_9MgBr	45
12	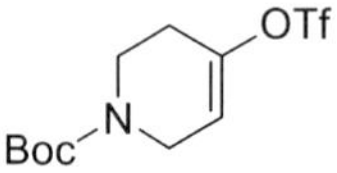 OTf, Boc, N	PhMgBr	47
13		$C_{14}H_{29}MgBr$	73
14[c]	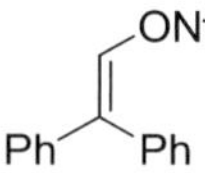 ONf, Ph, Ph	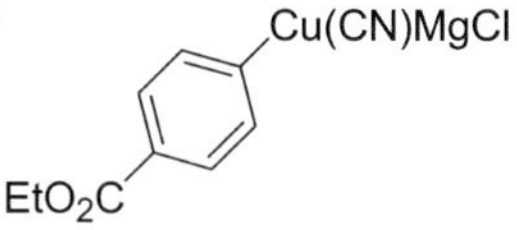 Cu(CN)MgCl, EtO_2C	77
15[c]	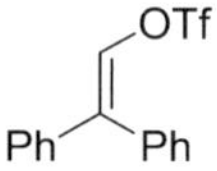 OTf, Ph, Ph	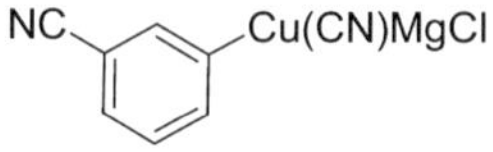 NC, Cu(CN)MgCl	56
16[d]	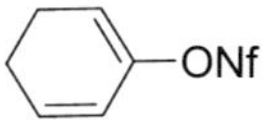 ONf	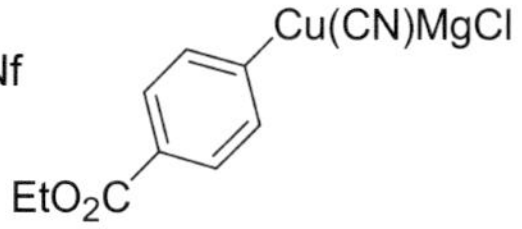 Cu(CN)MgCl, EtO_2C	86
17[e]	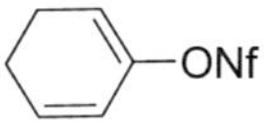 ONf	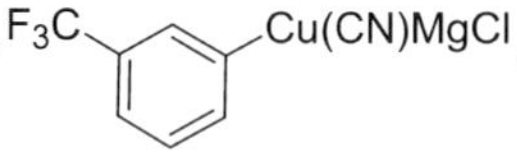 F_3C, Cu(CN)MgCl	90
18[e]	 ONf	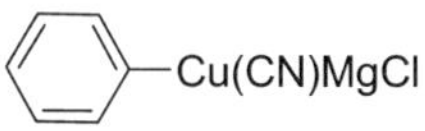 Cu(CN)MgCl	41

All reactions were performed using RMgX (1.1–1.3 equiv.) and $Fe(acac)_3$ (5 mol%) in THF-NMP at −30 °C unless stated otherwise.

[a] 1.8 equiv. of RMgX and 7 mol% of catalyst were used.

[b] 2.4 equiv. of RMgX and 10 mol% of catalyst were used.

[c] 2 equiv. arylcopper reagent was used.

[d] 1.4 equiv. arylcopper reagent was used.

[e] Reactions were carried out on a 1 mmol scale using 2.8 equiv. of arylcopper derivative and $Fe(acac)_3$ (10 mol%).

This is in line with the general tendency in cross-coupling chemistry that alkenyl halides are more reactive than aryl halides.

Alkenyl sulfones were also identified as suitable substrates by Julia and coworkers, but reduction and 1,4-addition were observed in notable quantities [19]. In some examples, sulfone electrophiles can give good yields and high stereoselectivities (Scheme 5.4).

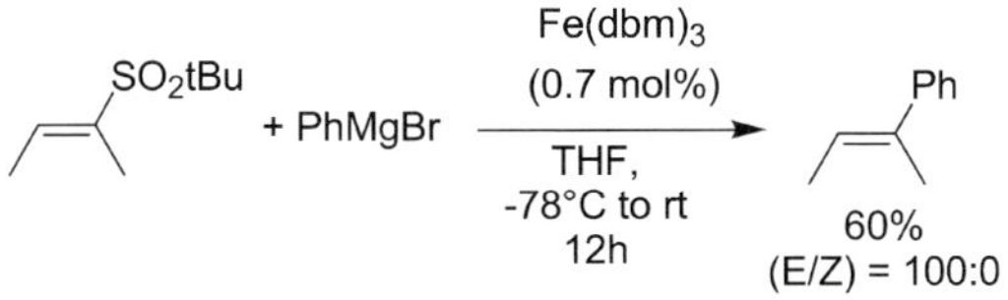

SO_2tBu + PhMgBr → [$Fe(dbm)_3$ (0.7 mol%); THF, -78°C to rt, 12h] → Ph product, 60%, (E/Z) = 100:0

Scheme 5.4 Iron-catalyzed cross-coupling reaction of alkenyl sulfones developed by Julia and coworkers.

Cahiez and Avedissian further reported the cross-coupling reaction of cheap and easily available enol phosphates [10]. These substrates are less reactive, which explains the higher catalyst loadings and 2 equiv. of Grignard reagent are required (Scheme 5.5).

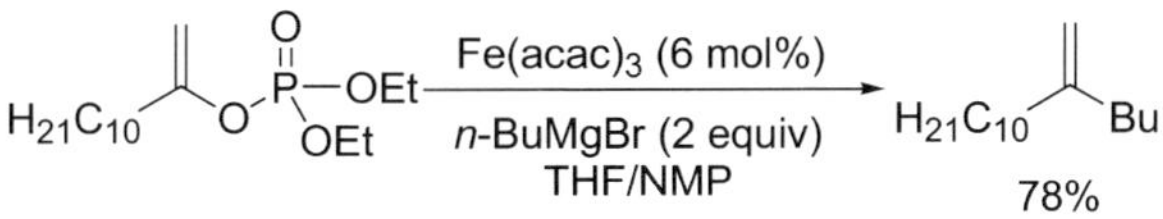

$H_{21}C_{10}$–C(=CH₂)–O–P(O)(OEt)₂ → [$Fe(acac)_3$ (6 mol%); *n*-BuMgBr (2 equiv); THF/NMP] → $H_{21}C_{10}$–C(=CH₂)–Bu, 78%

Scheme 5.5 Iron-catalyzed cross-coupling reaction of an enol phosphate reported by Cahiez and Avedissian.

Although bromothioethenes react chemoselectively at the halogen moiety (Table 5.1, entries 18 and 19), Itami *et al.* identified alkenyl sulfides as electrophiles for iron-catalyzed cross-coupling reactions with aryl- and alkyl-Grignard reagents (Scheme 5.6) [20]. When reactions were conducted with phenyl vinyl sulfide in the presence of catalytic amounts of $Fe(acac)_3$, the cross-coupling event occurred selectively in order to deliver the styrene product in good yield, whereas only tiny amounts of the biaryl product could be detected. The scope of this reaction is still rather limited, but the reactivity profile gives new appendages for mechanistic examinations of such transformations.

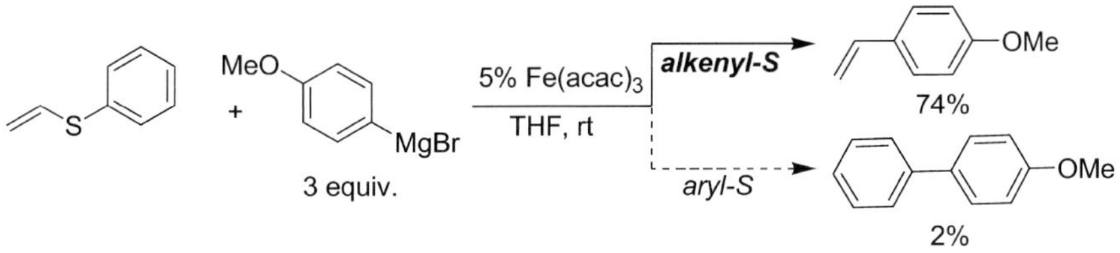

Phenyl vinyl sulfide + MeO–C₆H₄–MgBr (3 equiv.) → [5% $Fe(acac)_3$; THF, rt] → ***alkenyl-S***: 4-vinylanisole (OMe), 74%; *aryl-S*: 4-methoxybiphenyl (OMe), 2%

Scheme 5.6 Iron-catalyzed cross-coupling reaction of vinyl sulfides reported by Itami *et al.*

5.3
Cross-coupling Reactions of Aryl Electrophiles

Aryl halides and sulfonates emerged as excellent electrophiles in nickel- and palladium-catalyzed cross-coupling reactions [1]. Surprisingly, no comparable transformations have been reported employing iron catalysts. In 2000, Bogdanović's group reported, in one of a series of papers on the topic of inorganic Grignard reagents, $[Fe(MgX)_2]$ as a powerful catalyst to generate Grignard reagents from aryl chlorides and magnesium [21].

Such an inorganic Grignard reagent [22], in which magnesium has formally inserted into one or more metal–halide bonds, was generated by treatment of $FeCl_2$ with 4 equiv. of RMgX to give a cluster species with the formal composition $[Fe(MgX)_2]$. This consists of small magnesium and iron centers that are connected via intermetallic bonds. The reduction proceeded with release of the alkane R–H, the corresponding alkene and the homodimer R–R from the Grignard reagent (Scheme 5.7), similarly to the reduction process described by Smith and Kochi [7].

$$FeCl_2 + 4\,RCH_2CH_2MgX \xrightarrow{2\,(RCH_2CH_3\; +\; RCH{=}CH_2\; +\; RCH_2CH_2CH_2CH_2R)} [Fe(MgX)_2] + MgX_2$$

Scheme 5.7 Formation of an inorganic Grignard reagent described by Bogdanović's group.

Since the inorganic Grignard species consists of formally iron(−II) centers, the overall reaction does not stop once a zerovalent iron species is formed. Because such species are able to add oxidatively to aryl chlorides as shown by Bogdanović's group, Fürstner and Leitner re-evaluated iron-catalyzed cross-coupling reactions with aryl halides based on a mechanistic hypothesis as depicted in Scheme 5.8 [23]. Oxidative addition of the aryl halide to the inorganic Grignard reagent can result in an organometallic iron species with the formal oxidation state zero [Fe(0)], which, again, can be alkylated by the excess of Grignard reagent in analogy with the case of the elementary steps passed through during the initial formation of $[Fe(MgX)_2]$ from $FeCl_2$ and RMgX (Scheme 5.7). Reductive elimination of the organic ligands should form the desired product and regenerate the propagating iron species.

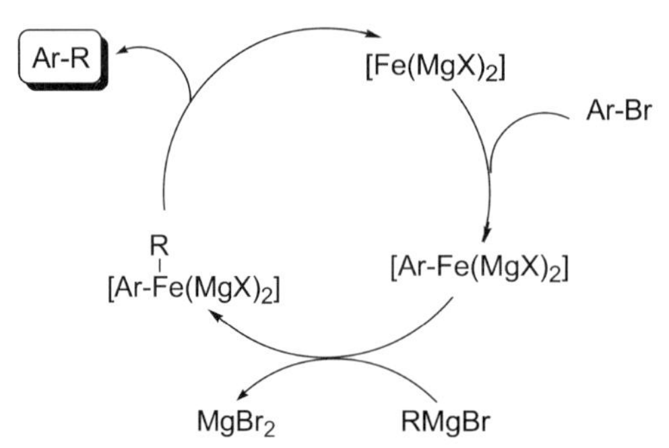

Scheme 5.8 Fürstner and Leitner's proposal for the iron-catalyzed cross-coupling reaction of aryl halides.

Table 5.4 Reactivity profile of aryl electrophiles in iron-catalyzed cross-coupling reactions.

n-$C_6H_{13}MgBr$, $Fe(acac)_3$ cat., THF/NMP, 0°C

A   B

Entry	X	A (GC) (%)	B (GC) (%)
1	I	27	46
2	Br	38	50
3	Cl	>95	—
4	OTf	>95	—
5	OTs	>95	—

In the event of a first substrate screening, aryl iodides and bromides yielded mainly the reduced compound **B** with n-$C_6H_{13}MgBr$ in presence of catalytic amounts of $Fe(acac)_3$ in THF–NMP at 0 °C (Table 5.4, entries 1 and 2). In contrast, the corresponding aryl chloride furnished the desired product in quantitative yield (entry 3); the triflate and even the tosylate, which turned out to be a difficult substrate in nickel- and palladium-catalyzed reactions [24], were converted in high yields (entries 4 and 5).

This reactivity profile turned out to be inverse to the general one of palladium- and nickel-catalyzed cross-coupling reactions, where the activity decreases in the order I > Br > Cl > OTs. Furthermore, activation of the C–Cl bond occurs with significant higher rates compared to the uncatalyzed nucleophilic attack on the polar ester group. These types of iron-catalyzed reactions occur with high rates within minutes at or below room temperature. The nature of the nucleophile emerged as crucial in this cross-coupling event. The iron-catalyzed cross-coupling of *n*- and *sec*-alkylmagnesium halides (with ≥2 carbon atoms), trialkyl zincates and all types of organomanganese reagents (RMnX, R_2Mn, $R_3MnMgCl$, R = alkyl, aryl) proceeds without incident, whereas methyl- and phenylmagnesium halides react only in special cases. Vinyl- and allyl-Grignard reagents, triethylaluminum and *n*-butyllithium failed to afford any product. This may arise from the inability of these organometallic compounds to reduce the iron pre-catalyst. Consistent with this notion is the fact that reactions with other "non-reducing" carbon nucleophiles such as boronic acids, stannanes and even RZnX and R_2Zn do not result in any product formation [27]. Fürstner and Leitner's mechanistic proposal is supported by a control experiment adopting a structurally well-defined, isolated, nucleophilic iron compound with a formal oxidation state of −II. $Li_2[Fe(C_2H_4)_4]$, which was prepared by Jonas and Schieferstein [25], afforded the cross-coupling reaction with comparatively high rates and yields (Scheme 5.9). This observation is of particular interest in the context that the highly nucleophilic $Na_2Fe(CO)_4$ [formally Fe(−II)] developed by Collman [26] did not pursue the cross-coupling reaction under the reported conditions.

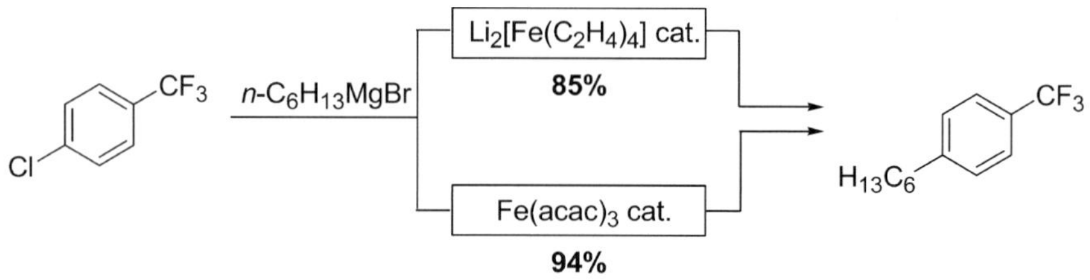

Scheme 5.9 $Li_2[Fe(C_2H_4)_4]$ as defined catalyst with a formal Fe(−II).

The iron-catalyzed cross-coupling reaction of aryl electrophiles with alkyl-Grignard reagents appears to be widely applicable [27]. As outlined in Table 5.5, moderately electron-deficient aryl and heteroaryl chlorides and tosylates reacted with good to excellent yields. The range comprises benzene derivatives substituted with electron-withdrawing substituents such as esters, nitriles, sulfonates, sulfonamides and $-CF_3$ groups (entries 1–7) and also many heterocyclic compounds such as pyridine [28], pyrimidine, triazine, quinoline, isoquinoline, carbazole, purine, pyridazine, pyrazine, quinoxaline, quinazoline, uracil, thiophene and benzothiophenes (entries 10–20). Most impressive is the replacement of a chloride function at C-6 of the per-*O*-acetylated purine-β-D-ribofuranoside by the Grignard reagent in presence of $Fe(acac)_3$ (entry 17). Uncatalyzed nucleophilic attack on the labile ester protecting groups did not occur. For the conversion of electron-rich arenes, the leaving group has to be changed from chloride to triflate to obtain the desired cross-coupling product (Table 5.5, entries 8 and 9) [29]. It is important to note that the reaction is, however, sensitive to steric hindrance and *ortho*-substituted substrates give lower yields than their *para*-substituted counterparts.

Furthermore, even the unprotected 6-chloropurine containing an N–H bond undergoes efficient cross-coupling reactions, although an extra equivalent of the Grignard reagent is necessary for the initial deprotonation and formation of the corresponding magnesium amide (entry 15). The cross-coupling reaction of 4-chloromethyl benzoate with nonylmagnesium bromide to generate the liquid crystalline material methyl 4-nonylbenzoate has been conducted even at multi-gram levels with high efficiency (Scheme 5.10) [30].

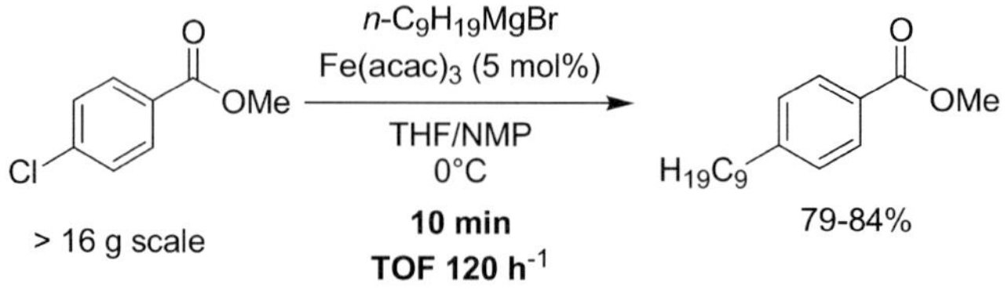

Scheme 5.10 Cross-coupling reaction on a multigram scale.

Kharasch and coworkers reported in a series of classical papers that aryl-Grignard reagents undergo catalytic decomposition with the formation of biaryls in presence

Table 5.5 Scope of aryl chlorides and sulfonates in iron-catalyzed cross-coupling reaction with RMgBr.

$$\text{Ar-X} + \text{RMgBr} \xrightarrow[\text{THF/NMP, 0°C}]{\text{Fe(acac)}_3\ (5\ \text{mol\%})} \text{Ar-R}$$

Entry	Ar–X	RMgBr	Yield (%)
1	CN, X	$C_{14}H_{29}MgBr$	91 (X = Cl)
2			87 (X = OTf)
3			83 (X = OTs)
4	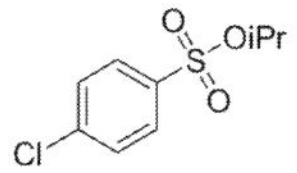 Cl, O, S, OiPr, O	$C_6H_{13}MgBr$	85
5	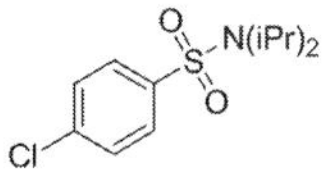 Cl, O, S, $N(iPr)_2$, O	$C_6H_{13}MgBr$	94
6	CF_3, X	$C_{14}H_{29}MgBr$	72 (X = OTf)
7			75 (X = OTs)
8	MeO, X, OMe	$C_{14}H_{29}MgBr$	0 (X = Cl)
9			90 (X = OTf)
10	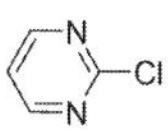 N, Cl, N	$C_{14}H_{29}MgBr$	93
11	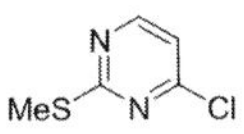 N, MeS, N, Cl	$C_{14}H_{29}MgBr$	89
12[a]		PhMgBr	53
13	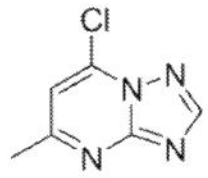 Cl, N, N, N, N	$C_{14}H_{29}MgBr$	67
14	N, N, H, Cl	n-C_2H_5MgBr	67
15	Cl, N, N, N, N, R	$C_{14}H_{29}MgBr$	85 (R = H)
16			90 (R = Me)
17	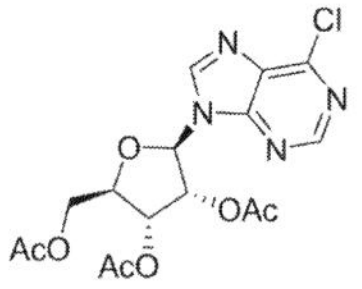 N, Cl, N, N, N, O, AcO, AcO, OAc	$C_{14}H_{29}MgBr$	72
18[a]	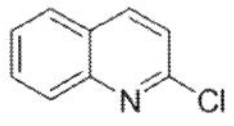 N, Cl	PhMgBr	71
19[a]	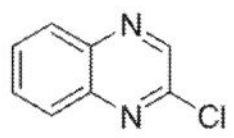 N, N, Cl	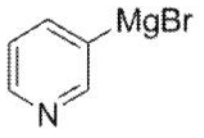 MgBr, N	82
20[a]	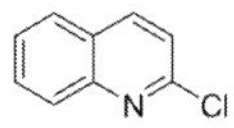 N, Cl	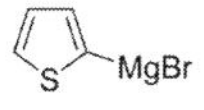 S, MgBr	63

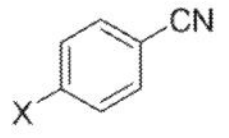

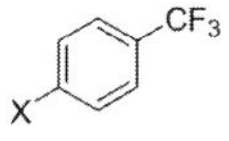

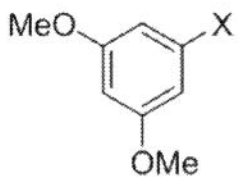

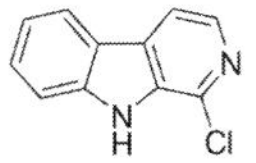

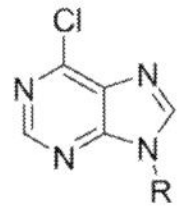

All reactions were carried using RMgBr (1.1–1.3 equiv.) and $Fe(acac)_3$ (5 mol%) in THF–NMP unless stated otherwise.

[a] Reactions were carried out in THF at −30 °C

of metal salts and aryl halides as stoichiometric oxidizing agents [31]. Although reactions with electron-deficient benzene derivatives under Fürstner *et al.*'s conditions showed mainly homodimerization of the magnesium reagent in analogy with the Kharasch reaction, π-electron-deficient heterocyclics undergo aryl–aryl cross-coupling reactions in pure THF at −30 °C with good yields, but varying amounts of Grignard dimerization product are inevitable (Table 5.5, entries 18–20). More recently, Knochel's group reported the first efficient iron-catalyzed aryl–aryl coupling with Grignard-derived functionalized copper reagents (Table 5.6) [32].

Table 5.6 Iron-catalyzed aryl–aryl coupling reaction developed by Knochel's group.

$$\text{Ar-I} + \text{Ar-I} + \text{Ar}^1\text{Cu(CN)MgCl} \xrightarrow[\text{25-80°C}]{\text{Fe(acac)}_3\ (10\ \text{mol\%})} \text{Ar-Ar}^1$$

Entry	Ar–I	R–M[a]	Yield (%)	Entry	Ar–I	R–M[a]	Yield (%)
1[b]	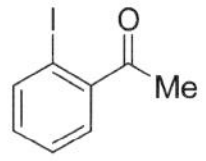 2-iodophenyl (C(O)Me)	Ph-Cu	86	6[c]	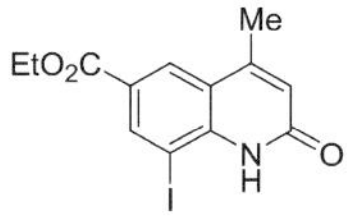 EtO$_2$C, Me, I, N-H, O (quinolinone)	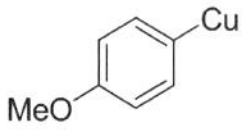 MeO-C$_6$H$_4$-Cu	91
2[b]	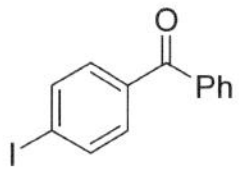 I-C$_6$H$_4$-C(O)Ph	Ph-Cu	80	7[c]	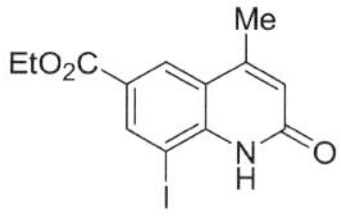 EtO$_2$C, Me, I, N-H, O (quinolinone)	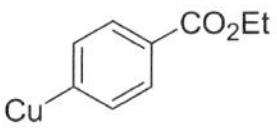 Cu-C$_6$H$_4$-CO$_2$Et	96
3[b]	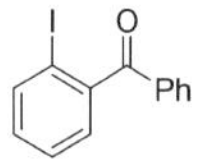 2-iodophenyl (C(O)Ph)	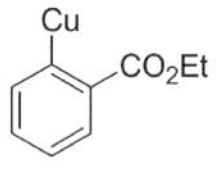 2-Cu-C$_6$H$_4$-CO$_2$Et	75	8[c]	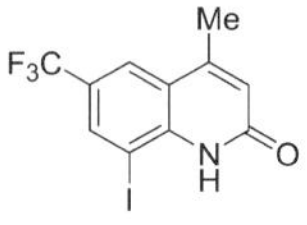 F$_3$C, Me, I, N-H, O (quinolinone)	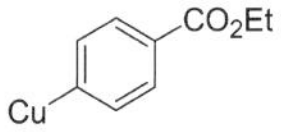 Cu-C$_6$H$_4$-CO$_2$Et	74
4[b]	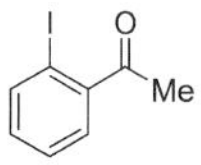 2-iodophenyl (C(O)Me)	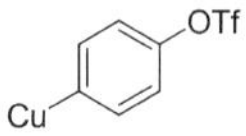 Cu-C$_6$H$_4$-OTf	62	9[c]	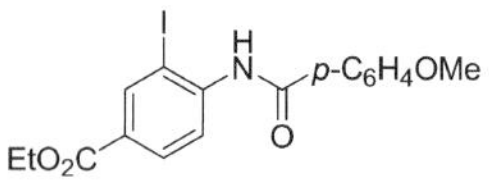 I, EtO$_2$C-C$_6$H$_3$-NHC(O)-*p*-C$_6$H$_4$OMe	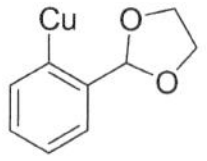 2-Cu-C$_6$H$_4$-(1,3-dioxolan-2-yl)	75
5[b]	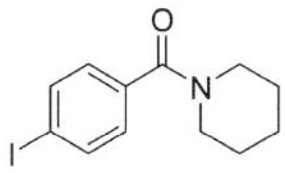 I-C$_6$H$_4$-C(O)N(piperidine)	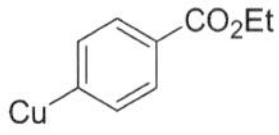 Cu-C$_6$H$_4$-CO$_2$Et	58	10[c]	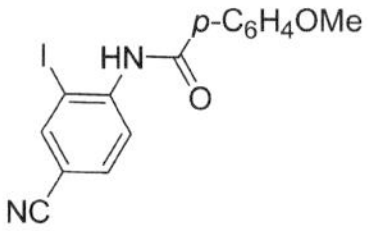 I, NC-C$_6$H$_3$-HN-C(O)-*p*-C$_6$H$_4$OMe	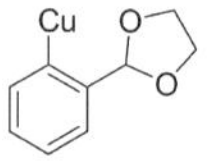 2-Cu-C$_6$H$_4$-(1,3-dioxolan-2-yl)	71

[a] The copper reagent is better represented as ArCu(CN)MgCl.
[b] Reactions were conducted at room temperature in presence of Fe(acac)$_3$ (10 mol%) in DME-THF.
[c] Reactions were performed at 80 °C.

It has been assumed that homo-coupling side-reactions may arise by the formation of ferrate complexes with the highly reactive organomagnesium compounds. Consequently, transmetallation of the aryl-Grignard to other metals should result in organometallic compounds with less tendency to dimerize. Applying organocopper reagents, however, the homodimerization reaction of the organometallic compound can be reduced and the aryl–aryl cross-coupling proceeds readily with good to excellent yields. Remarkably, the nature of the electrophile plays an important role and the reactivity profile appears to be inverse compared with the iron-catalyzed reaction described with organomagnesium compounds. Aryl iodides achieved full conversion to the desired biaryl with PhCu(CN)MgCl in presence of 10 mol% $Fe(acac)_3$ in DME–THF at room temperature after 30 min, whereas the corresponding aryl chloride and triflate react with significantly lower rates. No conversion was observed with tosylate as a leaving group. The iron-catalyzed aryl–aryl coupling with organocopper reagents tolerates a broad range of functional groups such as ketones, esters, nitriles, amides and acetals (Table 5.6). Remarkably, a methyl ketone, such as 2-iodophenyl methyl ketone, undergoes cross-coupling in high yields without formation of significant amounts of side-products by deprotonation of the methyl ketone or attack to the carbonyl group (entry 1). The copper reagent, bearing a triflate protected alcohol, reacts with ethyl 2-iodobenzoate to furnish the biaryl in 62% yield (entry 4). Functionalized quinolinones and secondary amides turned out to react with aryl cuprates, bearing electron-rich or electron-withdrawing subtitutents, with comparatively high yields (entries 6–10) [33].

The breakthrough towards the development of an efficient iron-catalyzed aryl–aryl cross-coupling reaction applying aryl-Grignard nucleophiles was accomplished by Hatakayama and Nakamura [34]. A broad evaluation of different iron sources and ligands afforded a novel combination of iron fluoride salts with an *N*-heterocyclic carbene ligand (NHC), which performs the conversion of aryl chlorides with aryl-Grignard reagents in high yields (Scheme 5.11). Competing homocoupling reaction of the Grignard reagent, as described by Kharasch's group, was suppressed, most likely due to addition of the fluoride anion. The catalyst is prepared by reduction of $FeF_3{\cdot}3H_2O$ with EtMgBr in the presence of the NHC ligand SIPr·HCl prior to its use in the catalytic reaction.

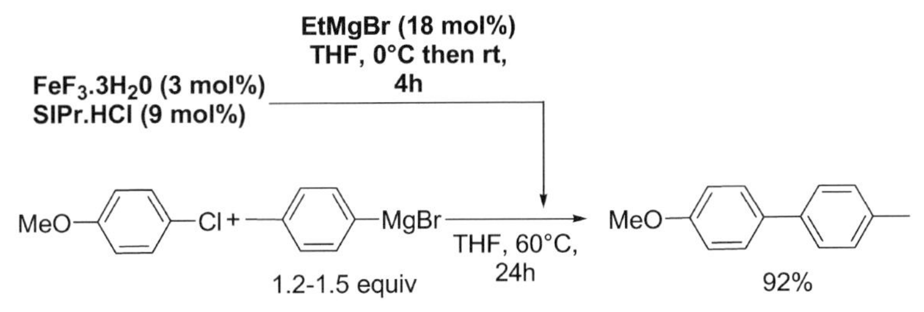

Scheme 5.11 Iron-catalyzed aryl–aryl cross-coupling reaction with Ar–MgX developed by Hatakayama and Nakamura.

The new coupling reaction is sensitive to the nature of the leaving group (Table 5.7, entries 1–5). Whereas aryl chlorides generally give high yields, aryl bromides and iodides show poor conversions and fluorobenzene did not react in any case. In general, this novel catalyst performs the cross-coupling reaction of electron-rich, -neutral and -poor aryl chlorides with high yields up to 98% (Table 5.7). Dimethylamino and methylthio groups did not interfere (entries 8 and 9). Reactions with sterically crowded substrates (entry 6) and even with mesitylmagnesium bromide (entry 7) were performed successfully. The mechanistic proposal suggests that the fluoride anion coordinating to the iron center suppresses the formation of a ferrate complex ($Ar^1Ar^2_2Fe$ and $Ar^1Ar^2_3Fe$) [35], which can undergo non-selective reductive elimination. From the mechanistic point of view it is important

Table 5.7 Iron-catalyzed aryl–aryl coupling with aryl-Grignard reagents developed by Hatakayama and Nakamura.

Entry	Ar–X	RMgBr	Yield (%)
1	X	MgBr	0 (X = F)
2			98 (X = Cl)
3			28 (X = Br)
4			23 (X = I)
5			27 (X = OTf)
6[a]	OMe, Cl	MgBr	92
7[b]	Cl	MgBr	93
8[c]	MeS, Cl	MgBr	80
9	Me$_2$N, Cl	MgBr	94
10	O, O, Cl	MgBr	88
11	Bu, Cl	F, MgBr	87
12[d]	N, Br	S, MgBr	74

Reactions were carried out under the conditions outlined in Scheme 5.11.
[a] 80 °C, 24h.
[b] 120 °C, 24h.
[c] 6 mol% of iron catalyst was used.
[d] 80 °C, 24h.

to note that the outlined aryl–aryl coupling displays much lower reaction rates (24 h at 60 °C) than the corresponding aryl-alkyl coupling (Table 5.4, 5 min at 0 °C).

Iron-catalyzed cross-coupling reactions of aryl electrophiles are not restricted to carbon nucleophiles such as Grignard reagents or cuprates. Recently, Taillefer *et al.* reported an iron–copper bimetallic catalyst which permits efficient cross-coupling reactions between aryl halides and nitrogen nucleophiles (Scheme 5.12) [36]. In the presence of 0.3 equiv. of $Fe(acac)_3$, 0.1 equiv. of CuO and base (2 equiv. Cs_2CO_3) in DMF at 100 °C, a broad variety of azoles such as pyrazole, imidazole, pyrrole, triazole, indole and a cyclic amide (pyrrolidin-2-one) undergo cross-coupling with aryl halides under mild reaction conditions with high yields up to 90%. In particular, this process is highly compatible with sensitive functional groups such as esters, aromatic amines, nitriles and nitro groups. Overall, the reaction proceeds cleanly and no obvious side-reactions, for example from the reduction of the aryl halide, were observed.

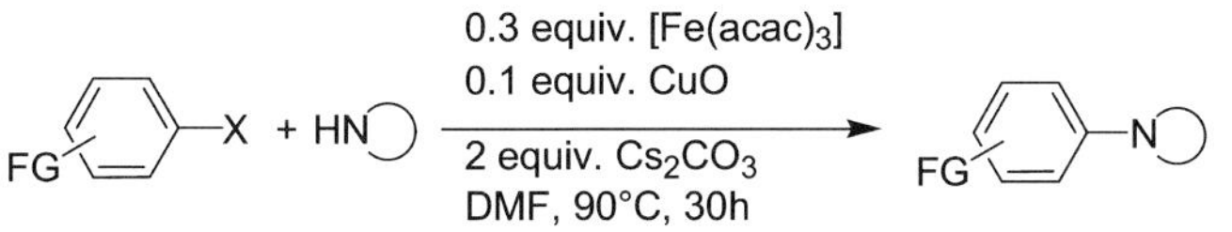

FG–C₆H₄–X + HN◯ → (0.3 equiv. $[Fe(acac)_3]$, 0.1 equiv. CuO / 2 equiv. Cs_2CO_3, DMF, 90°C, 30h) → FG–C₆H₄–N◯

Scheme 5.12 Iron-catalyzed C—N coupling reaction developed by Taillefer *et al.*

5.4 Cross-coupling Reactions of Alkyl Electrophiles

Transition metal-catalyzed cross-coupling reactions of alkyl halides and sulfonates have been an unsolved problem for a long time, although excellent ligand systems were known for aryl- and alkenyl electrophiles in Pd- and Ni-catalyzed processes [37]. In particular, the poor reactivity of alkyl electrophiles in order to add oxidatively to the metal center and the proclivity of the resulting metal fragment to undergo destructive β-hydrogen elimination rendered alkyl electrophiles especially challenging substrates [38]. The application and design of special ligands and careful fine tuning of the reaction conditions allowed the activation of primary and secondary alkyl electrophiles in nickel- and palladium-catalyzed reactions [39]. Taking these difficulties into account, it was remarkable when Nakamura *et al.* reported in 2004 the first iron-catalyzed cross-coupling reaction of secondary alkyl halides with aryl-Grignard reagents applying iron(III)chloride and TMEDA (*N,N,N′,N′*-tetramethylethylenediamine) as preferred catalyst [40]. Conducting the reactions at low temperature (−78 to 0 °C) and adding a solution of Grignard reagent and amine slowly via a syringe pump to the reaction mixture, cyclic, acyclic, primary and secondary alkyl halides were transformed in good to excellent yields in the range 45–99% and no side-reactions occurred with substrates bearing polar ester groups (Table 5.8, entries 1, 5, 9 and 10). At the same time, Nagano and Hayashi published similar work applying $Fe(acac)_3$ in boiling diethyl ether as the catalytic system (Table 5.8, entries 6–8) [41]. Slow addition of the nucleophile was not required but

Table 5.8 Scope of iron-catalyzed cross-coupling reactions with sp^3 electrophiles.

Entry	Electrophile	RMgX	Product	Yield (%)	Entry	Electrophile	RMgX	Product	Yield (%)
1[a] 2[c] 3[c] 4[c]	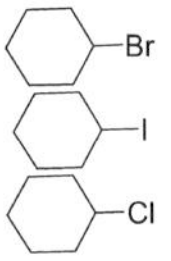 Br I Cl	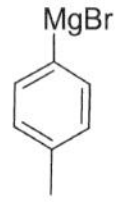 MgBr	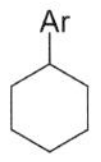 Ar	69 88 100 74	16[e]	EtOOC COOEt Br	Ph	EtOOC COOEt Ar	87
5[a]	n-$C_8H_{17}Cl$		n-$C_8H_{17}Ar$	70	17[e]	Br O OEt	Ph	Ar O OEt	87
6[b] 7 8	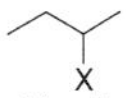 X X = I X = Br X = Cl	Ph	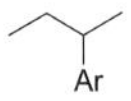 Ar	95 94 84	18[e]	Br	Ph	Ar	93
9[a]	O EtO I 5	Ph	O EtO Ar 5	88	19[e]	Br	Ph	Ar	84
10[a]	N I 3	Ph	N Ar	87	20[e]	Cl I	Ph	Cl Ph	86
11[c]	n-$C_8H_{17}Br$	Ph	n-$C_8H_{17}Ar$	62	21[e]	CN I	Ph	CN Ar	83
12[c]	Br	CH_3 MgBr	Ar	70	22[f]	Br	MgBr NMe_2	Ar	93
13[d]	Br	MgBr	Ar	78	23[f]	Br	MgBr MeO	Ar	83
14[d]	Br		Ar	91	24[f] 25[g]	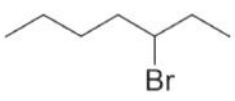 Br		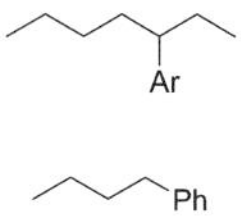 Ar	74 83
15[e]	I NCO	Ph	Ph NCO	90	26[g]	Br	Ph	Ph	75

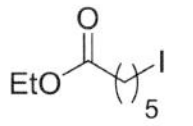

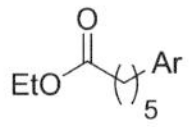

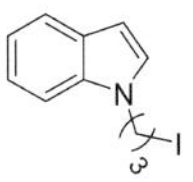

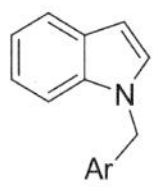

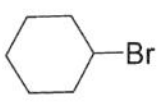

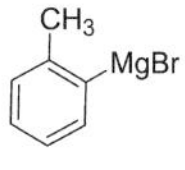

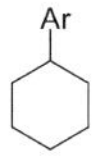

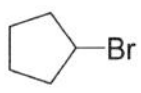

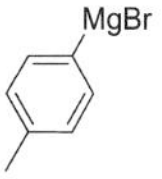

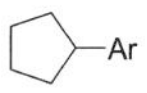

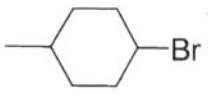

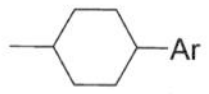

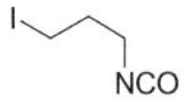

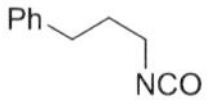

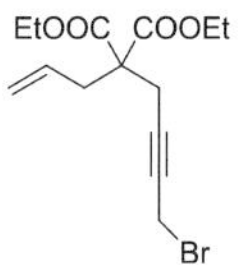

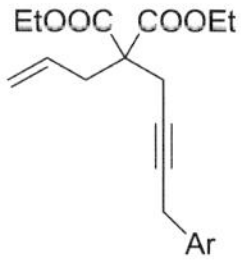

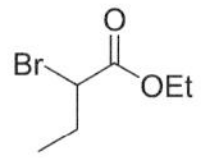

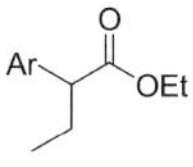

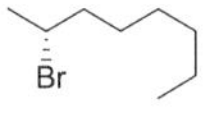

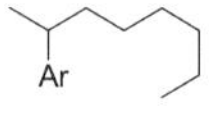

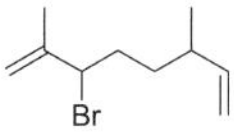

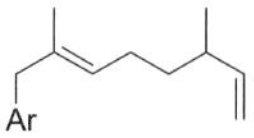

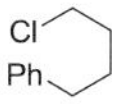

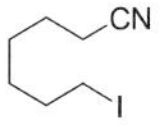

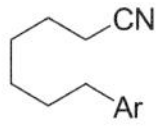

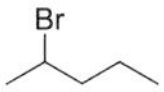

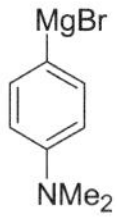

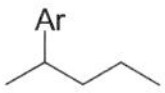

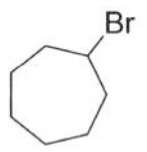

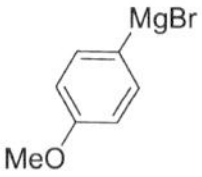

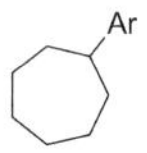

[a]**Nakamura** conditions: $FeCl_3$ (5 mol%), TMEDA (1.2 equiv.) in THF, −78 °C to 0 °C.
[b]**Hayashi** conditions: $Fe(acac)_3$ (5 mol%) in diethylether at reflux.
[c]**Bedford** conditions: $FeCl_3$ (10 mol%) in presence of DABCO.
[d]**Bedford** conditions: Iron nanoparticles (5 mol%) stabilized by PEG in diethyl ether at reflux.
[e]**Fürstner** conditions: $[Li(tmeda)]_2[Fe(C_2H_4)_4]$ (5 mol%), THF, −20 °C
[f]**Cahiez** conditions: $Fe(acac)_3$ (5 mol%) TMEDA (1.0 mmol), HMTA (0.5 mmol) in THF at 0 °C
[g]**Cahiez** conditions: $[(FeCl_3)_2(tmeda)_3]$ was used as catalyst.

lower yields (60–73%) were reported for similar substrates. Notably, the reaction conditions permit an intriguing reaction profile. A chemoselective conversion of a primary alkyl halide in the presence of an aryl triflate group could be accomplished. This is surprising, considering that Fürstner's group reported aryl triflates as valuable electrophiles for iron-catalyzed cross-coupling reactions. Subsequent treatment of the resulting aryl triflate with *n*-butylmagnesium bromide in THF–NMP under Fürstner's conditions delivered the final product in 90% yield (Scheme 5.13).

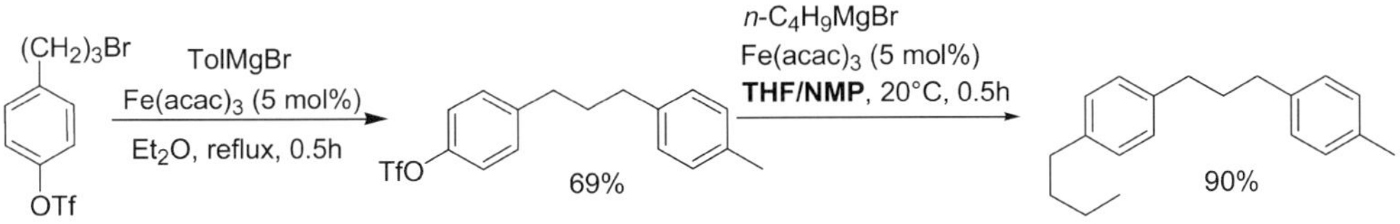

$(CH_2)_3Br$ OTf; TolMgBr, $Fe(acac)_3$ (5 mol%), Et_2O, reflux, 0.5h; TfO; 69%; n-C_4H_9MgBr, $Fe(acac)_3$ (5 mol%), **THF/NMP**, 20°C, 0.5h; 90%

Scheme 5.13 Chemoselectivity profile in iron-catalyzed cross-coupling reactions.

Bedford *et al.* demonstrated that pretreating of $FeCl_3$ with amines such as triethylamine, TMEDA or DABCO prior to its application in catalysis permits iron-catalyzed cross-coupling reactions, which can be performed in refluxing diethyl ether in high yields, and slow addition of the Grignard reagent is not required (Table 5.8, entries 2–4, 11 and 12) [42]. Optimum activities were obtained with either mono- or bidentate tertiary amines; chelating primary and secondary amines showed a lower performance. Under these conditions, no excess of amine is required and the Fe:N molar ratio was kept at 1:2. In a further intriguing report, Bedford *et al.* described polyethylene glycol (PEG)-stabilized iron nanoparticles as catalyst, which are generated *in situ* by reduction of $FeCl_3$ with the Grignard reagent [43]. Treatment of an ethereal solution of $FeCl_3$ and PEG with 5 equiv. of tolylmagnesium bromide gave a black solution. TEM analysis of this solution showed iron nanoparticles with a typical size range of about 7–13 nm in an MgX_2 matrix. This catalyst performed cross-coupling reactions of alkyl halides with aryl-Grignard reagents equally well with yields from 30 to 91% (Table 5.8, entries 13 and 14).

In line with the mechanistic proposal (Scheme 5.8) that iron species with a formal oxidation state of −II could be accountable for the iron-catalyzed cross-coupling of aryl halides, Fürstner's group further applied the isolated complex $[Li(tmeda)]_2[Fe(C_2H_4)_4]$ to the cross-coupling reaction of alkyl halides [44]. The tetrakis(ethylene) ferrate complex demonstrated a fascinating chemoselectivity profile. Apart from primary and secondary alkyl bromides, propargylic and allylic halides react smoothly with high yields of 67–98%. Under these mild reaction conditions (THF, −20 °C), a plethora of polar functional groups such as ketones, esters, enoates, chlorides, nitriles, isocyanates, ethers, acetals and trimethylsilyl functionalities remain intact (Table 5.8, entries 15–21). Despite allylic phosphates being described as substrates in iron-catalyzed reactions, efficient reaction with allylic bromides could be achieved for the first time with high selectivities in favor of the linear product (Table 5.8, entry 19). Note that complete loss of optical purity has been observed in the reaction of (*R*)-2-bromooctane (Table 5.8, entry 18). Tertiary alkyl halides and primary alkyl chlorides

(Table 5.8, entry 20) were found to be inert. This is contrary to the findings outlined above, where alkyl chlorides were reported as feasible substrates (Table 5.8, entries 4, 5 and 8).

Recently, Cahiez *et al.* reported hexamethylenetetramine (HMTA) in combination with TMEDA as an appropriate additive for reactions performed with $Fe(acac)_3$ (Table 5.8, entries 22–26) [45]. Although this catalyst showed no reaction with alkyl chlorides, primary and secondary alkyl bromides and iodides gave good to excellent yields of 39–94%. They also developed a more convenient method for the application of $FeCl_3$ as catalyst, which is highly hygroscopic and corrosive. Treatment of $FeCl_3$ with 1.5 equiv. of TMEDA gives a non- hygroscopic solid complex $[(FeCl_3)_2(tmeda)_3]$, which can be easily isolated quantitatively by simple filtration. The yields of the reactions conducted with this complex are slightly higher (5–10%) and the coupling can be performed at room temperature with slow addition of the Grignard reagent (Table 5.8, entries 25–26). Notably, $[(FeCl_3)_2(tmeda)_3]$ can be stored at room temperature without any special precautions and no additional additive has to be added to conduct the reaction.

Another interesting catalyst concept was reported by Bica and Gaertner, applying an iron-containing ionic liquid (bmim-$FeCl_4$) as catalyst (Scheme 5.14), which can be easily generated by treatment of commercially available bmim-Cl (butylmethylimidazolium chloride) and $FeCl_3{\cdot}6H_2O$ [46]. The hydrophobic ionic liquid can be isolated by phase separation.

$FeCl_4^{\ominus}$

0.1-10 mol%

-Br + -MgBr → Ether,0°C 30 min

89%

Scheme 5.14 bmim-$FeCl_4$ catalyst reported by Bica and Gaertner.

Cross-coupling reactions employing the ionic liquid catalyst were conducted under biphasic conditions without an inert atmosphere and proceeded with high yields in a range 60–89%. The product was removed by decantation and the catalyst trapped in the ionic liquid can be reused. This highly practicable work-up procedure can offer new possibilities for applying this reaction on a large scale in industrial processes. Since the yields and reaction rates of reactions with bmim-$FeCl_4$ are comparable to those in reactions conducted with $FeCl_3$ and an *N*-heterocyclic carbene as reported by Bedford and coworkers, one can conclude that the nature of the catalytically active species in both reactions is related [47].

Transition metal-catalyzed cross-coupling reactions between vinyl organometallic compounds and unactivated alkyl halides that can be usually performed with palladium, nickel and cobalt are of particular synthetic interest [37–39]. Recently, the groups of Cahiez [48] and Cossy [49] concurrently reported the first iron-catalyzed reaction of alkenyl Grignard compounds with primary and secondary alkyl halides (X = Br, I) (Scheme 5.15). The two protocols basically differ in the iron source

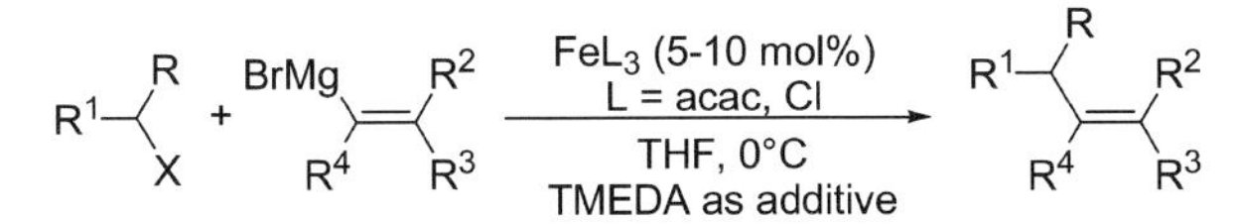

R¹–CH(R)–X + BrMg(R⁴)C=C(R²)(R³) → [FeL₃ (5-10 mol%), L = acac, Cl; THF, 0°C; TMEDA as additive] → R¹–CH(R)–C(R⁴)=C(R²)(R³)

R = Alkyl; R^1 = H, Alkyl; X = I, Br

R^2, R^3, R^4 = H, Alkyl, $SiMe_3$

Cossy: 35-98%
Cahiez: 48-84%

Scheme 5.15 Cross-coupling of alkenyl-Grignard reagents with alkyl electrophiles.

deployed, $Fe(acac)_3$ and $FeCl_3$, respectively, and in the quantitiy of amine additives applied. Whereas Cossy *et al.* reported 1.9 equiv. of TMEDA (with respect to the Grignard reagent) as an essential additive, Cahiez *et al.* transferred a similar catalyst, recently developed for the cross-coupling of alkyl halides with ArMgX [$Fe(acac)_3$–TMEDA–HMTA (1:2:1)] [45], successfully to perform this type of reaction.

The scope of this reaction includes α- or β-monosubsituted as well as α, β- or α, β-disubstituted alkenylmagnesium bromides, which undergo cross-coupling under the outlined reaction conditions with various cyclic and acyclic alkyl bromides and iodides. In general, this new method tolerates several functional groups such as ethyl esters, acetal groups, trimethylsilyl protecting groups, amides, ethers and nitrile groups (Table 5.9).

Table 5.9 Cross-coupling of alkenyl-Grignard reagents with sp^3 electrophiles.

Entry	Alkenyl halide	RMgX	Yield (%)	Entry	Alkenyl halide	RMgX	Yield (%)
1[a]	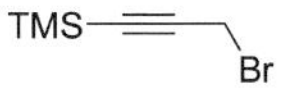 TMS–C≡C–CH₂Br	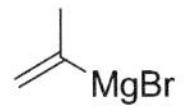 CH₂=C(CH₃)MgBr	80	7[a]	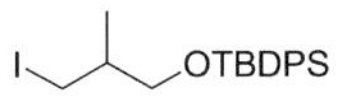 I–CH₂CH(CH₃)CH₂–OTBDPS	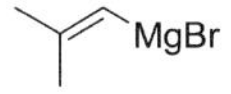 (CH₃)₂C=CH–MgBr	80
2[a]	 Ph–CH₂CH₂–I	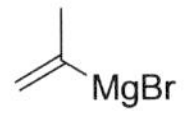 CH₂=C(CH₃)MgBr	98	8[a]	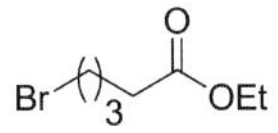 Br–(CH₂)₃–C(O)OEt	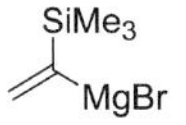 CH₂=C(SiMe₃)MgBr	40
3[a]	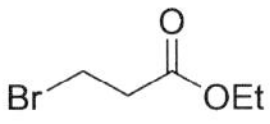 Br–CH₂CH₂–C(O)OEt	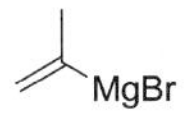 CH₂=C(CH₃)MgBr	0	9[b]	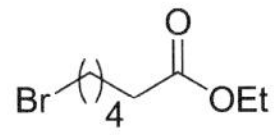 Br–(CH₂)₄–C(O)OEt	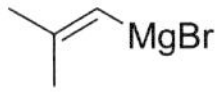 (CH₃)₂C=CH–MgBr	73
4[a]	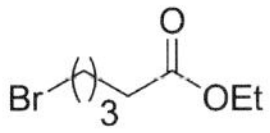 Br–(CH₂)₃–C(O)OEt	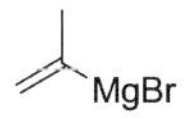 CH₂=C(CH₃)MgBr	62	10[b]	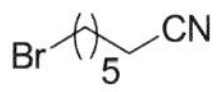 Br–(CH₂)₅–CN	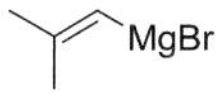 (CH₃)₂C=CH–MgBr	67
5[a]	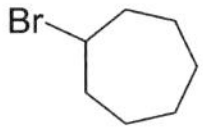 Br–cycloheptane	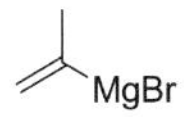 CH₂=C(CH₃)MgBr	94	11[b]	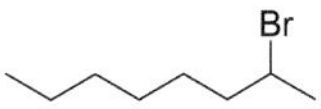 2-bromooctane	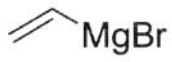 CH₂=CH–MgBr	55
6[a]	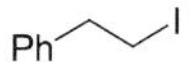 Ph–CH₂CH₂–I	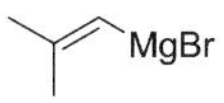 (CH₃)₂C=CH–MgBr	67	12[b]	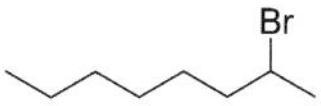 2-bromooctane	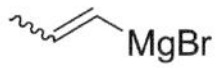 alkyl–CH=CH–MgBr	67 (*E/Z*: 85:15)

[a]**Cossy** conditions: $FeCl_3$ (5–10 mol%), THF, TMEDA (1.9 equiv.), 0 °C to room temperature, RMgX (2equiv.).
[b]**Cahiez** conditions: $Fe(acac)_3$–TMEDA–HMTA (1:2:1) (5 mol%), THF, 0 °C to room temperature, RMgX (1.5 equiv.).

Chai's group reported $Fe(OAc)_2$ in the presence of the bidentate phosphine ligand Xantphos in diethyl ether as catalyst for iron-promoted sp^3–sp^3 coupling reactions of alkyl bromides with alkyl-Grignard reagents (Scheme 5.16) [50]. The yields varied in the range 46–64%, since the formation of the alkene, which is formed by elimination of the alkyl halide in the presence of the organomagnesium compound, and also the disproportionation of the Grignard reagent, are difficult to suppress.

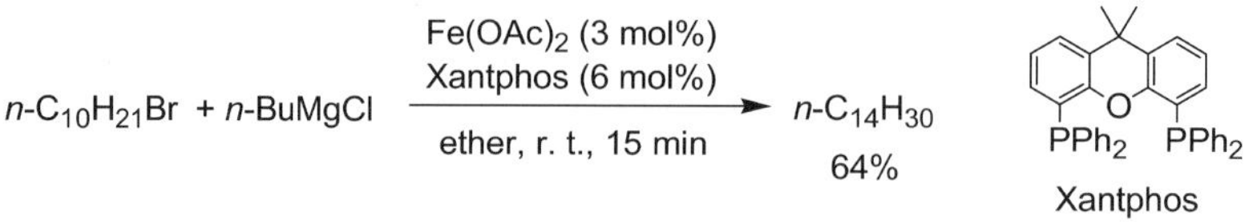

$n\text{-}C_{10}H_{21}Br$ + n-BuMgCl → [$Fe(OAc)_2$ (3 mol%), Xantphos (6 mol%); ether, r. t., 15 min] → $n\text{-}C_{14}H_{30}$ 64%

PPh_2 PPh_2 Xantphos

Scheme 5.16 sp^3–sp^3 coupling reactions reported by Chai's group.

Iron-catalyzed cross-coupling reactions of alkyl electrophiles are not limited to organomagnesium compounds [51]. Nakamura *et al.* reported the successful application of diarylzinc reagents employing $FeCl_3$ and TMEDA in THF at 50 °C as preferred catalyst [52]. To circumvent the dissipation of one aryl group of the diarylzinc reagent, Knochel's methodology to introduce the dummy ligand Me_3SiCH_2 has been applied successfully (Table 5.10, entries 2 and 3). Functionalized primary and secondary alkyl halides react with diorganozinc reagents in good to excellent yields. Note that heterocyclic organozinc reagents can also be employed. In general, arylzinc reagents bearing electron-withdrawing groups were found to be slightly less reactive that the corresponding phenyl reagent.

Table 5.10 Cross-coupling reaction of sp^3 electrophiles with diarylzinc reagents reported by Nakamura *et al.*

Entry	Electrophile	ArR'Zn	Product	Yield (%)
1	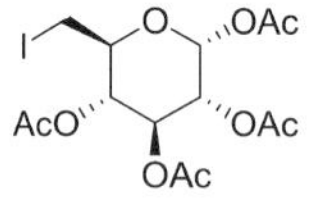 I, OAc, AcO, OAc, OAc (tetra-O-acetyl iodo sugar)	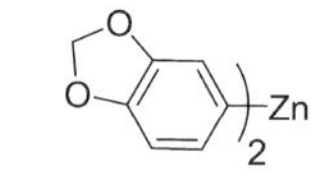 (benzodioxolyl)$_2$Zn	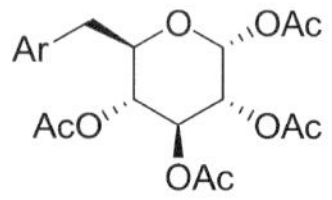 Ar, OAc, AcO, OAc, OAc	90
2	$n\text{-}C_{10}H_{21}I$	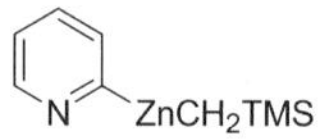 2-pyridyl-$ZnCH_2TMS$	$n\text{-}C_{10}H_{21}Ar$	98
3	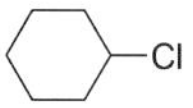 cyclohexyl–Cl	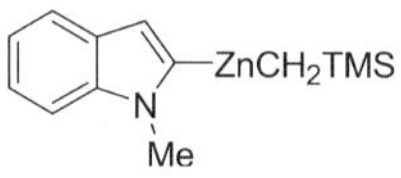 N-Me indol-2-yl-$ZnCH_2TMS$	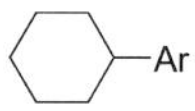 cyclohexyl–Ar	78
4	Me_3Si–≡–$(CH_2)_3I$	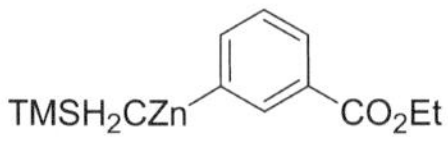 $TMSH_2CZn$–C_6H_4–CO_2Et	Me_3Si–≡–$(CH_2)_3Ar$	91

A controversial discussion about the mechanism of the iron-catalyzed cross-coupling reaction of alkyl electrophiles with aryl-Grignard reagents has appeared in the literature. Whereas Hayashi and coworkers proposed a classical catalytic cycle including the elementary steps of oxidative addition, transmetallation and reductive elimination, the groups of Nakamura, Bedford, Cossy and Cahiez suggested a radical related pathway. Scheme 5.17 delineates a highly simplified representation of a radical-based coupling reaction proposed by Bedford and coworkers. In the first step, the pre-catalyst is reduced to the oxidation state n. Subsequently the catalyst reacts with the alkyl halides by transfer of a single electron to generate an alkyl radical (via the intermediate formation of a radical anion) and an $[Fe^{(n+1)}]X]$ intermediate.

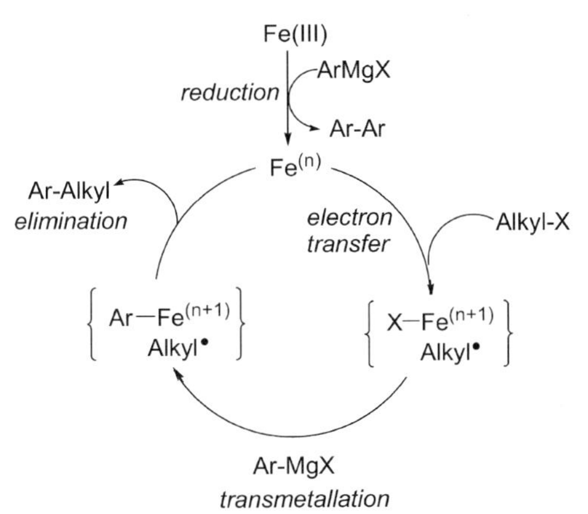

Scheme 5.17 Simplified radical coupling pathway proposed by Bedford and coworkers.

Nakamura's and Bedford's groups propose a metal-bound radical intermediate rather than a free radical. Transmetallation with the Grignard reagent generates an iron–aryl complex which reacts further with the alkyl radical to give the cross-coupling product with regeneration of the catalyst. This mechanistic proposal could be supported by following findings: (1) enantiomerically pure alkyl halides react with complete loss of optical purity [53] and (2) radical cyclization and ring opening reactions occur prior to the cross-coupling reaction [Scheme 5.18, reactions (1) and (2)].

Considering the racemization in the cross-coupling process of enantiomerically pure alkyl halides (Table 5.8, entry 18), Fürstner *et al.* partially support a theory delineating a metal-bound radical. However, they also call attention to several compounds set up for analogous 5-*exo-trig* cyclizations but that do not cyclize under identical conditions of the cross-coupling reaction [Scheme 5.18, reaction (3)]. Consequently, care has to be taken in generalizing this mechanistic proposal.

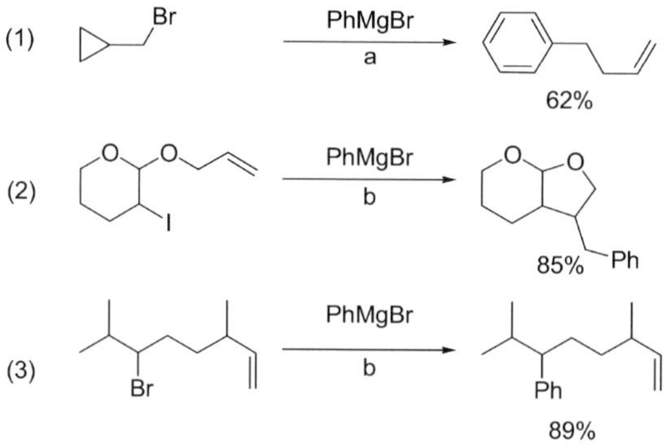

(1) [cyclopropylmethyl bromide] —PhMgBr, a→ [4-phenylbut-1-ene] 62%

(2) —PhMgBr, b→ 85% Ph

(3) Br —PhMgBr, b→ Ph 89%

[a]Reaction conducted with Fe-nano particles (5 mol%) in ether, reflux.
[b]Reaction was conducted with
$[Li(tmeda)]_2[Fe(C_2H_4)_4]$ in THF, -20°C.

Scheme 5.18 Radical cyclication *versus* direct cross-coupling.

5.5
Cross-coupling Reactions of Acyl Electrophiles

Functionalized ketones display valuable synthetic building blocks and their functionality is present in various natural products, pharmaceuticals and materials science target molecules [54]. In general, ketones are generated by acylation of organometallic intermediates and organomanganese reagents have proved to be especially useful [55]. Organomagnesium compounds cause an inherent reactivity problem and stoichiometric reactions generate significant amounts of tertiary alcohols through over-reaction of the ketone with another Grignard molecule. This destructive side-reaction can be suppressed by implementing a transition metal catalyst. Especially iron has been shown to be a highly efficient catalyst in acylation reactions of Grignard reagents with acyl chlorides. Based on early studies of iron-catalyzed acylation reactions by Kraus and coworkers [56], Marchese and coworkers found general conditions for the acylation of aryl- and alkyl-Grignard reagents applying $Fe(acac)_3$ as catalyst in THF at 0 °C or room temperature (Table 5.11, entries 1–5) [57].

Thioesters were also identified as suitable substrates (Table 5.11, entry 3) and the reaction can also be performed with di-Grignard reagents in high yields (Table 5.11, entry 5) [58]. Fürstner's group further improved the ketone synthesis by applying a lower reaction temperature of −78 °C, which opened up the possibility of employing functionalized substrates with a view to application in total synthesis (Table 5.11, entries 6–9) [17, 59]. Likewise, Reddy and Knochel reported the application of diorganozinc reagents in a similar reaction with acid chlorides (Table 5.11, entry 10) [60].

Because the iron-catalyzed formation of diaryl ketones oftengives low yields, Knochel's group tackled this problem by introducing aroyl cyanides as alternative acylation agents [61]. These types of compounds are more powerful in acylation

Table 5.11 Range of acyl chlorides and sulfonates in iron-catalyzed cross-coupling reactions.

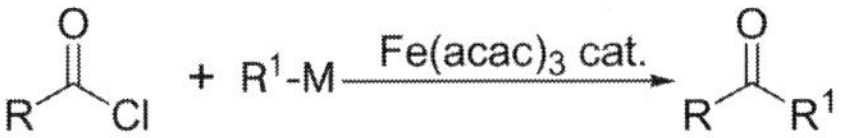

$RCOCl + R^1\text{-}M \xrightarrow{Fe(acac)_3\ cat.} RCOR^1$

Entry	Electrophile	R^1-M	Yield (%)	Entry	Electrophile	R^1-M	Yield (%)
1[a]	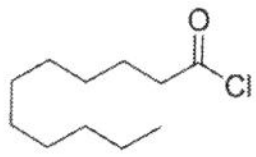 O, Cl	CH_3MgCl	84	6[c]	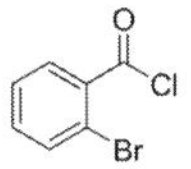 O, Cl, Br	CH_3MgBr	85
2[a]	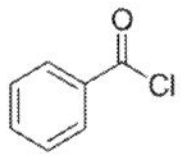 O, Cl	CH_3MgCl	80	7[c]	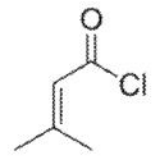 O, Cl	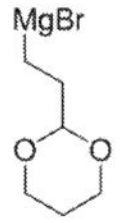 MgBr, O, O	80
3[a]	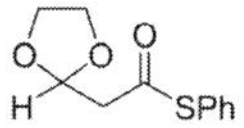 O, O, O, H, SPh	n-$C_6H_{13}MgBr$	85	8[c]	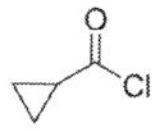 O, Cl	n-$C_6H_{13}MgBr$	92
4[a]	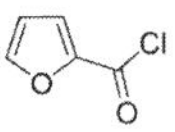 O, Cl, O	n-C_4H_9MgBr	83	9[c]	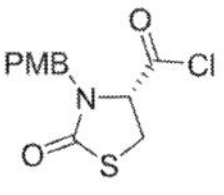 PMB, N, O, Cl, O, S	CH_3MgBr	80
5[b]	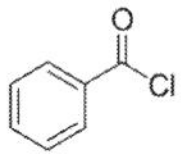 O, Cl	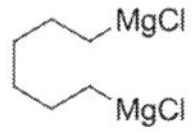 MgCl, MgCl	65	10[d]	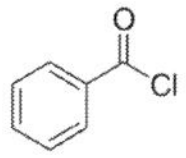 O, Cl	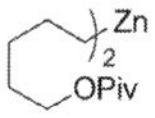 Zn, 2, OPiv	82

[a] Reactions conducted with 3 mol% $Fe(acac)_3$ at room temperature in THF.
[b] 6 mol% $Fe(acac)_3$ applied in THF at 0°C.
[c] Reactions proceed at −78 °C in THF in the presence of $Fe(acac)_3$ (3 mol%).
[d] $FeCL_3$ (10 mol%) in THF at −10 °C.

reactions, because the adjacent cyano group enhances the reactivity of the carbonyl group. In THF at −10 °C in thepresence of $Fe(acac)_3$ (5 mol%), a broad variety of functionalized aroyl cyanides can be transformed to the corresponding diaryl ketones with high yields of up to 98% and functional groups such as esters, nitriles and methyl ethers remain intact under the described reaction conditions (Scheme 5.19).

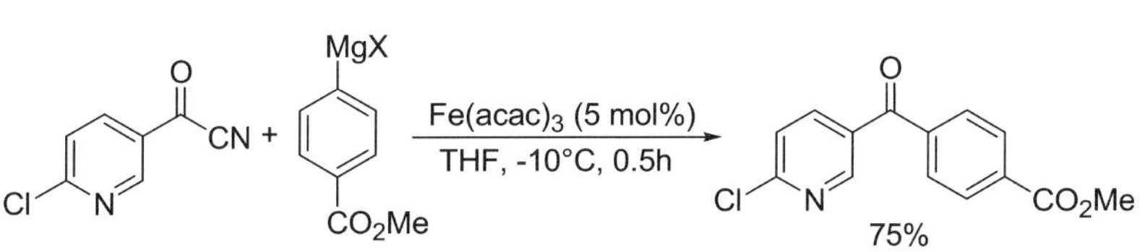

O, CN + MgX, CO_2Me, Cl, N

$Fe(acac)_3$ (5 mol%)
THF, -10°C, 0.5h

O, Cl, N, CO_2Me
75%

Scheme 5.19 Ketone synthesis *via* aroyl cyanides developed by Knochel's group.

5.6 Iron-catalyzed Carbometallation Reactions

The iron-catalyzed addition reaction of organometallic compounds to alkynes and alkenes, generally refered as carbometallation reaction [62], constitutes a further, but less investigated although still very important, carbon–carbon bond formation method. Early results of Lardicci and coworkers indicated that the addition reaction of trialkylaluminum reagents to alkynes in presence of $FeCl_3$ results in a complex mixture of products containing minor amounts of the desired compound [63]. In contrast, Hosomi's group reported efficient carbometallation reactions of alkyllithium reagents with alkynes bearing alkoxy or a tertiary amine group in presence of 10 mol% of an iron catalyst with high yields up to 97% (Scheme 5.20) [64].

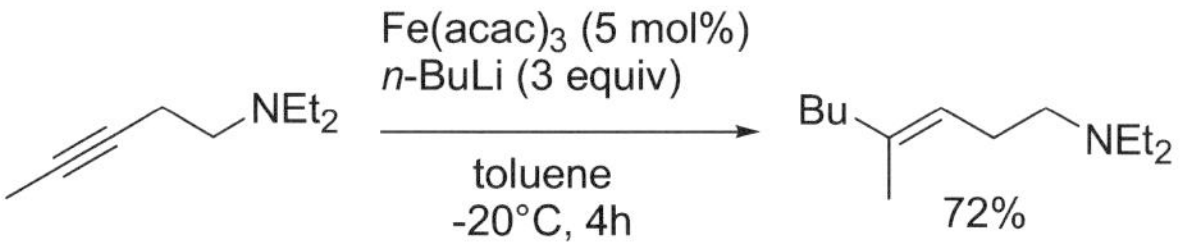

NEt2 → [Fe(acac)3 (5 mol%), n-BuLi (3 equiv); toluene, -20°C, 4h] → Bu, NEt2, 72%

Scheme 5.20 Carbometallation reaction employing alkyllithium reagents reported by Hosomi's group.

Since an excess of *n*-BuLi has been inevitable, iron-ate complexes were proposed as catalytic relevant species, which generate a vinyllithium intermediate by carbometallation. The latter was verified by a quench with aldehydes and ketones with generation of the corresponding allylic alcohol. It is important to note that these reactions took place in low yields with alkyl-Grignard reagents.

Arylmagnesation of unfunctionalized alkynes with aryl-Grignard reagents employing a cooperative catalyst that consists of a copper and an iron salt was reported by Hayashi's group (Scheme 5.21) [65]. In general, these reactions occur with good yields from 36 to 90% and high stereoselectivity ($E/Z = 18–99:1$), which indicates that the arylmagnesation proceeds with a high *syn* selectivity.

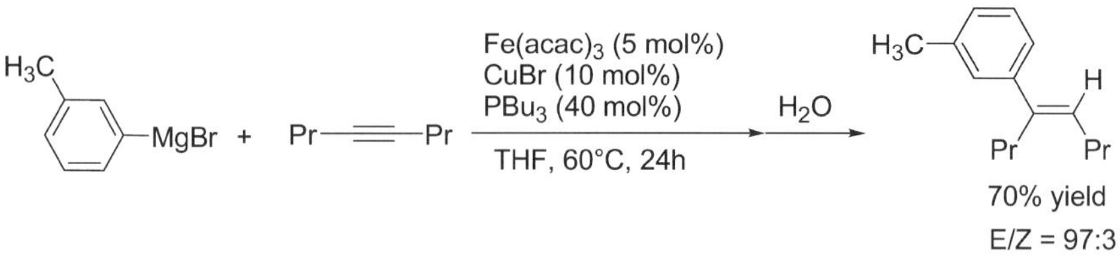

H3C, MgBr + Pr—≡—Pr → [Fe(acac)3 (5 mol%), CuBr (10 mol%), PBu3 (40 mol%); THF, 60°C, 24h] → H2O → H3C, H, Pr, Pr, 70% yield, E/Z = 97:3

Scheme 5.21 Arylmagnesation developed by Hayashi's group.

Note that no product formation could be obtained in absence of the iron or copper catalyst. Without copper, the iron catalyst generates only stoichiometric amounts (with respect to iron) of an alkenyliron species, which is not reactive towards further transformations such as transmetallation or polymerization. A cuprate, generated by CuBr and ArMgBr, did not undergo carbometallation with the alkyne. These results indicate that the main role of the copper catalyst is most likely to promote the metal exchange between the alkenyliron species and the aryl-Grignard reagent. As depicted in Scheme 5.22, addition of an aryliron species generates an alkenyliron intermediate, which transfers the alkenyl group to copper by transmetallation with the diaryl cuprate with regeneration of the iron catalyst. Further, transmetallation between the alkenyl (aryl) cuprate and ArMgBr releases alkenylmagnesium bromide with formation of $Ar_2CuMgBr$ to complete the catalytic cycle.

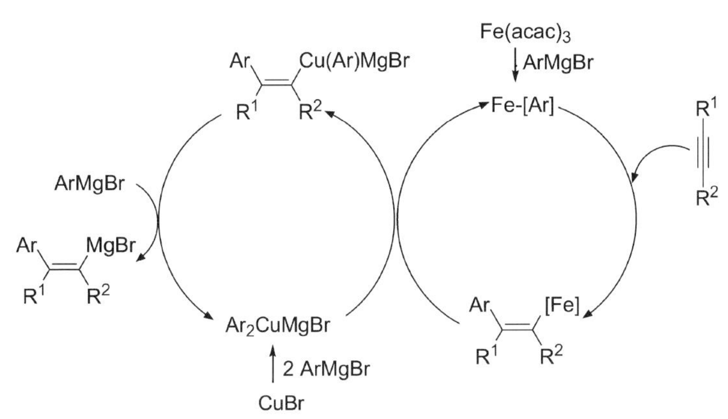

Scheme 5.22 Mechanistic proposal for the arylmagnesation developed by Hayashi's group.

Recently, Zhang and Ready reported an iron-catalyzed carbomagnesation reaction of primary and secondary (homo)propargylic alcohols with *n*-alkyl-Grignard reagents which generates highly regioselective *Z*-configured tri- and tetrasubstituted alkenes (Scheme 5.23) [66].

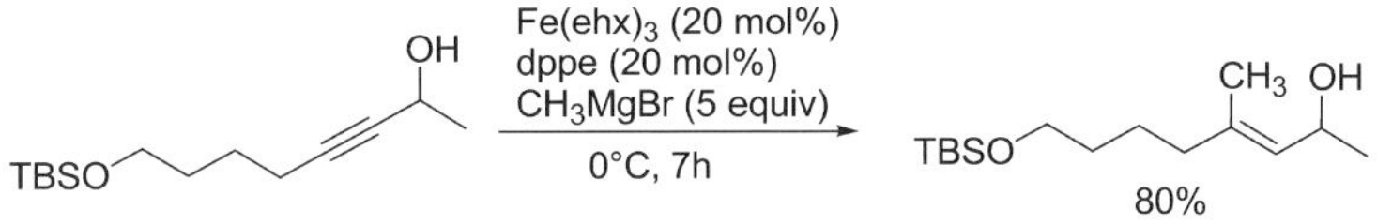

TBSO OH → TBSO CH3 OH

$Fe(ehx)_3$ (20 mol%), dppe (20 mol%), CH_3MgBr (5 equiv), 0°C, 7h; 80%

Scheme 5.23 Carbometallation of (homo)propargylic alcohols reported by Zhang and Ready.

In contrast to the carbometallation with aryl-Grignard reagents to unfunctionalized alkynes, this reaction does not require a co-catalyst such as copper. Only in one example, applying PhMgBr, was CuBr added in accord with the protocol reported by Hayashi's group. Although the nature of the catalytically active species remains unclear, an alkoxide-directed carbometallation which yields a (vinyl)iron intermediate is proposed.

Although the iron-catalyzed synthesis of allenes from propargylic halides was reported by Pasto and coworkers in 1978 [67], little progress was achieved in this field until recently [68]. In 2003, Fürstner and coworkers discovered propargylic epoxides as valuable substrates for the reaction with Grignard reagents in presence of catalytic amounts of $Fe(acac)_3$ to generate 2,3-allenol derivatives (Scheme 5.24) [69].

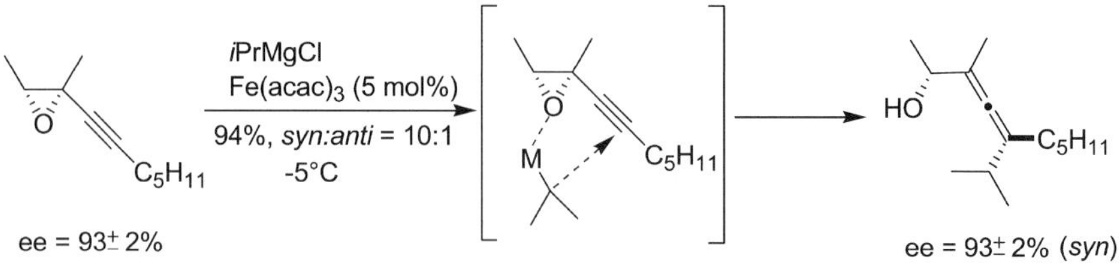

Scheme 5.24 Synthesis of 2,3-allenol derivatives developed by Fürstner and coworkers.

The reaction can be conducted under mild reaction conditions and the major isomer is *syn* configured, most likely due to directed delivery of the nucleophile to the alkyne by coordination of the catalyst to the oxygen atom of the substrate. Thereby, the central chirality of the substrate is transferred to the axial chirality of the resulting 2,3-allenol. This method complements established copper-catalyzed reactions with propargyl epoxides, which furnish the *anti*-configured allenol. In particular, these reactions occur under ligand-free conditions and gives high yields up to 98% [70].

Nakamura *et al.* reported iron-catalyzed carbometallation reactions of cyclopropane acetal derivatives (Scheme 5.25) and bicyclic alkenes with Grignard and organozinc reagents [71]. Of particular interest is the application of the chiral diphosphine (*R*)-*p*-Tol-BINAP in combination with the diamine TMEDA in enantioselective addition where an *ee* of up to 92% could be obtained. These findings are of importance considering that most iron-catalyzed cross-coupling reactions can be performed under *ligand-less* conditions. The high *ee* value clearly demonstrates that the phosphine ligand remains on the iron center during the reaction since no racemic by-products are formed by competing side-reactions.

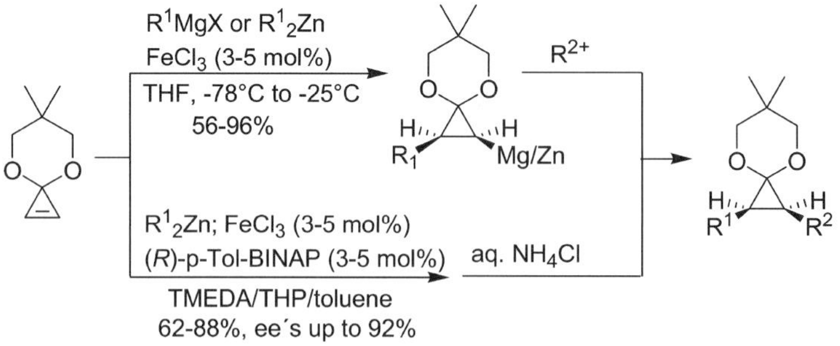

Scheme 5.25 Carbometallation of cyclopropane acetals reported by Nakamura *et al.*

5.7
Conclusion

Since the discovery of transistion metal-catalyzed cross-coupling reactions by the groups of Kochi, Kumada and Corriu in the early 1970s, impressive progress has been achieved in the development of these methods into a synthetically indispensable tool. Whereas nickel- and palladium-catalyzed processes have frequently been applied in academia and industry, iron-catalyzed transformations seemed to have been largely overlooked. Increasing production costs based on high prices of energy and raw materials and also high standards in environmental protection are forcing chemical research to develop clean, selective, highly efficient, environmentally benign and sustainable chemical processes. In this context, chemical reactions-catalyzed by cheap, non-toxic iron salts are highly desirable. In recent years, important discoveries have been accomplished in the field of iron-catalyzed cross-coupling chemistry. Although by far less general compared with their nickel and palladium counterparts, a detailed study of the rather unclear mechanism should open up new opportunities to develop new reactions and improve turnover numbers, space–time yields and selectivity profiles of existing transformations. The renaissance of iron as a homogeneous catalyst for novel chemical reactions is reflected in the rapidly increasing number of publications, and exciting discoveries within this fast-growing field of research can be expected in the near future.

References

1 (a) N. Miyaura (ed.), *Topics in Current Chemistry*, Vol. 219, Springer, New York, 2002; (b) A. de Meijere, F. Diederich (eds.), *Metal-catalyzed Cross-coupling Reactions*, Wiley-VCH, Weinheim, 2004; (c) J. Tsuji (ed.), *Palladium Reagents and Catalysis, Innovations in Organic Synthesis*, Wiley, New York, 1995; (d) M. Beller, A. Zapf, W. Mägerlein, *Chem. Eng. Techn.* 2001, **24**, 575; (e) K. C. Nicolaou, R. Hanko, W. Hartwig (eds.), *Cross-coupling Reactions. Handbook of Combinatorial Chemistry: Drugs, Catalysts, Materials*, Wiley-VCH, Weinheim, 2002.

2 (a) J.-P. Corbet, G. Mignani, *Chem. Rev.* 2006, **106**, 2651–2710; (b) E. I. Neghishi (ed.), *Handbook of Organopalladium Chemistry for Organic Synthesis*, 2002, Wiley, New York.

3 V. V. Grushin, H. Alper, in S. Murai (ed.), *Activation of Unreactive Bonds and Organic Synthesis*, Springer, Berlin, 1999, p. 203.

4 Reviews: (a) H. Shinokubo, K. Oshima, *Eur. J. Org. Chem.* 2004, 2081–2091; (b) C. Bolm, J. Legros, J. Le Paih, L. Zani, *Chem. Rev.* 2004, **104**, 6217–6254; (c) A. Fürstner, R. Martin, *Chem. Lett.* 2005, **34**, 624–629;

5 (a) R. J. P. Corriu, J. P. Masse, *J. Chem. Soc., Chem. Commun.* 1972, 144; (b) K. Tamao, K. Sumitani, M. Kumada, *J. Am. Chem. Soc.* 1972, **94**, 4374.

6 (a) M. Tamura, J. K. Kochi, *J. Am. Chem. Soc.* 1971, **93**, 1487; (b) M. Tamura, J. K. Kochi, *Synthesis* 1971, 303.

7 R. S. Smith, J. K. Kochi, *J. Org. Chem.* 1976, **41**, 502.

8 (a) J. K. Kochi, *Acc. Chem. Res.* 1974, **7**, 351; (b) S. M. Neumann, J. K. Kochi, *J. Org. Chem.* 1975, **40**, 599; (c) J. K. Kochi, *J. Organomet. Chem.* 2002, **653**, 11.

9 G. A. Molander, B. J. Rahn, D. C. Shubert, S. E. Bonde, *Tetrahedron Lett.* 1983, **24**, 5449.

10 G. Cahiez, H. Avedissian, *Synthesis* 1998, 1199–1205.

11 M. A. Fakhakh, X. Franck, R. Hocquemiller, B. Figadère, *J. Organomet. Chem.* 2001, **624**, 131–135.

12 (a) G. Cahiez, S. Marquais, *Tetrahedron Lett.* 1996, **37**, 1773–1776; (b) A. Fürstner, H. Brunner, *Tetrahedron Lett.* 1996, **37**, 7009.

13 V. Fiandanese, G. Miccoli, F. Naso, L. Ronzini, *J. Organomet. Chem.* 1986, **312**, 343–348.

14 N. Ostergaard, B. T. Pedersen, N. Skjaerbaek, P. Vedso, M. Begtrup, *Synlett* 2002, **11**, 1889–1891.

15 M. Seck, X. Franck, R. Hocquemiller, B. Figadère, J.-F. Peyrat, O. Provot, J.-D. Brion, M. Alami, *Tetrahedron Lett.* 2004, **45**, 1881–1884; see also M. A. Fakhfakh, X. Franck, R. Hocquemiller, B. Figadère, *J. Organomet. Chem.* 2001, **624**, 131–135.

16 W. Dohle, F. Kopp, G. Cahiez, P. Knochel, *Synlett* 2001, **12**, 1901–1904.

17 B. Scheiper, M. Bonnekessel, H. Krause, A. Fürstner, *J. Org. Chem.* 2004, **69**, 3943–3949.

18 G. Dunet, P. Knochel, *Synlett* 2006, 407–410.

19 (a) J. L. Fabre, M. Julia, J. N. Verpeaux, *Tetrahedron Lett.* 1982, **23**, 2469–2472; (b) E. Alvarez, T. Cuvigny, C. Herve du Penhoat, M. Julia, *Tetrahedron* 1988, **44**, 111–118; (c) E. Alvarez, T. Cuvigny, C. Herve du Penhoat, M. Julia, *Tetrahedron* 1988, **44**, 119–126; (d) J. L. Fabre, M. Julia, J. N. Verpeaux, *Bull. Soc. Chim. Fr.* 1985, **5**, 772–778.

20 K. Itami, S. Higashi, M. Mineno, J-i. Yoshida, *Org. Lett.* 2005, **7**, 1219–1222.

21 B. Bogdanović, M. Schwickardi, *Angew. Chem.* 2000, **112**, 4788–4790.

22 (a) B. Bogdanović, N. Janken, H. G. Kinzelmann, *Chem. Ber.* 1990, **123**, 1507; (b) L. E. Aleandri, B. Bogdanović, P. Bons, C. Dürr, A. Gaidies, T. Hartwig, S. C. Huckett, M. Lagarden, U. Wilczok, R. A. Brand, *Chem. Mater.* 1995, **7**, 1153.

23 A. Fürstner, A. Leitner, *Angew. Chem. Int. Ed.* 2002, **41**, 609–612.

24 M. E. Limmert, A. H. Roy, J. F. Hartwig, *J. Org. Chem.* 2005, **12**, 9364–9370.

25 (a) K. Jonas, L. Schieferstein, *Angew. Chem.* 1979, **91**, 590; *Angew. Chem. Int. Ed. Engl.* 1979, **18**, 549; (b) K. Jonas, *Angew. Chem.* 1975, **87**, 809; *Angew. Chem. Int. Ed. Engl.* 1975, **14**, 752.

26 M. F. Semmelhack, "Organoiron and Organochromium Chemistry", in M. Schlosser (ed.), *Organometallics in Synthesis*, Wiley-VCH, Weinheim, 2002, 1003–1113.

27 A. Fürstner, A. Leitner, M. Méndez, H. Krause, *J. Am. Chem. Soc.* 2002, **124**, 13856–13863.

28 Application in total synthesis: (a) A. Fürstner, A. Leitner, *Angew. Chem. Int. Ed.* 2003, **42**, 308–311; (b) B. Scheiper, F. Glorius, A. Leitner, A. Fürstner, *Proc. Natl. Acad. Sci. USA* 2004, **101**, 11960–11965.

29 G. Seidel, D. Laurich, A. Fürstner, *J. Org. Chem.* 2004, **69**, 3950–3952.

30 A. Fürstner, A. Leitner, G. Seidel, *Org. Synth.* 2005, **81**, 33–41.

31 M. S. Kharasch, E. K. Fields, *J. Am. Chem. Soc.* 1941, **63**, 2316.

32 I. Sapountzis, W. Lin, C. C. Kofink, C. Despotopoulou, P. Knochel, *Angew. Chem.* 2005, **117**, 1682–1685; *Angew. Chem. Int. Ed.* 2005, **44**, 1654–1658.

33 C. C. Kofink, B. Blank, S. Pagano, N. Götz, P. Knochel, *Chem. Commun.* 2007, 1954–1956.

34 T. Hatakeyama, M. Nakamura, *J. Am. Chem. Soc.* 2007, **129**, 9844–9845.

35 (a) A. Fürstner, H. Krause, C. W. Lehmann, *Angew. Chem.* 2006, **118**, 454–458; *Angew. Chem. Int. Ed.* 2006, **45**, 440–444; (b) T. Kauffmann, *Angew. Chem.* 1996, **108**, 401–418; *Angew. Chem. Int. Ed.* 1996, **35**, 386–403.

36 M. Taillefer, N. Xia, A. Quali, *Angew. Chem.* 2007, **119**, 952–954; *Angew. Chem. Int. Ed.* 2007, **46**, 934–936.

37 (a) A. C. Frisch, M. Beller, *Angew. Chem. Int. Ed.* 2005, **44**, 674–688; (b) M. R. Netherton, G. C. Fu, in J. Tsuji (ed.), *Topics in Organometallic Chemistry: Palladium in Organic Synthesis*, Springer, New York, 2005, 85–108.

38 (a) D. J. Cárdenas, *Angew. Chem. Int. Ed.* 2003, **42**, 384–387; (b) S. R. Chemler, D. Trauner, S. J. Danishefsky, *Angew. Chem.* 2001, **113**, 4676; *Angew. Chem. Int. Ed.*2001, **40**, 4544.

39 R. Netherton, G. C. Fu, *Adv. Synth. Catal.* 2004, **346**, 1525–1532.

40 M. Nakamura, K. Matsuo, S. Ito, E. Nakamura, *J. Am. Chem. Soc.* 2004, **126**, 3686–3687.

41 T. Nagano, T. Hayashi, *Org. Lett.* 2004, **6**, 1297–1299.

42 R. B. Bedford, D. W. Bruce, R. M. Frost, M. Hird, *Chem. Commun.* 2005, 4161–4163.

43 R. B. Bedford, M. Betham, D. W. Bruce, S. A. Davis, M. Hird, *Chem. Commun.* 2006, 1398–1400.

44 (a) R. Martin, A. Fürstner, *Angew. Chem. Int. Ed.* 2004, **43**, 3955–3957; for further applications of the Jonas complex in cycloisomerization reactions, see A. Fürstner, R. Martin, K. Majima, *J. Am. Chem. Soc.* 2005, **127**, 12236–12237.

45 G. Cahiez, V. Habiak, C. Duplais, A. Moyeux, *Angew. Chem. Int. Ed.* 2007, **46**, 4364–4366.

46 K. Bica, P. Gaertner, *Org. Lett.* 2006, **8**, 733–735.

47 (a) R. B. Bedford, M. Betham, D. W. Bruce, A. A. Danopoulos, R. M. Frost, M. Hird, *J. Org. Chem.* 2006, **71**, 1104–1110; for the application of iron (III) salen-type catalysts, see R. B. Bedford, D. W. Bruce, R. M. Frost, J. W. Goodby, M. Hird, *Chem. Commun.* 2006, 2822–2823.

48 G. Cahiez, C. Duplais, A. Moyeux, Org. Lett. 2007, 9, 3253–3254.

49 A. Guérinot, S. Reymond, J. Cossy, *Angew. Chem.* 2007, 119, 6641–6644; *Angew. Chem. Int. Ed.* 2007, **47**, 6521–6524.

50 K. G. Dongol, H. Koh, M. Sau, C. L. L. Chai, *Adv. Synth. Catal* 2007, **349**, 1015–1018.

51 For an iron catalyzed Reformatsky-type reaction, see M. Durandetti, J. Périchon, *Synthesis* 2006, 1542–1548.

52 M. Nakamura, S. Ito, K. Matsuo, E. Nakamura, *Synlett* 2005, **11**, 1794–1798.

53 See also: B. Hölzer, R. W. Hoffmann, *Chem. Commun.* 2003, 732–733.

54 (a) R. K. Dieter, *Tetrahedron* 1999, **55**, 4177; (b) N. J. Lawrence, *J. Chem. Soc., Perkin Trans. 1* 1998, 1739;

55 (a) G. Cahiez, B. Laboue, *Tetrahedron Lett.* 1989, **30**, 7369; (b) G. Cahiez, B. Laboue, *Tetrahedron Lett.* 1989, **30**, 3545.

56 (a) J. Cason, K. W. Kraus, *J. Org. Chem.* 1961, **26**, 1768–1772; (b) J. Cason, K. W. Kraus, *J. Org. Chem.* 1961, **26**, 1772–1779.

57 (a)V. Fiandanese, G. Marchese, V. Martina, L. Ronzini, *Tetrahedron Lett.* 1984, **25**, 4805; (b) M. M. Dell'Anna, P. Mastrorilli, C. F. Nobile, G. Marchese, M. R. Taurino, *J. Mol. Catal. A* 2000, **161**, 239; (c) K. Ritter, M. Hanack, *Tetrahedron Lett.* 1985, **26**, 1285; (d) C. Cardellicchio, V. Fiandanese, G. Marchese, L. Ronzini, *Tetrahedron Lett.* 1985, **26**, 3595; (e) C. Cardellicchio, V. Fiandanese, G. Marchese, L. Ronzini, *Tetrahedron Lett.* 1987, **28**, 2053.

58 F. Babudri, A. D'Ettole, V. Fiandanese, G. Marchese, F. Naso, *J. Organomet. Chem.* 1991, **405**, 53–58.

59 For a novel application in total synthesis, see A. Fürstner, D. Kirk, M. D. B. Fenster, C. Aïssa, D. De Souza, C. Nevado, C. T. T. Tuttle, W. Thiel, O. Müller, *Chem. Eur. J.* 2007, **13**, 135–149; (b) A. Fürstner, D. De Souza, L. Turet, M. D. B. Fenster, L. Parra-Rapado, C. Wirtz, R. Mynott, C. W. Lehmann, *Chem. Eur. J.* 2007, **13**, 115–134.

60 C. K. Reddy, P. Knochel, *Angew. Chem. Int. Ed.* 1996, **34**, 1700–1701.

61 C. Duplais, F. Bures, I. Sapountzis, T. J. Korn, G. Cahiez, P. Knochel, *Angew. Chem. Int. Ed.* 2004, **43**, 2968–2970.

62 E. I. Negishi, *Acc. Chem. Res.* 1987, **20**, 65–72.

63 (a) A. M. Caporusso, L. Lardicci, G. Giacomelli, *Tetrahedron Lett.* 1977, 4351; (b) A. M. Caporusso, G. Giacomelli, L. Lardicci, *J. Chem. Soc., Perkin Trans. 1* 1979, **1**, 3139.

64 M. Hojo, Y. Murakami, H. Aihara, R. Sakuragi, Y. Baba, A. Hosomi, *Angew. Chem.* 2001, **113**, 641; *Angew. Chem. Int. Ed.* 2001, **40**, 621.

65 E. Shirakawa, T. Yamagami, T. Kimura, S. Yamaguchi, T. Hayashi, *J. Am. Chem. Soc.* 2005, **127**, 17164–17165.

66 D. Zhang, J. M. Ready, *J. Am. Chem. Soc.* 2006, **128**, 15050–15051.

67 (a) D. J. Pasto, G. F. Hennion, R. H. Shults, A. Waterhouse, S.-K. Chou, *J. Org. Chem.* 1976, **41**, 3496; (b) D. J. Pasto, S.-K. Chou,

A. Waterhouse, R. H. Shults, G. F. Hennion, *J. Org. Chem.* 1978, **43**, 1385.

68 A. S. K. Hashmi, G. Szeimis, *Chem. Ber.* 1994, **127**, 1075–1089.

69 A. Fürstner, M. Méndez, *Angew. Chem. Int. Ed.* 2003, **42**, 5355–5357.

70 For application in the total synthesis of Amphidinolide X, see O. Lepage, E. Kattnig, A. Fürstner, *J. Am. Chem. Soc.* 2004, **126**, 15970–15971.

71 M. Nakamura, A. Hirai, E. Nakamura, *J. Am. Chem. Soc.* 2000, **122**, 978.

6
Iron-catalyzed Aromatic Substitutions

Jette Kischel, Kristin Mertins, Irina Jovel, Alexander Zapf, and Matthias Beller

6.1
General Aspects

Functionalization of C−H bonds via aromatic substitution is an important means of adding functional groups to all kinds of arenes and as such is of significant relevance to many areas of chemistry. Notably, the key step in several industrially important reactions is an electrophilic attack on aromatic π-systems by carbocations or other strong electrophiles.

Typically, in electrophilic aromatic substitutions, the electrophilic attack results from the high electron concentration on both sides of the aromatic ring [1]. For the large majority of such reactions the mechanism can be described as an arenium ion mechanism (Scheme 6.1): The attacking electrophile forms a π-complex **(1)** first. Then, a σ-bond to an aromatic carbon atom is formed, resulting in a change of hybridization from sp^2 to sp^3. The positive charge of this σ-complex **(2)** is delocalized by the corresponding mesomerically stabilized cyclohexadienyl cation. Finally, the proton is released from the sp^3 C atom with concomitant re-aromatization of the ring system. Here, also a π-complex occurs as a transition state [2, 3]. In addition to protons, sulfur, nitrogen, oxygen, halogen and carbocations are the most common electrophiles, resulting in Friedel–Crafts reactions [4], formylations and carboxylations.

Another possible but rarely seen mechanism is the unimolecular S_E1 mechanism (Scheme 6.2). It is observed in certain cases in the presence of strong bases.

Nucleophilic aromatic substitutions are more difficult in principle. Here, three types of mechanisms are discussed (Scheme 6.3): addition–elimination mechanism (S_NAr), elimination–addition mechanism (S_N1) and aryne mechanism. In addition, radical-type mechanisms are also possible.

With respect to iron catalysts, iron(III) chloride is one of the most common catalysts known for electrophilic aromatic substitutions and has been widely used in the past. In general, it is an inexpensive and eco-friendly reagent featuring a higher catalytic activity than other metal chlorides [5, 6].

Iron Catalysis in Organic Chemistry. Edited by Bernd Plietker

ISBN: 978-3-527-31927-5

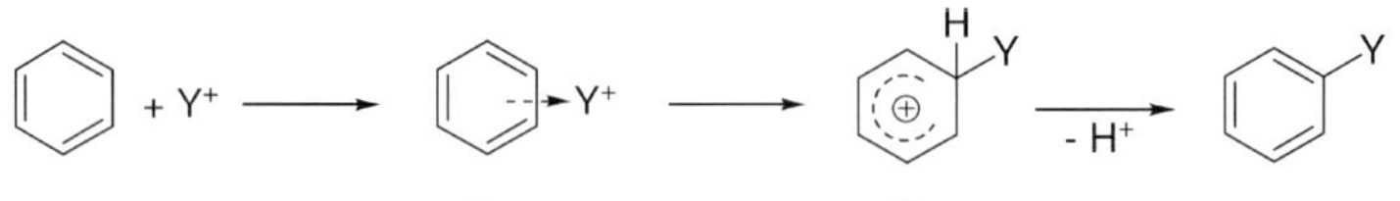

Scheme 6.1 Arene-mechanism of electrophilic aromatic substitution.

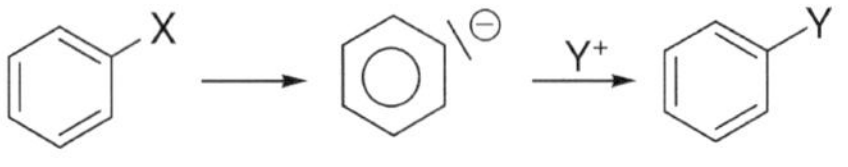

Scheme 6.2 S_E1 mechanism (unimolecular electrophilic substitution).

S_NAr-mechanism:

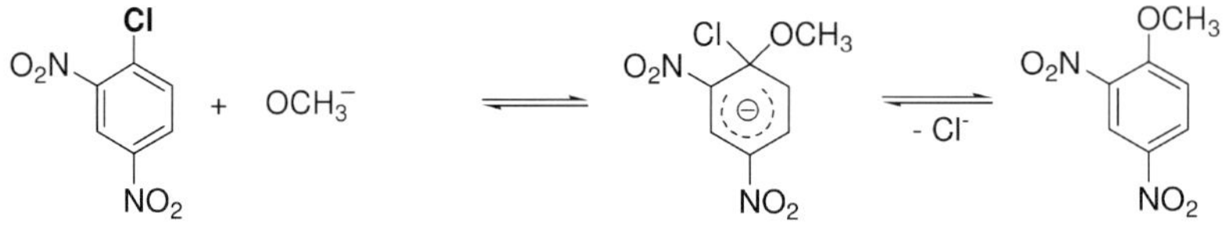

S_N1-mechanism (of interest only in reactions of diazonium salts):

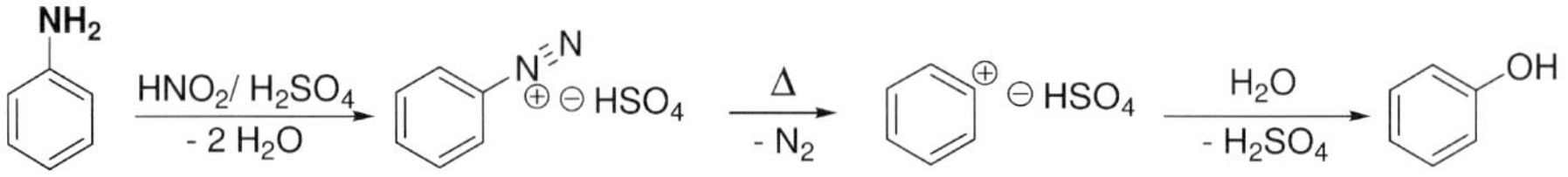

aryne-mechanism (cine substitution):

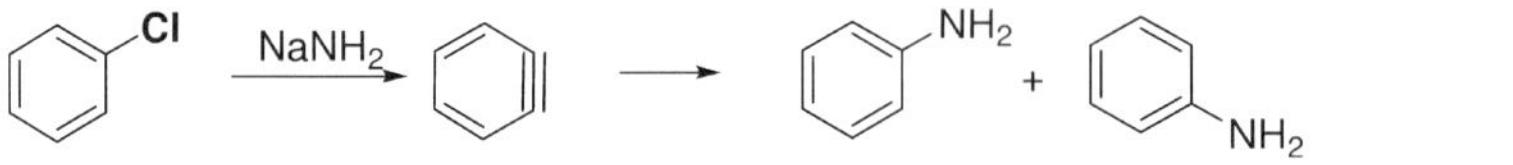

Scheme 6.3 Examples of the mechanism of nucleophilic aromatic substitutions.

6.2 Electrophilic Aromatic Substitutions

6.2.1 Halogenation Reactions

One of the most important reactions in arene synthesis is the halogenation of arenes. Conventionally, these reactions are performed directly by bromine or chlorine [7]. However, on the laboratory scale chlorine is not easily manageable and is a toxic gas. Therefore, it is not often used in academic research. For iodinations normally a strong oxidizing agent is required. In halogenations, often unintentional side-chain

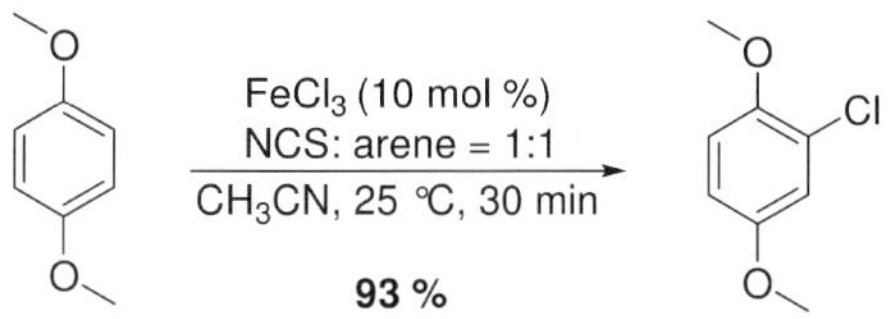

$FeCl_3$ (10 mol %)
NCS: arene = 1:1
CH_3CN, 25 °C, 30 min
93 %
Cl

Scheme 6.4 Halogenation of 1,4-dimethoxybenzene with *N*-chlorosuccinimide catalyzed by $FeCl_3$.

halogenations take place. Hence the development of more practical and selective halogenation procedures is still an important topic. Tanemura *et al.* [8] presented an iron(III) chloride-catalyzed halogenation of arenes using *N*-chloro-, *N*-bromo- and *N*-iodosuccinimide to give the corresponding core-substituted products in good yields. Whereas electron-rich substrates can be halogenated by catalytic amounts (10 mol%) of $FeCl_3$ (Scheme 6.4), for electron-poor substrates stoichiometric amounts of a Lewis acid are necessary. This simple and mild method represents a useful alternative to classical halogenation reactions, but selectivity issues are not improved, unfortunately.

6.2.2 Nitration Reactions

Aromatic nitrations, classically performed by the use of large quantities of nitric or sulfonic acid [9], are notorious for their corrosiveness, potential risk of explosion, low regioselectivity, oxidative degradations and low efficiency. Recent alternative methodologies, such as the Kyodai nitration [10], utilize nitrogen oxides instead of acid systems. For instance, Mori and coworkers [11] presented an iron(III)-catalyzed nitration procedure with nitrogen dioxide and molecular oxygen under neutral conditions. Both non-activated and moderately activated arenes were nitrated successfully, giving the corresponding nitro derivatives in fair to good yield and selectivity. Two examples are shown in Scheme 6.5.

Other viable alternatives are organic nitration agents and solid acid catalysts. Improved regioselectivities of nitration reactions and in certain cases even reversal of

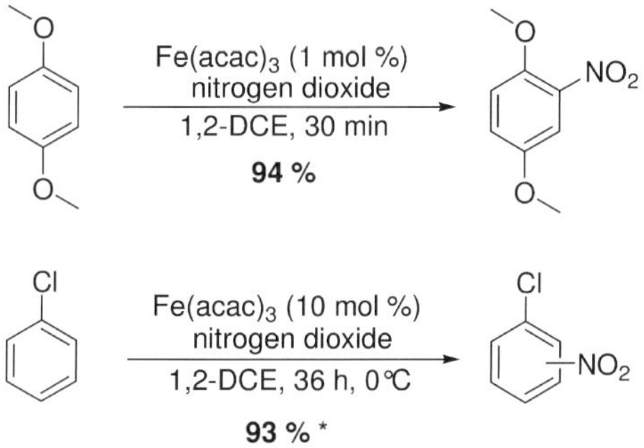

Fe(acac)$_3$ (1 mol %)
nitrogen dioxide
1,2-DCE, 30 min
94 %
NO_2

Cl
Fe(acac)$_3$ (10 mol %)
nitrogen dioxide
1,2-DCE, 36 h, 0 °C
93 % *
Cl
NO_2

Scheme 6.5 Iron-catalyzed nitration of arenes. *$o:m:p = 32:<1:68$.

the isomer ratio of nitration products are observed, for instance in zeolite-induced nitrations [12].

6.2.3
Sulfonylation Reactions

Organosulfones are versatile synthons and of great interest as agrochemicals [13], polymers [14] and pharmaceuticals [15]. They can be effectively synthesized by using sulfonyl chlorides in a modified Friedel–Crafts acylation. Aluminum chloride is widely used as a Lewis acid catalyst in the laboratory and industry but generates an enormous amount of solid waste. Therefore, a really catalytic sulfonylation method is highly desirable. As catalysts for the sulfonylation of arenes some metal halides, zeolites [16] and Brønsted acids [17] (polyphosphoric acid [18]) have been reported. Traditionally, $FeCl_3$ is used as a Lewis acid in this type of reaction and it has been employed at least in stoichiometric amounts [19]. However, several interesting new catalysts have been reported lately [20]. Among them, an Fe(III)-exchanged montmorillonite catalyst [21], which catalyzes the sulfonylation of benzene with *p*-tolylsulfonyl chloride, produced 84% of the corresponding aryl sulfone product within 12 h at 80 °C. This example indicates a novel inexpensive methodology for Friedel–Crafts sulfonylation, because the catalyst can be reused for several cycles. Additional advantages are operational simplicity, environmental acceptability, non-corrosiveness, mild reaction conditions and high yields. Another $FeCl_3$-catalyzed sulfonylation [22] of aromatic substrates utilizes microwave (MW) irradiation under solvent-free conditions. With this method, developed by Dubac and coworkers, even electron-poor aryl halides can be sulfonylated (Scheme 6.6). Benzene and its derivatives such as toluene also gave effective sulfonylation with reaction times of a few minutes.

Friedel–Crafts sulfonylation of benzene and its derivatives can also be carried out using iron(III) chloride-based ionic liquids, as was shown by Samant and coworkers [23]. For example, the synthesis of diphenyl sulfone, an important intermediate for the anti-leprosy drug Dapsone, was achieved in 78% yield catalyzed by 1-butyl–3-methylimidazolium chloride combined with $FeCl_3$ (Scheme 6.7).

This method was further improved by utilizing catalytic amounts of ionic liquids under MW irradiation. The reactions were extremely clean, giving good yields of the corresponding sulfones, and a substantial reduction in reaction time was observed (Scheme 6.8).

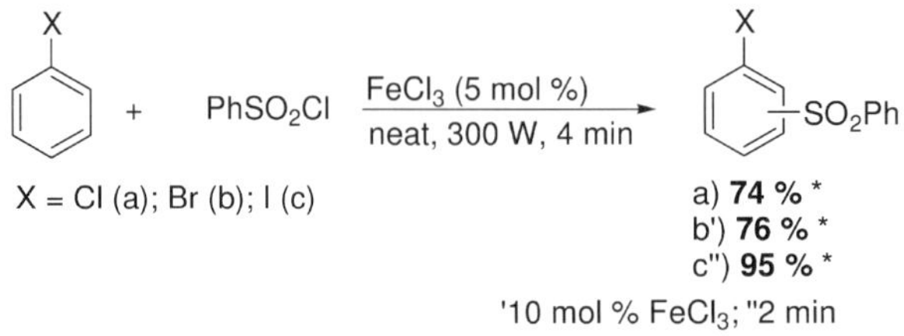

X

\+ $PhSO_2Cl$ —$FeCl_3$ (5 mol %) / neat, 300 W, 4 min→ X, SO_2Ph

X = Cl (a); Br (b); I (c)

a) **74 %** *
b') **76 %** *
c") **95 %** *

'10 mol % $FeCl_3$; "2 min

Scheme 6.6 $FeCl_3$-catalyzed sulfonylation of aromatics.
*$o:m:p$ = (a) 2 : 0 : 98; (b) 0 : 0 : 100; (c) 2 : 0 : 98.

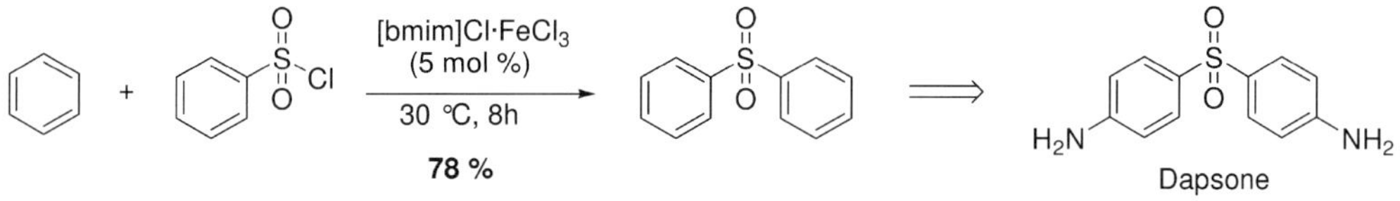

[bmim]Cl·FeCl3 (5 mol %)
30 °C, 8h
78 %
Dapsone

Scheme 6.7 Sulfonylation of benzene catalyzed by $FeCl_3$-based ionic liquid.

[bmim]Cl·FeCl3
microwave irradiation

a) R = CH3, 5 mol % cat., 160 °C, 3 min : **90 %** *
b) R = Cl, 10 mol % cat., 150 °C, 15 min: **64 %** *

Scheme 6.8 [bmim]Cl·$FeCl_3$-catalyzed sulfonylation of toluene and chlorobenzene under MW irradiation. *$o:m:p$ = (a) 11 : 0 : 89; (b) 0 : 0 : 100.

6.2.4 Friedel–Crafts Acylations

The main industrial route to aromatic ketones is via the Friedel–Crafts acylation, nowadays classified as an environmentally hostile process with gaseous effluents and mineral wastes. Typically catalyzed by the Lewis acid $AlCl_3$, this is a self-blocking reaction due to the complexation of the Lewis acid with substrates with alkoxy substituents or/and with the aryl ketone product. Stoichiometric amounts of the non-regenerable catalyst $AlCl_3$ are required, accompanied by high temperatures. Further, the work-up steps involving hydrolysis of the intermediate complex produces copious amounts of inorganic waste materials.

Therefore, the search for better and "really" catalytic Friedel–Crafts acylations has been an ongoing task during recent decades. Numerous protocols aimed at this target were published, including the application of different Lewis acids ($FeCl_3$ [24], $ZnCl_2$ [25], $BiCl_3$ [26]) (Scheme 6.9) and various metal chlorides [27], Brønsted acids, such as superacidic systems [28] and sulfonic acids [29], more specifically trifluoromethanesulfonic acid (triflic acid) and zeolites [30] (Scheme 6.10). Recently, water-stable metal triflates [31] and their analogues, the bis(trifluoromethane) sulfonimides [32], have also been reported as catalysts for acylation reactions.

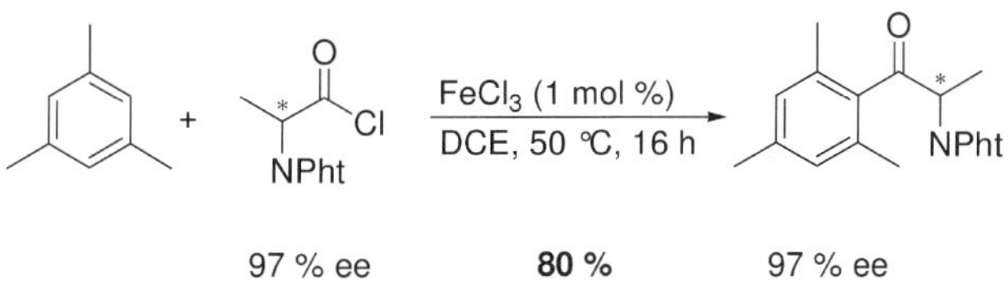

FeCl3 (1 mol %)
DCE, 50 °C, 16 h

97 % ee　**80 %**　97 % ee

Scheme 6.9 $FeCl_3$-catalyzed acylation with an optically active acyl chloride.

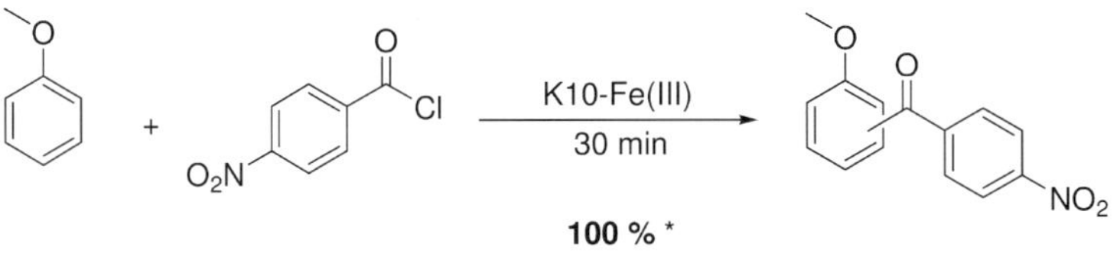

K10-Fe(III)
30 min
100 % *

Scheme 6.10 Montmorillonite K10–Fe(III) as reusable catalyst for the quantitative acylation of anisole by *p*-nitrobenzoyl chloride. *$o:m:p = 20:0:80$.

In most instances these catalysts were not satisfactory in terms of yield, range of substrates and turnover numbers of the catalyst. Additionally, they were mostly efficient only in the case of activated aromatics. In order to develop also Friedel–Crafts acylations of less activated or even deactivated arenes, a new generation of catalysts had to be developed. Here, particularly bismuth(III) triflate [33] and hafnium(IV) triflate in the presence of lithium perchlorate [34] or triflic acid [35] are effective, alternative catalysts actually acylating benzene, toluene and halobenzenes.

In 2000, Dubac's group reported the microwave-assisted Friedel–Crafts acylation of slightly activated and deactivated arenes under solvent-free conditions with $FeCl_3$ as catalyst. Here, for the acylation of toluene a 90% product yield is obtained after 5 min of irradiation and an overall reaction time of 30 min in the presence of only 5 mol% of $FeCl_3$. A sequential MW irradiation at 300 W afforded the acylation of fluorobenzene with 2-chlorobenzoyl chloride, with a surprisingly high yield of 92% of 2-chloro–4′-fluorobenzophenone (Scheme 6.11).

Although these reactions operate under solvent-free conditions (with an excess of the arene), many Friedel–Crafts acylations utilize volatile and hazardous halogenated solvents. Here, their replacement by ionic liquids can considerably lower the environmental risks and provide a "greener chemistry". Ionic liquids with their unique miscibility properties, high thermal stability and miniscule vapor pressure are valuable alternatives for the wide range of traditional solvents available.

Recently, it was proven by *in situ* spectroscopic studies [36] that the mechanism of the Friedel–Crafts acetylation is not altered by using ionic liquids. Experimental data

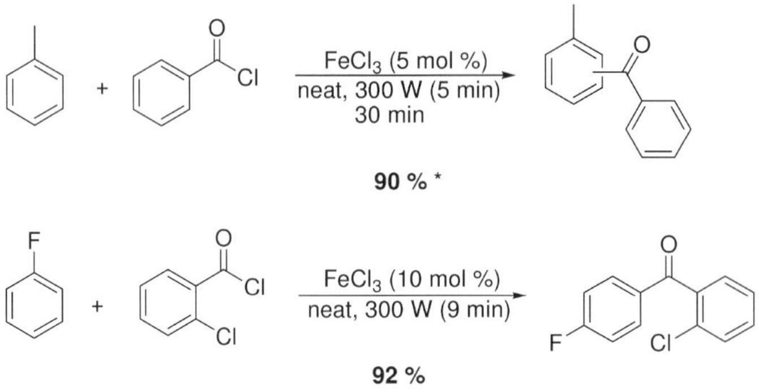

$FeCl_3$ (5 mol %)
neat, 300 W (5 min)
30 min
90 % *

$FeCl_3$ (10 mol %)
neat, 300 W (9 min)
92 %

Scheme 6.11 Arylation of arenes catalyzed by $FeCl_3$ under microwave irradiation. *$o:m:p = 12:3:85$.

and theoretical calculations indicate the acylium ion $[CH_3CO]^+[MCl_4]^-$ as key intermediate in these reactions and an ionic mechanism is assumed.

In 2001, Hölderich's group [37] presented 1-methyl–3-butylimidazolium chloroferrate (Fe-IL) in addition to Al-IL and Sn-IL as a catalyst for Friedel–Crafts acylations. In the acetylation of anisole with acetic anhydride, full conversion of the acylating agent was observed using Fe-IL. The immobilization of these catalysts, however, led to some serious problems such as catalyst leaching.

Xiao and Malhotra [38] reported investigations with pyridinium-based ionic liquids. Here, especially $[EtPy]^+[CF_3COO]^-$ was very effective in the acylation reaction of benzene (81% conversion of acylating agent; 75 °C) even in the absence of any catalyst. On employing additional $FeCl_3$ (stoichiomerically) with this ionic liquid, the conversion of acetic anhydride improved up to 97% at 50 °C. Higher conversions were seen with toluene compared with benzene, but bromobenzene was less reactive. The combination of $[EtPy]^+[CF_3COO]^-$ with $FeCl_3$ was found to be an excellent system for this type of acetylation since the ionic liquid can be recovered quantitatively with negligible loss of activity.

6.2.5 Friedel–Crafts Alkylations

One of the most common examples of an electrophilic aromatic substitution is Friedel–Crafts alkylation [40]. These days, many important industrial processes are based on this type of Friedel–Crafts-chemistry [41]. The manufacture of high-octane gasoline, ethylbenzene, synthetic rubber, plastics and detergent alkylates are examples. Moreover, the Friedel–Crafts alkylation is among the most fundamental and convenient processes for C—C bond formation on arenes, especially for the synthesis of fine chemicals and agrochemicals containing functionalized arenes and heteroarenes.

Through the interaction of an alkylating agent, a hydrogen atom of the aromatic core, in the case of alkylation of arenes, is replaced by an alkyl group driven by a Friedel–Crafts catalyst. A variety of alkylating agents (e.g. alkyl halides, alcohols,

Figure 6.1 Charles Friedel and James Mason Crafts [39].

alkenes, alkynes and many more) can be used in this reaction. The most common ones are shown in Equations (6.1)–(6.3).

$$PhH + RX \xrightarrow{\text{catalyst}} PhR + HX \tag{6.1}$$

$$PhH + ROH \xrightarrow{\text{catalyst}} PhR + H_2O \tag{6.2}$$

$$PhH + RCH{=}CH_2 \xrightarrow{\text{catalyst}} Ph(R)C{=}CH_2 \tag{6.3}$$

Equation (6.1) shows the conventional alkylation method using alkyl halides and a Lewis acid catalyst such as anhydrous $AlCl_3$. However, as stated before, this catalyst poses several problems, such as its need for stoichiometric amounts, difficulty in its separation and recovery, disposal of spent catalyst, corrosion, high toxicity and moisture sensitivity. Beyond this, the reaction has other significant drawbacks such as drastic reaction conditions (high temperature, strongly acidic conditions), low regioselectivities, undesired side-reactions (polyalkylations, skeletal rearrangements) and, above all, large amounts of waste (salt) byproducts. Additionally, often the application of protecting group chemistry or the introduction of activating groups is needed to build up defined C–C bonds. For these reasons, nowadays the development of new direct C–C coupling reactions of arenes has become an important topic in organometallic chemistry, which demands organometallic catalysis. In recent years, a great deal of effort has been directed towards the promotion of this reaction, mainly through replacement of common Lewis acids and the development of alternative alkylating agents. Selected examples are discussed below.

6.2.5.1 Alkylation with Alcohols, Ethers and Esters

Alcohols are preferred alkylating agents rather than alkyl halides because only water occurs as a byproduct in these reactions instead of hydrogen halides. However, alkylations with alcohols and related agents (ethers, esters, etc.) typically require considerably larger quantities of catalyst due to its interaction with the alkylating agent. Another reason is the deactivation of the catalyst by water formed during the course of the reaction.

Besides traditional Brønsted and Lewis acids, Friedel–Crafts reactions specifically with benzyl alcohols have been studied primarily with scandium triflate [42], but also with triflates of various lanthanides [43]. Advantageously, most of these catalysts are efficient and stable also in the presence of water and can be easily recovered. Equally, heterogeneous catalysts [44] were successfully employed for Friedel–Crafts alkylations with alcohols, offering an easy set-up and work-up. Until today, surprisingly few other transition metal-catalysts have been reported for this type of reaction. Here, until recently only stoichiometric [45] palladium-mediated benzylation of arenes have been described, although a multitude of related allylations [46] using allylic alcohols and palladium catalysts are known. More recently, also ruthenium [47], rhenium [48], gold [49] and iron [50] complexes have attracted attention for arene alkylations but mainly by propargylic alcohols and their derivatives (Scheme 6.12). In 1989, the ruthenium-catalyzed benzylation of arenes with benzyl formates [51] was described, requiring high temperatures (200 °C). A combination of an iridium complex and tin(II) chloride employed in catalytic

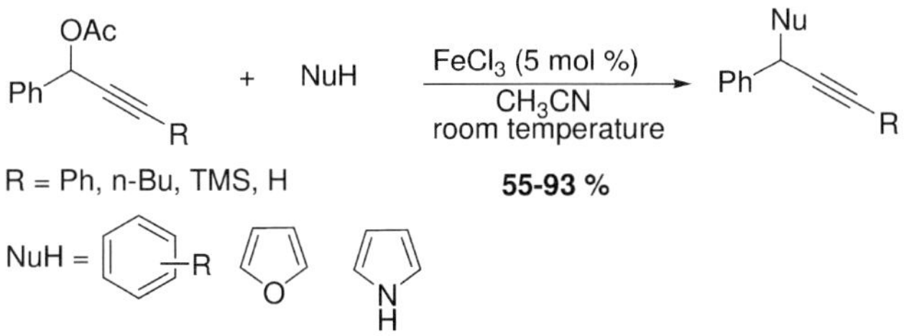

OAc
Ph — R + NuH → ($FeCl_3$ (5 mol %), CH_3CN, room temperature) Nu, Ph — R

R = Ph, n-Bu, TMS, H

55-93 %

NuH = —R, O, N H

Scheme 6.12 $FeCl_3$-catalyzed alkylation of arenes with propargylic acetates.

amounts allows the benzylation of toluene with 4-methylbenzyl alcohol with a maximum product yield of 95% [52]. More recently, Baba's group published an indium trichloride-catalyzed alkylation of indoles [53] and other nucleophiles such as active methylene compounds and alkoxy ketones.

We demonstrated that in the presence of catalytic amounts of several transition metal compounds [54], arenes and heteroarenes could be easily benzylated using benzylic acetates as benzylating agents. Interestingly, this method allows also the benzylation of non-activated arenes. Here, $IrCl_3$ and $H_2[PtCl_6]$ were the best catalysts. Later $HAuCl_4$ [55] also proved to be a very effective catalyst in this type of C–C coupling reaction.

The first general iron-catalyzed arylation of benzyl alcohols and benzyl carboxylates was published by our group in 2005 [56]. A practical synthesis of a multitude of

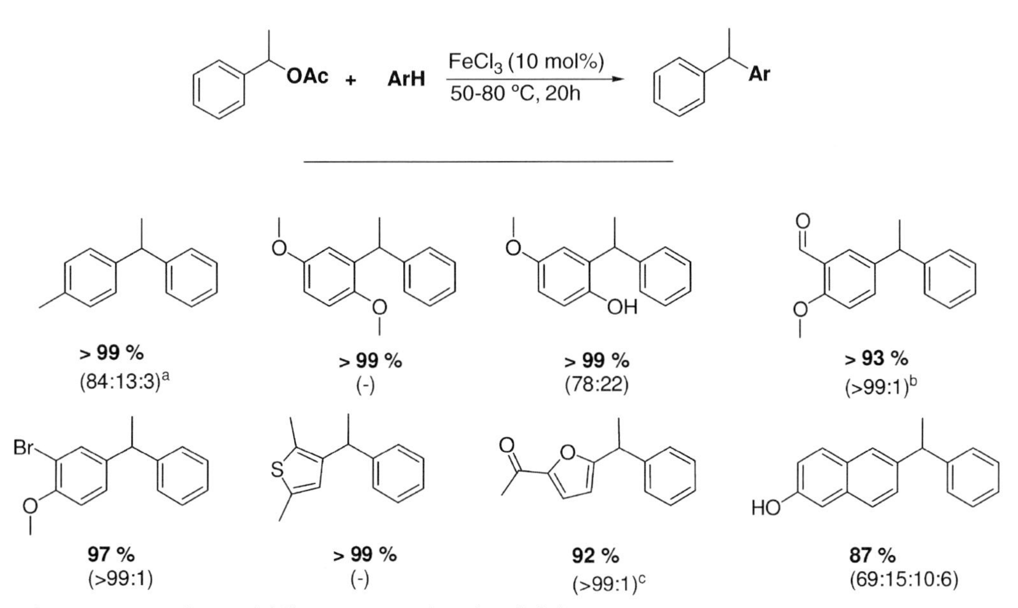

Scheme 6.13 Benzylation of different arenes with 1-phenylethyl acetate; GC yield (regioselectivity). Reaction conditions: 0.5 mmol 1-phenylethyl acetate, 10 mol% $FeCl_3$, 2.0 mmol arene, 5 mL CH_2Cl_2, 50 °C, 20 h; (a) 5 mL arene, no solvent; (b) 5 mL $MeNO_2$, 120 °C; (c) 5 mL $MeNO_2$.

Table 6.1 Various alcohols in the reactions with *o*-xylene[a].

Alcohol	Yield (%)	Regioselectivity[b]
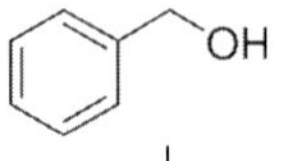 OH	99	99 : 1
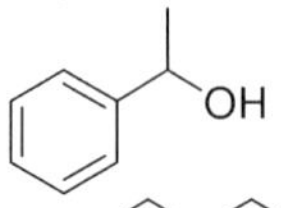 OH	>99	89 : 11
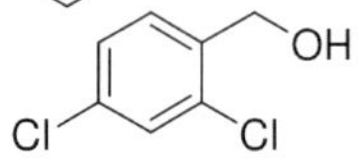 OH Cl Cl	>99	69 : 31
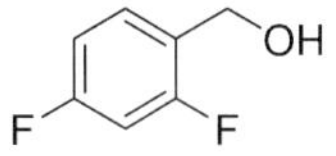 OH F F	99	69 : 31
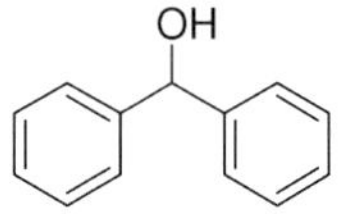 OH	>99	93 : 7
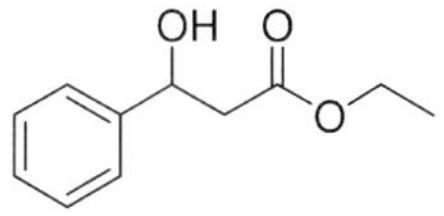 OH O O	98	99 : 1
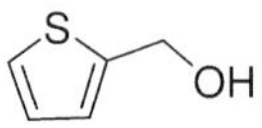 S OH	61	75 : 25
 O O OH	37	62 : 38

[a] Reaction conditions: 0.5 mmol alcohol, 5 mL *o*-xylene, 10 mol% $FeCl_3$, 80 °C, 24 h.
[b] 4-alkylated *o*-xylene:3-alkylated *o*-xylene.

diarylmethanes and arylheteroarylmethanes under mild conditions (50–80 °C, without strong acid and base) was described. Several examples are given in Scheme 6.13.

Notably, this reaction shows a high tolerance towards a wide variety of functional groups, such as CHO, CO_2R, I, Br, Cl, F, OH and OMe, and also thiophene and furan derivatives, which is a rare feature in Friedel–Crafts-type chemistry. In the reaction with *o*-xylene, various alcohols were successfully used as benzylation agents, forming the corresponding products in high yield and selectivity (Table 6.1).

In these transformations, water is the only by-produced, so this arylation method is a state-of-the-art "green" route to diarylmethanes via Friedel–Crafts chemistry.

6.2.5.2 **Alkylation with Alkenes**

An even more attractive version of Friedel–Crafts alkylations is the utilization of alkenes as electrophiles. These reactions profit on the one hand from the versatility and the availability of the starting materials and on the other from the high atom economy [57]. In this respect, in 2006 we presented a convenient iron-catalyzed protocol for the hydroarylation of styrenes (Scheme 6.14) [58].

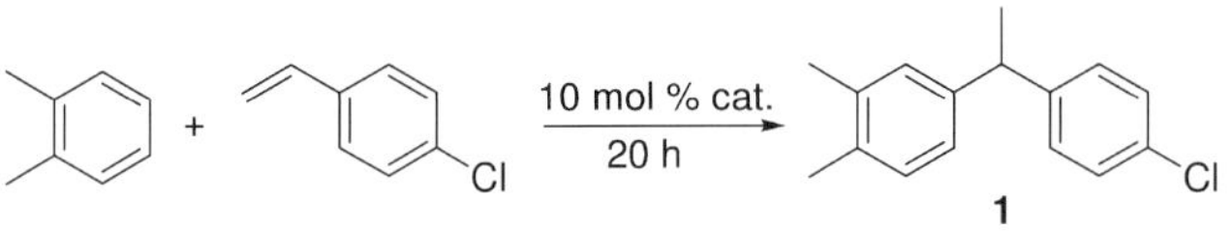

10 mol % cat.
20 h

1

Scheme 6.14 Reaction of *o*-xylene with 4-chlorostyrene.

Again $FeCl_3 \cdot 6H_2O$ is the catalyst of choice for the benzylation of *o*-xylene by 4-chlorostyrene (Table 6.2). This iron-catalyzed arylation reaction of styrenes allows the synthesis of a wide variety of 1,1-diarylalkanes employing various styrenes and diverse arenes. Scheme 6.15 illustrates selected examples of successfully synthesized 1,1-diarylalkanes and Table 6.3 shows the diversity of different substituted styrenes employed as benzylating agents.

Apart from our work, few publications have appeared so far concerning this type of hydroarylation of styrene. For example, Periana's group described in 2005 the hydroarylation of styrene with benzene using a bis-tropolonato–iridium(III) organometallic complex [59]. However, in contrast to the results above, the linear product (anti-Markovnikov product) was preferentially generated.

Table 6.2 Catalyst screening for the reaction of *o*-xylene with 4-chlorostyrene[a].

Entry	Catalyst	Temperature (°C)	Conversion (%)[b]	Yield (%)[c]	Regioselectivity[d]
1	HCl	80	8	0	–
2	TFA	80	10	0	–
3	pTSA	80	20	11	93 : 7
4	HOAc	80	24	0	–
5	$FeCl_3$	80	100	87	>99 : 1
6	$FeCl_3 \cdot 6H_2O$	80	100	87	>99 : 1
7[e]	$FeCl_3 \cdot 6H_2O$	80	93	84	98 : 2
8	$ZnCl_2$	80	2	0	–
9	$CuCl_2 \cdot 2H_2O$	80	6	0	–
10	$NiCl_2$	80	1	0	–
11	$Co(OAc)_2 \cdot 4H_2O$	80	16	0	–
12	$PdCl_2$	80	5	0	–
13	$MesW(CO)_3$	80	3	0	–
14	$RhCl_3$	80	2	0	–
15	$IrCl_3$	80	19	9	96 : 4
16	H_2PtCl_6	80	80	57	>99 : 1
17	$CeCl_3$	80	5	0	–
18	$La(OTf)_3$	80	3	0	–
19	$Sc(OTf)_3$	80	40	30	94 : 6
20	$Ag(OTf)_3$	80	4	0	–
21	$Y(OTf)_3$	80	7	0	–
22	$Yb(OTf)_3$	80	11	0	–

[a] Reaction conditions: 0.5 mmol 4-chlorostyrene, 5 mL *o*-xylene, 20 h.
[b] GC conversion of 4-chlorostyrene.
[c] GC yield of arylated products with 4-[1-(4-chlorophenyl)ethyl]-1,2-dimethylbenzene as main product.
[d] 4-/3-substitution.
[e] 1 h.

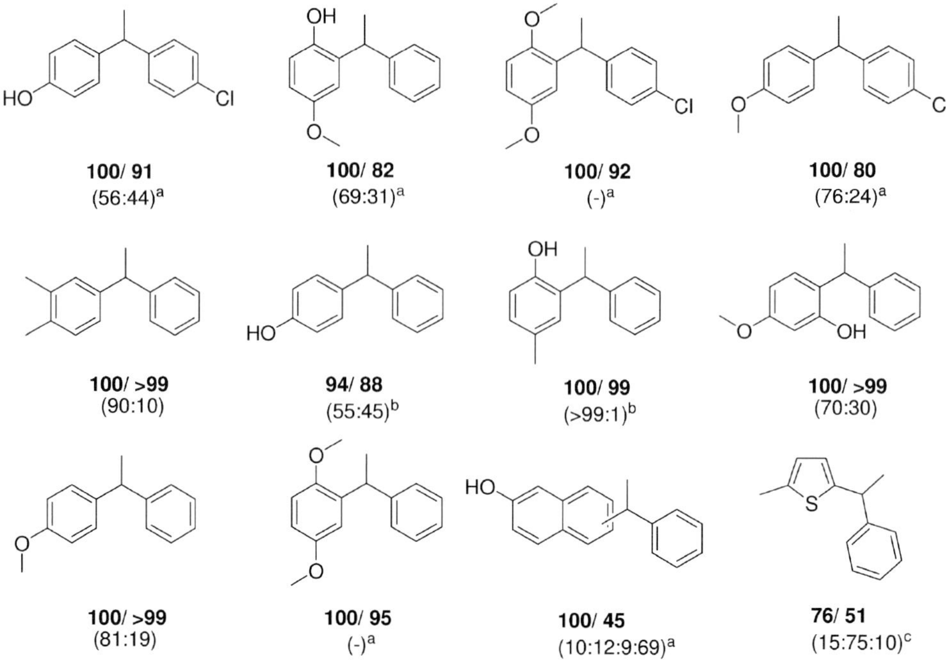

Scheme 6.15 Reaction of styrene and 4-chlorostyrene with various (hetero)arenes; GC conversion of styrene/GC product yield (regioselectivity). Reaction conditions: 0.5 mmol styrene/ 4-chlorostyrene, 10 mol% $FeCl_3$, 5 mL arene, 80 °C, 4 h; (a) 2 mmol arene, 5 mL CH_2Cl_2; (b) 2 mmol arene, 5 mL cyclohexane, 20 h; (c) 2 mL arene, 32 h.

During the same year, both an acid-catalyzed [60] and a $TiCl_4$-catalyzed [61] *ortho*-alkylation of anilines with styrene were published, but offering only a very low selectivity. The reaction of anisole and styrene catalyzed by $Mo(CO)_6$ [62] afforded the corresponding branched diaryl product with a comparable yield and selectivity as shown by our protocol. Furthermore, Michelet and coworkers successfully employed substituted styrenes in a gold-catalyzed tandem Friedel–Crafts-type addition–carbocyclization reaction [63].

Interestingly, enantioselective alkylation reactions [64] were also developed using, for instance, $Cu(OTf)_2$, [65], $[Cu(SbF_6)_2$, $Zn(OTf)_2]$ [66], $Cu(ClO_4)_2{\cdot}6H_2O$ [67] or Sc $(OTf)_3$ [68] in combination with diverse chiral ligands. Remarkably, organocatalytic alkylations of pyrroles, indoles and anilines by 3-phenylpropenal have been also developed [69].

6.3 Nucleophilic Aromatic Substitutions

For reasons of completeness, nucleophilic aromatic substitution reactions [70] will also be mentioned here, but they are less important and therefore not discussed in detail.

Table 6.3 $FeCl_3$-catalyzed reaction of *o*-xylene with different styrenes[a].

Entry	Styrene	Product[b]	Conversion (%)[c]	Yield (%)[d]	Regioselectivity[e]
1	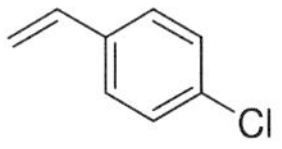 Cl	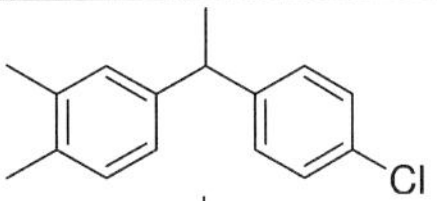 Cl	100	89	>99 : 1
2	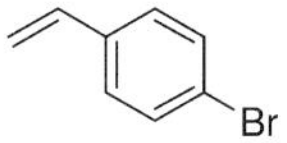 Br	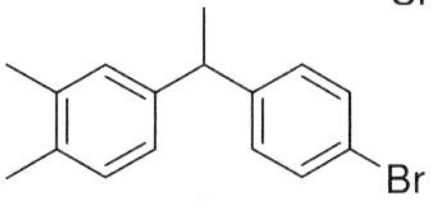 Br	100	98	90 : 10
3	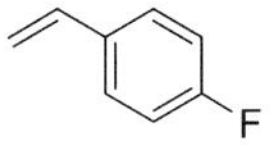 F	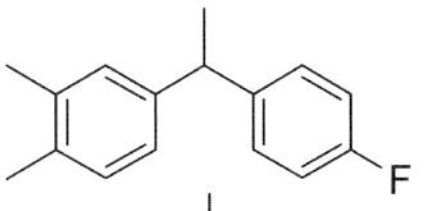 F	100	85	86 : 14
4	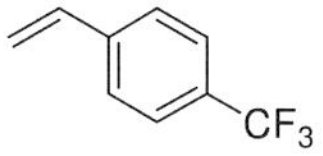 CF_3	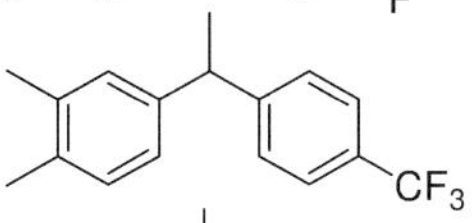 CF_3	100	99	84 : 16
5	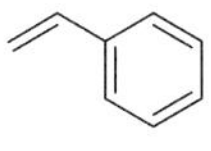	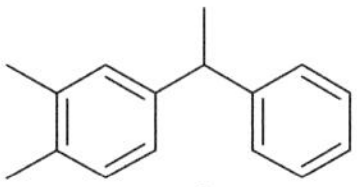	100	>99	90 : 10
6	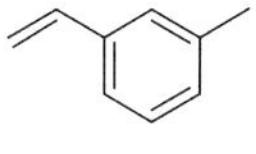	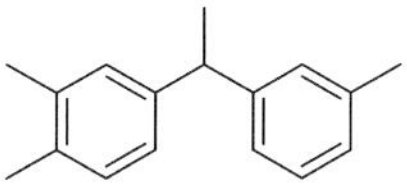	100	77	90 : 4
7[f]	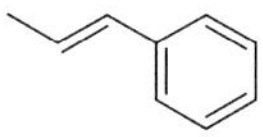	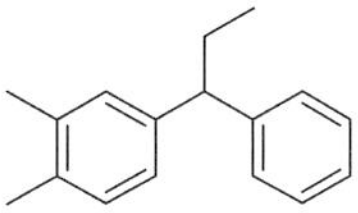	100	95	72 : 28
8[f]	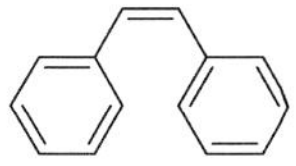	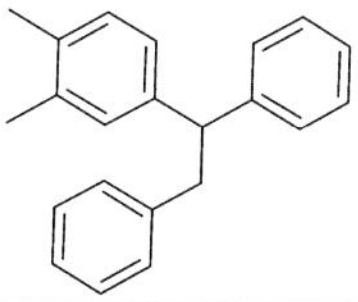	67	65	>99 : 1

[a] Reaction conditions: 0.5 mmol (substituted) styrene, 10 mol% $FeCl_3$, 5 mL *o*-xylene, 80 °C, 4 h.
[b] Main product.
[c] GC conversion of styrene.
[d] GC yield of arylated products.
[e] 4-/3-substitution.
[f] 20 h.

Both iron- [71] and ruthenium-activated [72] nucleophilic aromatic substitutions have been intensively studied in solution. These metal-assisted S_NAr reactions are employed for diverse applications such as the synthesis of monomers for liquid crystal production [73], macrocyclic ethers [74], unsymmetrical diaryl ethers [75], novel *p*-phenylenediamines [76] and biologically active heterocycles [77] and also for the α-arylation of carbonyl and heterocarbonyl compounds [78]. A general process of heteroaromatic substitution by catalytic amounts of an iron(II) salt was described by Minisci's group in 1983 [79]. Here, hydroxylamine-*O*-sulfonic acid is involved in a radical decomposition initiated by $FeSO_4$. Employing various hydrogen donor molecules, selective substitutions of protonated 4-methylquinoline were achieved (Scheme 6.16).

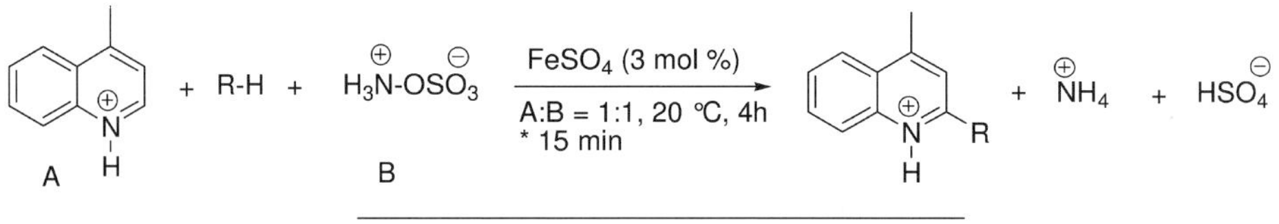

$FeSO_4$ (3 mol %)
A:B = 1:1, 20 °C, 4h
* 15 min

R-H	Radical	Product yield
a) MeOH	• CH_2OH	90 %*
b) 1,4-dioxane	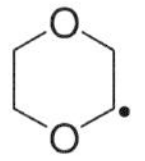	95 %
c) $HCONH_2$	• $CONH_2$	84 %

Scheme 6.16 Radical functionalization of heteroarenes.

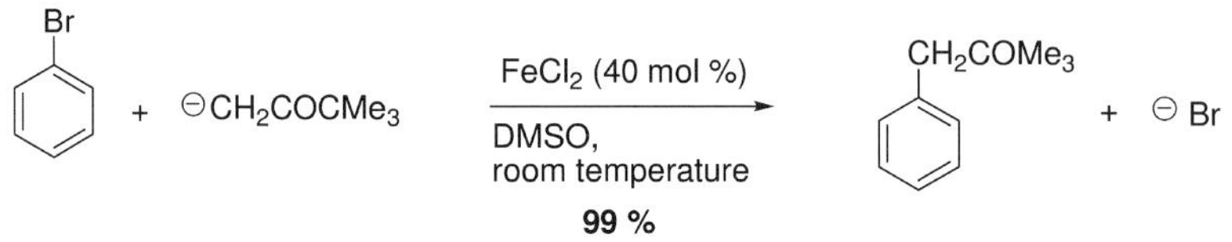

$FeCl_2$ (40 mol %)
DMSO,
room temperature
99 %

Scheme 6.17 Cation-induced S_NAr reaction.

In the following years, this method was improved to a regioselective acylation [80] and also an alkylation reaction [81].

Iron sulfate [82] and iron chloride [83] have also been reported as catalysts for the nucleophilic aromatic substitution of unactivated aryl halides. As solvents, liquid ammonia and DMSO at room temperature are used (Scheme 6.17).

Systematic studies [84] revealed later that a radical $S_{RN}1$ pathway occurs in the case of conjugated substrates.

In order to investigate the scope and limitation of this iron-catalyzed substitution reaction, various nucleophiles [85] and electrophiles were employed in subsequent years [86]. $FeBr_2$ [87] was found to promote this reaction, also, but stoichiometric amounts were necessary.

In 2001, $FeBr_2$ was described as an active catalyst in the synthesis of α-aryl- and α-alkylthioacetylamides even in very low quantities (0.6–1.6 mol%) [88]. An impressive example is shown in Scheme 6.18.

In 2002, an iron-activated nucleophilic aromatic substitution on a solid phase (resin-bound) was published by Ruhland *et al.* [89]. They demonstrated its first application by the synthesis of a library of *o*-, *m*- and *p*-heteroatom-substituted

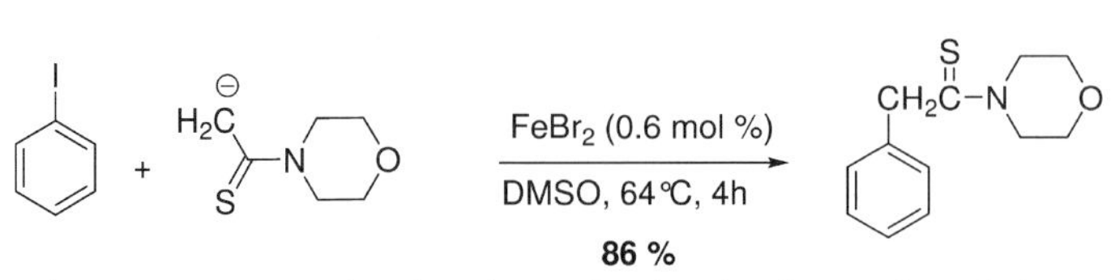

$FeBr_2$ (0.6 mol %)
DMSO, 64 °C, 4h
86 %

Scheme 6.18 $FeBr_2$-catalyzed synthesis of α-phenyl-*N*-thioacetylmorpholine.

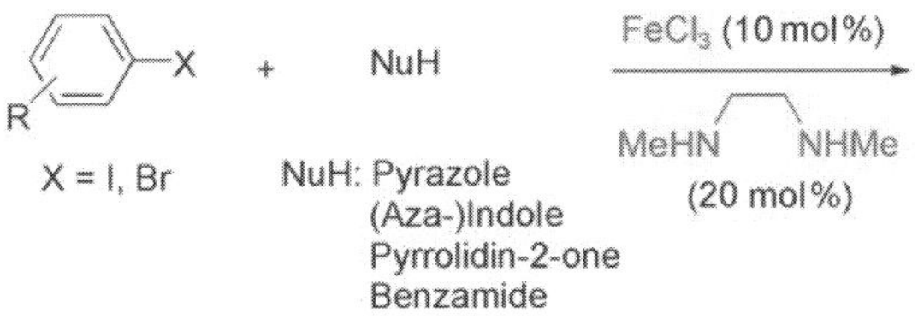

Scheme 6.19 Fe-catalyzed *N*-arylation according to Correa and Bolm.

N-phenylpiperazines and related compounds. Noteworthily, in contrast to methods in solution with covalently bound activating groups such as nitro or carbonyl groups, here the cyclopentadienyl iron moiety is removable.

Recently, Taillefer *et al.* reported an Fe/Cu cooperative catalysis in the assembly of *N*-aryl heterocycles by C–N bond formation [90]. Similarly, Wakharkar and co-workers described the *N*-arylation of various amines with aryl halides in the presence of Cu–Fe hydrotalcite [91]. Interestingly, Correa and Bolm developed a novel and promising ligand-assisted iron-catalyzed *N*-arylation of nitrogen nucleophiles without any Cu co-catalysts (Scheme 6.19) [92]. Differently substituted aryl iodides and bromides react with various amides and *N*-heterocycles. The new catalyst system consists of a mixture of inexpensive $FeCl_3$ and *N,N′*-dimethylethylenediamine (dmeda). Clearly, this research established a useful starting point for numerous future applications of iron-catalyzed arylation reactions.

References

1 Taylor, R. (ed.) *Electrophilic Aromatic Substitution*, Wiley: Chichester, 1990.

2 March, J. *Advanced Organic Chemistry*, 3rd edn., Wiley: New York, 1985.

3 Barton, S. D.; Ollis, V. D. *Comprehensive Organic Chemistry – The Synthesis and Reactions of Organic Compounds*, Vol. 1, Pergamon Press: Oxford, 1979.

4 (a) Olah, G. A. (ed.) *Friedel–Crafts and Related Reactions*, Vols. I–IV, Wiley-Interscience: New York 1963–65; (b) Olah, G. A. (ed.) *Friedel–Crafts Chemistry*, Wiley: New York, 1973; (c) Haeney, H. "The Bimolecular Aromatic Friedel–Crafts Reaction", in Trost, B. M. (ed.) *Comprehensive Organic Synthesis*, Vol. 2, Pergamon Press: Oxford, 1991, pp. 733–752.

5 Review of iron-catalyzed reactions in organic synthesis: Bolm, C.; Legros, J.; Le Paih, J.; Zani, L. *Chem. Rev.* 2004, **104**, 6217–6254.

6 Review particularly of iron(III) chloride in organic synthesis: Diaz, D. D.; Miranda, P. O.; Padrón, J. I.; Martin, V. S. *Curr. Org. Chem.* 2006, **10**, 457–476.

7 Bromination promoted by zeolite: Smith, K.; Bahzad, D. *J. Chem. Soc., Chem. Commun.* 1996, 467–468.

8 Tanemura, K.; Suzuki, T.; Nishida, Y.; Satsumabayashi, K.; Horaguchi, T. *Chem. Lett.* 2003, **32**, 932–933.

9 (a) Feuer, H. (ed.) *Chemistry of the Nitro and Nitroso Groups*, Wiley: New York, 1969; see also: (b) Patai, S. (ed.) *Chemistry of Amino, Nitroso and Nitro Compounds, Supplement F*, Wiley: Chichester, 1982; (c) Topchiev, A. V. *Nitration of Hydrocarbons*, Pergamon Press: Oxford, 1959; (d) De La Mare, P. B. D.; Ridd, J. H. *Aromatic Substitution, Nitration and Halogenation*, Butterworths: London, 1959; (e) for nitrations catalyzed by lanthanide(III) triflates, see Waller, F. J.; Barrett, A. G. M.; Braddock, D. C.;

Ramprasad, D. *J. Chem Soc., Chem. Commun.* 1997, 613–614; (f) Barrett, A. G. M.; Braddock, D. C.; Ducray, R.; McKinnell, R. M.; Waller, F. J. *Synlett* 2000, 57–58; (g) Waller, F. J.; Barrett, A. G. M.; Braddock, D. C.; McKinnell, R. M.; Ramprasad, D. *J. Chem. Soc., Perkin Trans. 1*, 1999, 867–872; (h) Waller, F. J.; Barrett, A. G. M.; Braddock, D. C.; McKinnell, R. M.; White, A. J. P.; Williams, D. J.; Ducray, R. *J. Org. Chem.* 1999, **64**, 2910–2913.

10 For a general survey, see: (a) Nonoyama, N.; Mori, T.; Suzuki, H. *Russ. J. Org. Chem.* 1998, **34**, 1521–1531; (b) Suzuki, T.; Noyori, R. *Chemtracts* 1997, **10**, 813–817; (c) Mori, T.; Suzuki, H. *Synlett* 1995, 383–392; (d) Mori, T.; Suzuki, H. *J. Chem. Soc., Perkin Trans. 2* 1995, 41–44.

11 Suzuki, H.; Yonezawa, S.; Nonoyama, N.; Mori, T. *J. Chem. Soc., Perkin Trans.* 1996, 1, 2385–2389.

12 (a) Peng, X.; Fukui, N.; Mizuta, M.; Suzuki, H. *Org. Biomol. Chem.* 2003, **1**, 2326–2335; (b) Smith, K.; Musson, A.; DeBoos, G. A. *J. Chem. Soc., Chem. Commun.* 1996, 469–470.

13 Michaely, W. J.; Kraatz, G. W. *US Patent 4 780 127*, 1988; *Chem. Abstr.* 1989, **111**, P129017a.

14 Robsein, R. L.; Straw, J. J.; Fahey, D. R. *US Patent 5 260 489*, 1993; *Chem. Abstr.* 1994, **120**, P165200z.

15 Padwa, A.; Bullock, W. H.; Dyszlewski, A. D. *J. Org. Chem.* 1990, **55**, 955.

16 Smith, K.; Ewart, G. M.; Randles, K. R. *J. Chem. Soc., Perkin Trans. 1*, 1997, 1085–1086.

17 Using triflic acid: (a) Effenberger, F.; Huthmacher, K. *Chem. Ber.* 1976, **109**, 2315–2326; (b) Ono, M.; Nakamura, Y.; Sato, S.; Itoh, I. *Chem. Lett.* 1988, 395–398.

18 (a) Graybill, B. M. *J. Org. Chem.* 1967, **32**, 2931–2933; (b) Sipe, H. J., Jr.; Clary, D. W.; White, S. B. *Synthesis* 1984, 283–284; (c) Ueda, M.; Uchiyama, K.; Kano, T. *Synthesis* 1984, 323–325.

19 Furusho, Y.; Okada, Y.; Takata, *T. Bull. Chem. Soc. Jpn.* 2000, **73**, 2827–2828.

20 Reviews and recent articles about FC sulfonylation: (a) F. R. Jensen, Goldman, G. in: Olah, G. A. (ed.) *Friedel–Crafts and Related Reactions*, Vol. III, Wiley-Interscience: New York, 1964, pp. 1319–1367; (b) Taylor, R. in: Banford, C. H. Tipper, C. F. H. (eds.) *Comprehensive Chemical Kinetics*, Elsevier: New York, 1972, pp. 77–83; (c) Taylor, R. *Electrophilic Aromatic Substitution*, Wiley: Chichester, 1990, pp. 334–337; (d) using triflates: Répichet, S.; Le Roux, C.; Hernandez, P.; Dubac, J.; Desmurs, J. R. *J. Org. Chem.* 1999, **64**, 6479–6482; (e) utilizing a combination of $BiCl_3$ and TfOH: Répichet, S.; Le Roux, C.; Dubac, J. *Tetrahedron Lett.* 1999, **40**, 9233–9234.

21 Chaudary, B. M.; Sreenivasa Chowdari, N.; Lakshmi Kantam, M.; Kannan, R. *Tetrahedron Lett.* 1999, **40**, 2859–2862.

22 Marquié, J.; Laporterie, A.; Dubac, J. *J. Org. Chem.* 2001, **66**, 421–425.

23 Alexander, M. V.; Khandekar, A. C.; Samant, S. D. *J. Mol. Catal. A* 2004, **223**, 75–83.

24 (a) Pearson, D. E.; Buehler, C. A. *Synthesis* 1972, 533–542, and references therein; (b) Effenberger, F.; Steegmüller, D. *Chem. Ber.* 1988, **121**, 117–123; (c) Effenberger, F.; Steegmüller, D.; Null, V.; Ziegler, T. *Chem. Ber.* 1988, **121**, 125–130.

25 (a) Cornélis, A.; Laszlo, P.; Wang, S. *Tetrahedron Lett.* 1993, **34**, 3849–3852; (c) Clark, J. H.; Cullen, S. R.; Barlow, S. J.; Bastock, T. W. *J. Chem. Soc., Perkin Trans. 2*, 1994, 1117–1130.

26 Desmurs, J. R.; Labrouillère, M.; Dubac, J.; Laporterie, A.; Gaspard, H.; Metz, F. *Bismuth(III) Salts in Friedel–Crafts Acylation. Industrial Chemistry Library, Vol. 8, The Roots of Organic Development;* Elsevier: Amsterdam, 1996, pp. 15–28.

27 Pivsa-Art, S.; Okuro, K.; Miura, M.; Murata, S.; Nomura, M. *J. Chem. Soc., Perkin Trans. 1*, 1994, 1703–1707.

28 (a) Olah, G. A. *Angew. Chem.* 1993, **105**, 805–827; *Angew. Chem. Int. Ed. Engl.* 1993, **32**, 767–788; (b) Yato, M.; Ohwada, T.; Shudo, K. *J. Am. Chem. Soc.* 1991, **113**,

691–692; (c) Sato, Y.; Yato, M.; Ohwada, T.; Saito, S.; Shudo, K. *J. Am. Chem. Soc.* 1995, **117**, 3037–3043.

29 (a) Effenberger, F.; Epple, D. *Angew. Chem.* 1972, **84**, 295–296; *Angew. Chem. Int. Ed. Engl.* 1972, **11**, 300–301; (b) Effenberger, F.; Sohn, E.; Epple, G. *Chem. Ber.* 1983, **116**, 1195–1208; (c) Effenberger, F.; Eberhard, J. K.; Maier, A. H. *J. Am. Chem. Soc.* 1996, **118**, 12572–12579; (d) Izumi, J.; Mukaiyama, T. *Chem. Lett.* 1996, 739–740.

30 (a) Hölderich, W.; Hesse, M.; Näumann, F. *Angew. Chem.* 1988, **100**, 232–251; *Angew. Chem. Int. Ed. Engl.* 1988, **27**, 226–246; (b) Akporiaye, D. E.; Daasvatn, K.; Solberg, J.; Stöcker, M.; in M. Guisnet *et al.* (eds.) *Heterogeneous Catalysis and Fine Chemicals III*, Elsevier: Amsterdam, 1993, pp. 521–526; (c) Spagnol, M.; Gilbert, L.; Alby, D.; in J. R. Desmurs; S. Ratton (eds.) *Industrial Chemistry Library, Vol. 8, The Roots of Organic Development;* Elsevier: Amsterdam, 1996, pp. 29–38; (d) Cornélis, A.; Gerstmans, A.; Laszlo, P.; Mathy, A.; Zieba, I. *Catal. Lett.* 1990, **6**, 103–109; (e) Das, D.; Cheng, S. *Appl. Catal. A* 2000, **201**, 159–168, and references therein.

31 (a) Olah, G. A.; Farook, O.; Morteza, S.; Farnia, F.; Olah, J. *J. Am. Chem. Soc.* 1988, **110**, 2560–2565; (b) Kawada, A. Mitamura, S.; Kobayashi, S. *J. Chem. Soc., Chem. Commun.* 1993, 1157–1158; (c) Kawada, A. Mitamura, S.; Kobayashi, S. *Synlett*, 1994, 545–546.

32 (a) Mikami, K.; Kotera, O.; Motoyama, Y.; Sakaguchi, H.; Maruta, M. *Synlett* 1996, 171–172; (b) Nie, J.; Xu, J.; Zhou, G. *J. Chem. Res. (S)* 1999, 446–447.

33 (a) Desmurs, J. R.; Labrouillère, M.; Le Roux, C.; Laporterie, A.; Dubac, J. *Tetrahedron Lett.* 1997, **38**, 8871–8874; (b) Répichet, S.; Le Roux, C.; Dubac, J.; Desmurs, J. R. *Eur. J. Org. Chem.* 1998, 2743–2746.

34 (a) Hachiva, I.; Moriwaki, M.; Kobayashi, S. *Tetrahedron Lett.* 1995, **36**, 409–412; (b) Hachiva, I.; Moriwaki, M.; Kobayashi, S. *Bull. Chem. Soc. Jpn.* 1995, **68**, 2053–2060.

35 Kobayashi, S.; Iwamoto, S. *Tetrahedron Lett.* 1998, **39**, 4697–4700.

36 (a) Csihony, S.; Mehdi, H.; Horváth, I. T. *Green Chem.* 2001, **3**, 307–309; (b) Csihony, S.; Mehdi, H.; Homonnay, Z.; Vértes, A.; Farkas, Ö.; Horváth, I. T. *J. Chem. Soc., Dalton Trans.* 2002, 680–685.

37 Valkenberg, M. H.; de Castro, C.; Hölderich, W. F. *Appl. Catal. A* 2001, **215**, 185–190.

38 Xiao, Y.; Malhotra, S. V. *J. Organomet. Chem.* 2005, **690**, 3609–3613.

39 Pictures taken from Olah, G. A. (ed.) *Friedel–Crafts and Related Reactions*, Vols. I–IV, Wiley-Interscience: New York, 1963–1965.

40 Trost, B. M.; Flemming, I. *Comprehensive Organic Synthesis*, 1st edn., Vol.3, Pergamon Press: Oxford, 1991, p. 293.

41 (a) Weissermel, K.; Arpe, H.-J. *Industrielle Organische Chemie,* 5th edn., Wiley-VCH: Weinheim, 1998; (b) Franck, H. G.; Stadelhofer, J. W. *Industrielle Aromatenchemie: Rohstoffe, Verfahren, Produkte*, Springer: Berlin, 1987.

42 (a) Tsuchimoto, T.; Tobita, K.; Hiyama, T.; Fukuzawa, S. *Synlett* 1996, 557–559; (b) El Gihani, M. T.; Heaney, H.; Shuhaibar, K. F. *Synlett*, 1996, 871–872; (c) Tsuchimoto, T.; Tobita, K.; Hiyama, T.; Fukuzawa, S. *J. Org. Chem.* 1997, **62**, 6997–7005; (d) Mukaiyama, T.; Kamiyama, H.; Yamanaka, H. *Chem. Lett.* 2003, **32**, 814–815; (e) see also review: Kobayashi, S. *Eur. J. Org. Chem.* 1999, 15–27.

43 (a) Shiina, I.; Suzuki, M. *Tetrahedron Lett.* 2002, **43**, 6391–6394; (b) Noji, M.; Ohno, T.; Fuji, K.; Futaba, N.; Tajima, H.; Ishii, K. *J. Org. Chem.* 2003, **68**, 9340–9347.

44 (a) Transition metal-doped montmorillonite: Laszlo, P.; Mathy, A. *Helv. Chim. Act.* 1987, **70**, 577–586; (b) Nafion-H: Yamato, T.; Hideshima, C.; Prakash, G. K. S.; Olah, G. A. *J. Org. Chem.* 1991, **56**, 2089–2091; (c) iron-promoted sulfated zirconia: Suja, H.; Deepa, C. S.; Sreejarani, K.; Sugunan, S. *React. Kinet. Catal. Lett.* 2003, **79**, 373–379;

(d) K10-supported iron oxides: Shrigadi, N. B.; Shinde, A. B.; Samant, S. D. *Appl. Catal. A* 2003, **252**, 23–35.

45 Mincione, E.; Bovicelli, P. *Gazz. Chim. Ital.* 1982, **112**, 437–440.

46 (a) Tada, Y.; Satake, A.; Shimizu, I.; Yamamoto, A. *Chem. Lett.* 1996, 1021–1022; (b) Kimura, M.; Futamata, M.; Mukai, R.; Tamura, Y. *J. Am. Chem. Soc.* 2005, **127**, 4592–4593; (c) Tamura, Y. *Eur. J. Org. Chem.* 2005, 2647–2656; (d) Kimura, M.; Fukasaka, M.; Tamura, Y. *Synthesis* 2006, **21**, 3611–3616; (e) enantioselective allylation of indoles: Trost, B. M.; Quancard, J. *J. Am. Chem. Soc.* 2006, **128**, 6314–6315; (f) Kimura, M.; Fukasaka, M.; Tamura, Y. *Heterocycles* 2006, **67**, 535–542; (f) for molybdenum- and tungsten-catalyzed allylation of aromatic compounds, see also Shimizu, I.; Sakamoto, T.; Kawaragi, S.; Maruyama, Y.; Yamamoto, A. *Chem. Lett.* 1997, 137–138.

47 (a) Nishibayashi, Y.; Inada, Y.; Hidai, M.; Uemura, S. *J. Am. Chem. Soc.* 2002, **124**, 7900–7901; (b) Nishibayashi, Y.; Yoshikawa, M.; Inada, Y.; Hidai, M.; Uemura, S. *J. Am. Chem. Soc.* 2002, **124**, 11846–11847; (c) Nishibayashi, Y.; Inada, Y.; Yoshikawa, M.; Hidai, M.; Uemura, S. *Angew. Chem.* 2003, **115**, 1533–1536; *Angew. Chem. Int. Ed.* 2003, **42**, 1495–1498; (d) for ruthenium-catalyzed propargylations with other nucleophiles, see also Nishibayashi, Y.; Milton, M. D.; Inada, Y.; Yoshikawa, M.; Wakiji, I.; Hidai, M.; Uemura, S. *Chem. Eur. J.* 2005, **11**, 1433–1451.

48 Kennedy-Smith, J. J.; Young, L. A.; Toste, F. D. *Org. Lett.* 2004, **6**, 1325–1327.

49 (a) Georgy, M.; Boucard, V.; Campagne, J.-M. *J. Am. Chem. Soc.* 2005, **127**, 14180–14181; (b) Liu, J.; Muth, E.; Flörke, U.; Henkel, G.; Merz, K.; Sauvageau, J.; Schwake, E.; Dyker, G. *Adv. Synth. Catal.* 2006, **348**, 456–462.

50 (a) Propargylic acetates: Zhan, Z.-P.; Cui, Y.-Y.; Liu, H.-J. *Tetrahedron Lett.* 2006, **47**, 9143–9146; (b) propargylic alcohols: Zhan, Z.-P.; Yu, J.-L.; Liu, H.-J.; Cui, Y.-Y.; Yang, R.-F.; Yang, W.-Z.; Li, J.-P. *J. Org. Chem.* 2006, **71**, 8298–8301.

51 Kondo, T.; Tantayanon, S.; Tsuji, Y.; Watanabe, Y. *Tetrahedron Lett.* 1989, **30**, 4137–4140.

52 Choughury, J.; Podder, S.; Roy, S. *J. Am. Chem. Soc.* 2005, **127**, 6162–6163.

53 Yasuda, M.; Somyo, T.; Baba, A. *Angew. Chem.* 2006, **118**, 807–810; *Angew. Chem. Int. Ed.* 2006, **45**, 793–796.

54 Mertins, K.; Jovel, I.; Kischel, J.; Zapf, A.; Beller, M. *Angew. Chem.* 2005, **117**, 242–246; *Angew. Chem. Int. Ed.* 2005, **44**, 238–242.

55 Mertins, K.; Jovel, I.; Kischel, J.; Zapf, A.; Beller, M. *Adv. Synth. Catal.* 2006, **348**, 691–995.

56 Jovel, I.; Mertins, K.; Kischel, J.; Zapf, A.; Beller, M. *Angew. Chem.* 2005, **117**, 3981–3985; *Angew. Chem. Int. Ed.* 2005, **44**, 3913–3917.

57 selected examples of ethylene arylation: (a) Matsumoto, T.; Taube, D. J.; Periana, R. A.; Taube, H.; Yoshida, H. *J. Am. Chem. Soc.* 2000, **122**, 7414–7415; (b) Ritleng, V.; Sirlin, C.; Pfeffer, M. *Chem. Rev.* 2002, **102**, 1731–1769; (c) Kakiuchi, F.; Murai, S. *Acc. Chem. Res.* 2002, **35**, 826–834; (d) Lail, M.; Arrowood, B. N.; Gunnoe, T. B. *J. Am. Chem. Soc.* 2003, **125**, 7506–7507; (e) Guari, Y.; Castellanos, A.; Sabo-Etienne, S.; Chaudret, B. *J. Mol. Catal. A* 2004, **212**, 77–82; examples of diverse alkene arylations: (f) Matsui, M.; Yamamoto, H. *Bull. Chem. Soc. Jpn.* 1995, **68**, 2657–2661; (g) Thalji, R. K.; Ahrendt, K. A.; Bergmann, R. G.; Ellman, J. A. *J. Am. Chem. Soc.* 2001, **123**, 9692–9693; (h) Dorta, R.; Togni, A. *Chem Commun.* 2003, 760–761; (i) Ahrendt, K. A.; Bergmann, R. G.; Ellman, J. A. *Org. Lett.* 2003, **5**, 1301–1303; (j) Youn, S. W.; Pastine, S. J.; Sames, D. *Org. Lett.* 2004, **6**, 581–584; (k) Karshtedt, D.; Bell, A. T.; Tilley, T. D. *Organometallics* 2004, **23**, 4169–4171.

58 Kischel, J.; Jovel, I.; Mertins, K.; Zapf, A.; Beller, M. *Org. Lett.* 2006, **8**, 19–22.

59 Bhalla, G.; Oxgaard, J.; Goddard, W. A.; Periana, R. A. *Organometallics* 2005, **24**, 3229–3232.

60 Cherian, A. E.; Domski, G. J.; Rose, J. M.; Lobkovski, E. B.; Coates, G. W. *Org. Lett.* 2005, **7**, 5135–5137.

61 Kaspar, L. T.; Fingerhut, B.; Ackermann, L. *Angew. Chem.* 2005, **117**, 6126–6128; *Angew. Chem. Int. Ed.* 2005, **44**, 5972–5974.

62 Shimizu, I.; Khien, K. M.; Nagatomo, M.; Nakajima, T.; Yamamoto, A. *Chem. Lett.* 1997, 851–852.

63 Toullec, P. Y.; Genin, E.; Leseurre, L.; Genêt, J.-P.; Michelet, V. *Angew. Chem.* 2006, **118**, 7587–7590; *Angew. Chem. Int. Ed.* 2006, **45**, 7427–7430.

64 (a) for a recent review on stereoselective Friedel–Crafts type reactions with unsaturated compounds, see Bandini, M.; Melloni, A.; Umani-Ronchi, A. *Angew. Chem.* 2004, **116**, 560–566; *Angew. Chem. Int. Ed.* 2004, **43**, 550–556; (b) for stereoselective alkylations of indoles, see also Bandini, M.; Melloni, A.; Tommasi, S.; Umani-Ronchi, A. *Synlett* 2005, **8**, 1199–1222.

65 (a) Zhuang, W.; Hansen, T.; Jørgensen, K. A. *Chem. Commun.* 2001, 347–348; (b) Jørgensen, K. A. *Synthesis* 2003, **7**, 1117–1125; (c) Palomo, C.; Oiarbide, M.; Kardak, B. G.; García, J. M.; Linden, A. *J. Am. Chem. Soc.* 2005, **127**, 4154–4155.

66 Jensen, K. B.; Thorhauge, J.; Hazell, R. G.; Jørgensen, K. A. *Angew. Chem.* 2001, **113**, 164–167; *Angew. Chem. Int. Ed.* 2001, **40**, 160–163.

67 (a) Zhou, J.; Tang, Y. *J. Am. Chem. Soc.* 2002, **124**, 9030–9031; (b) Zhou, J.; Ye, M.-C.; Huang, Z.-Z.; Tang, Y. *J. Org. Chem.* 2004, **69**, 1309–1320.

68 Evans, D. A.; Fandrick, K. R.; Song, H.-J. *J. Am. Chem. Soc.* 2005, **127**, 8942–8943.

69 (a) Paras, N. A.; MacMillan, D. W. C. *J. Am. Chem. Soc.* 2001, **123**, 4370–4371; (b) Austin, J. F.; MacMillan, D. W. C. *J. Am. Chem. Soc.* 2002, **124**, 1172–1173; (c) Paras, N. A.; MacMillan, D. W. C. *J. Am. Chem. Soc.* 2002, **124**, 7894–7895.

70 (a) Bunnett, J. F. *Acc. Chem. Res.* 1978, **11**, 413; (b) Rossi, R. A.; Rossi, R. H., *Aromatic Substitution by the $S_{RN}1$ Mechanism*, ACS Monograph 178, American Chemical Society: Washington, DC, 1983.

71 (a) Astruc, D. *Tetrahedron* 1983, **39**, 4027–4095; (b) Astruc, D. *Top. Curr. Chem.* 1992, **160**, 47–95.

72 Moriarty, R. M.; Gill, U. S.; Ku, Y. Y. *J. Organomet. Chem.* 1988, **350**, 157–190.

73 (a) Pearson, A. J.; Gelormini, A. M. *Macromolecules* 1994, **27**, 3675–3677; (b) Pearson, A. J.; Sun, L. *J. Polym. Sci., Part A* 1997, **35**, 447–453.

74 Pearson, A. J.; Park, J. G.; Zhu, P. Y. *J. Org. Chem.* 1992, **57**, 3583–3589.

75 Sawyer, J. S. *Tetrahedron* 2000, **56**, 5045–5065.

76 (a) Pearson, A. J.; Gelormini, A. M. *J. Org. Chem.* 1996, **61**, 1297–1305; (b) Pearson, A. J.; Gelormini, A. M. *Tetrahedron Lett.* 1997, **38**, 5123–5126.

77 Sutherland, R. G.; Piórko, A.; Gill, U. S.; Lee, C. C. *J. Heterocycl. Chem.* 1982, **19**, 801–803; (b) Sutherland, R. G.; Piórko, A.; Lee, C. C. *J. Heterocycl. Chem.* 1988, **25**, 1911–1916; (c) Storm, J. P.; Ionescu, R. D.; Martinsson, D.; Andersson, C.-M. *Synlett* 2000, **7**, 975–978; (d) *Encyclopedia of Reagents for Organic Synthesis*, Vol. 4, ed.: L. A. Paquette, Wiley: Chichester, 1995.

78 (a) Abd-El-Aziz, A. S.; Lee, C. C.; Piórko, A.; Sutherland, R. G. *Synth. Commun.* 1988, **18**, 291–300; (b) Abd-El-Aziz, A. S.; de Denus, C. R. *J. Chem. Soc., Perkin Trans. 1*, 1993, 293–298; (c) Artamkina, G. A.; Sazonov, P. K.; Beletskaya, I. P. *Tetrahedron Lett.* 2001, **42**, 4385–4387.

79 Citterio, A.; Gentile, A.; Minisci, F.; Serravalle, M.; Ventura, S. *J. Chem. Soc., Chem. Commun.* 1983, 916–917.

80 Minisci, F.; Vismara, E.; Fontana, F.; Radaelli, D. *Gazz. Chim. Ital.* 1987, **117**, 363.

81 Minisci, F.; Vismara, E.; Fontana, F. *J. Org. Chem.* 1989, **54**, 5224–5227.

82 Galli, C.; Bunnett, J. F. *J. Org. Chem.* 1984, **49**, 3041–3042.

83 Galli, C.; Gentili, P. *J. Chem. Soc., Perkin Trans. 2,* 1993, 1135–1140.

84 Galli, C.; Gentili, P.; Rappoport, Z. *J. Org. Chem.* 1994, **59**, 6786–6795.

85 Galli, C.; Gentili, P.; Guarnieri, A. *Gazz. Chim. Ital.* 1997, **127**, 159.

86 (a) Rossi, R. A.; Alonso, R. A. *J. Org. Chem.* 1980, **45**, 1239–1241; (b) van Leeuwen, P. W. M. N.; McKillop, A. J. *Chem. Soc., Perkin Trans. 1,* 1993, 2433; (c) Murguía, M. C.; Rossi, R. A. *Tetrahedron Lett.* 1997, **38**, 1355–1358.

87 Nazareno, M. A.; Rossi, R. A. *J. Org. Chem.* 1996, **61**, 1645–1649.

88 Murguía, M. C.; Ricci, C. G.; Cabrera, M. I.; Luna, J. A.; Grau, R. J. *J. Mol. Catal. A.* 2001, **165**, 113–120.

89 Ruhland, T.; Bang, K. S.; Andersen, K. *J. Org. Chem.* 2002, **67**, 5257–5268.

90 Taillefer, M.; Xia, N.; Oualli, A. *Angew. Chem.* 2007, **119**, 952; *Angew. Chem. Int. Ed.* 2007, **46**, 934.

91 Jadhav, V. H.; Dumbre, D. K.; Phapale, V. B.; Borate, H. B.; Wakharkar, R. D. *Catal. Commun.* 2007, **8**, 65.

92 Correa, A.; Bolm, C. *Angew. Chem.* 2007, **119**, 9018; *Angew. Chem. Int. Ed.* 2007, **46**, 8862.

7
Iron-catalyzed Substitution Reactions

Bernd Plietker

7.1
Introduction

Substitution reactions, i.e. the replacement of an atom or molecular fragment for another, represent one of the fundamental transformations in organic chemistry. Depending on the character of the leaving group, nucleophiles or electrophiles can be introduced at a defined position in the starting material. This Section will focus on iron-catalyzed nucleophilic substitutions as a mean for the chemo-, regio- and stereoselective construction of new C-atom bonds. Due to their increasing importance, nucleophilic substitutions involving reaction between alkyl halides or sulfonates and aryl or alkyl metal species in the presence of catalytic amounts of iron are summarized in a separate section on cross-coupling chemistry (Chapter 5). Reactions involving an Fe-catalyzed electrophilic substitution of an aromatic C—H bond by any other atom are presented in Chapter 6.

7.2
Iron-catalyzed Nucleophilic Substitutions

7.2.1
Nucleophilic Substitutions of Non-activated C—X Bonds

7.2.1.1 Introduction

Several methods for iron-catalyzed nucleophilic substitution reactions have been developed in recent years. In general, these reactions can be subdivided into two main categories depending on the catalytic character of the iron source employed. Traditionally, several formal substitutions are catalyzed by Fe^{3+} salts. In these reactions, the metal salt acts as a Lewis acid and coordinates to the leaving group X in the starting material. The resulting increase in positive charge at the carbon atom facilitates the reaction with an incoming nucleophile. Depending on the steric and

Iron Catalysis in Organic Chemistry. Edited by Bernd Plietker

ISBN: 978-3-527-31927-5

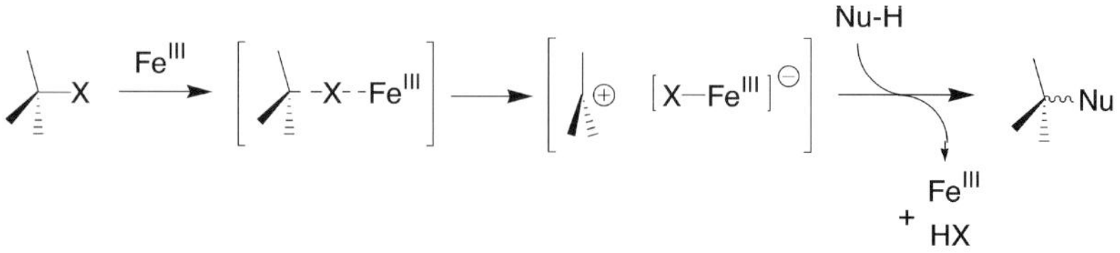

Scheme 7.1 Nucleophilic substitutions catalyzed by Lewis acidic Fe salts.

electronic properties of the starting material and the reaction conditions, this classical reaction pathway can follow an S_N1- or S_N2-type mechanism (Scheme 7.1).

Within the past 20 years, ferrates, i.e. anions possessing iron as the center atom, have found increasing application as nucleophilic complexes in substitution chemistry. In these reactions, the ferrate replaces the leaving group X in a first nucleophilic substitution event. A transfer of one ligand from the metal atom (i.e. a reductive elimination, *path A*, Scheme 7.2) or substitution of the metal atom via external attack of the nucleophile (*path B*) concludes this mechanistic scenario. However, the exact mechanism in ferrate-catalyzed nucleophilic substitutions is still under debate. Apart from the ionic mechanism, radical processes are also discussed in the literature.

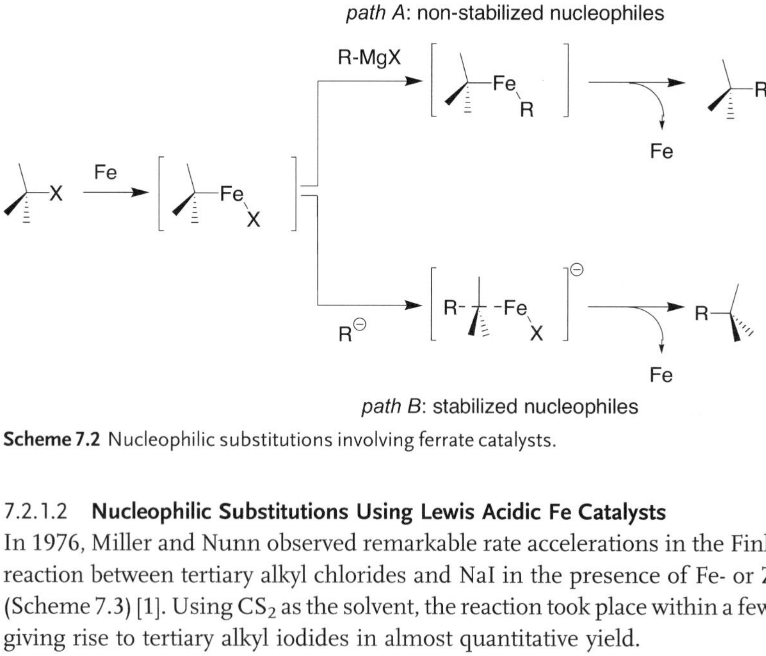

path A: non-stabilized nucleophiles

path B: stabilized nucleophiles

Scheme 7.2 Nucleophilic substitutions involving ferrate catalysts.

7.2.1.2 Nucleophilic Substitutions Using Lewis Acidic Fe Catalysts

In 1976, Miller and Nunn observed remarkable rate accelerations in the Finkelstein reaction between tertiary alkyl chlorides and NaI in the presence of Fe- or Zn salts (Scheme 7.3) [1]. Using CS_2 as the solvent, the reaction took place within a few hours, giving rise to tertiary alkyl iodides in almost quantitative yield.

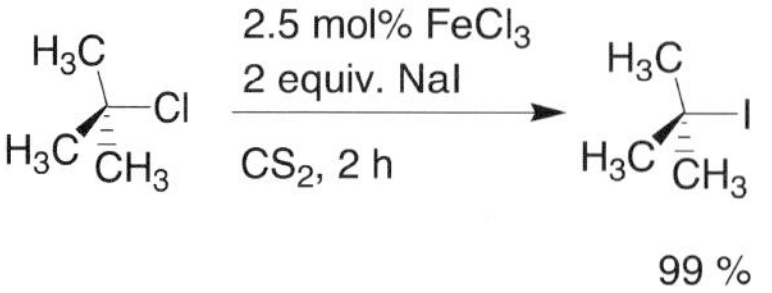

2.5 mol% $FeCl_3$, 2 equiv. NaI, CS_2, 2 h

99 %

Scheme 7.3 Fe-catalyzed Finkelstein reaction of tertiary alkyl chlorides.

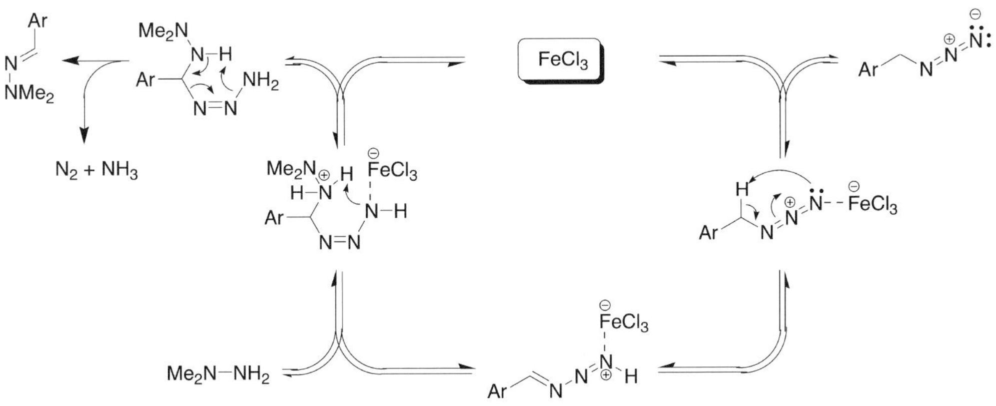

Scheme 7.4 Mechanism of the preparation of hydrazones from azides.

Using azides as pseudohalide substituents, Fe salts open up new reaction pathways. In the presence of catalytic amounts of $FeCl_3$ and stoichiometric amounts of dimethylhydrazine, benzylic azides can be transformed into aryl-substituted hydrazones [2]. The reaction is believed to involve a Lewis acid coordination to the terminal nitrogen followed by a proton shift and subsequent addition of an aryl-substituted hydrazine. The following skeletal rearrangement is driven by the extrusion of nitrogen while generating the desired products in good yields (Scheme 7.4).

Apart from these examples, the activated substrate–Fe complex is more frequently used as an electrophilic reagent, for example in electrophilic aromatic substitutions (Chapter 6).

7.2.1.3 **Substitutions Catalyzed by Ferrate Complexes**

Whereas iron-based Lewis acidic catalysts have found frequent use in organic chemistry, the nucleophilic displacement of a leaving group by an iron–ate complex to give organo-iron complexes (Scheme 7.2) has attracted comparatively less attention in catalysis. This is even more surprising if one considers the beginning of the Fe-catalyzed cross-coupling chemistry in the early 1980s. At that time, Kocchi presented the first example of an Fe-catalyzed cross-coupling reaction [3]. The ferrate ion needed for this reaction was prepared by reacting an Fe salt with an excess of a Grignard reagent. Since then, the stoichiometric organometallic chemistry of ferrate ions formed *in situ* has been investigated in detail; however, catalytic applications using these complexes have been reported only occasionally [4]. The recent search for new, environmentally benign and inexpensive transition metal catalysts has led to a revival of this chemistry. Based upon Kocchi's seminal contributions [3], a variety of research groups were able to develop sophisticated and improved protocols for iron-catalyzed cross-coupling reactions (Chapter 5) and nucleophilic substitutions. In most of these reactions, the active iron species is formed *in situ* by reacting an Fe salt with a Grignard or zinc reagent to form ferrates via a formal ligand exchange event [5]. Mechanistic studies on catalytic reactions involving tetraalkylferrates indicate that a radical mechanism takes place with a double SET [6]. Two structurally characterized,

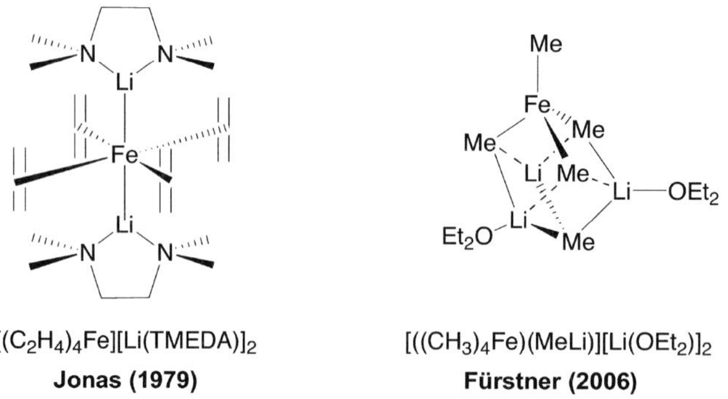

$[(C_2H_4)_4Fe][Li(TMEDA)]_2$
Jonas (1979)

$[((CH_3)_4Fe)(MeLi)][Li(OEt_2)]_2$
Fürstner (2006)

Figure 7.1 Structures of isolated catalytically active low-valent Fe complexes.

catalytically active low-valent Fe complexes are shown in Figure 7.1, with the latter complex most likely being involved in reactions catalyzed by a combination of an Fe salt and an alkylmetal reagent [7].

Although the exact mechanism still remains elusive and may be dependent upon the reaction conditions, various substitution processes are efficiently catalyzed. The most prominent area of application of these complexes is the cross-coupling chemistry. The dehalogenation of aromatic halides using catalytic amounts of Fe salts and a reducing agent such as lithium powder is mechanistically related and can be applied to a variety of different halogenated aromatic compounds [8]. Alkyl-Grignard reagents can also be used as reducing agents in this type of chemistry; however, this observation is in sharp contrast to the results obtained in Fe-catalyzed cross-coupling reactions (Scheme 7.5) [9].

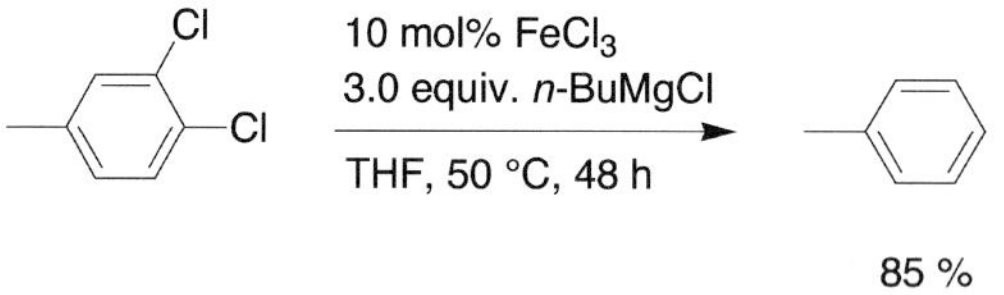

10 mol% $FeCl_3$
3.0 equiv. *n*-BuMgCl
THF, 50 °C, 48 h
85 %

Scheme 7.5 Fe-catalyzed dechlorination of aryl chlorides.

A formally related yet most likely Ullman type of arylation of nitrogen nucleophiles was reported most recently. A bimetallic catalytic system of $Fe(acac)_3$–CuO was shown to be highly active in the simple arylation of a variety of *N*-heterocycles with aryl iodides and bromides in the presence of $CsCO_3$. Although a deeper mechanistic understanding awaits further investigations, the broad scope plus its operational simplicity make this procedure especially amenable to applications on an industrial scale (Scheme 7.6) [10].

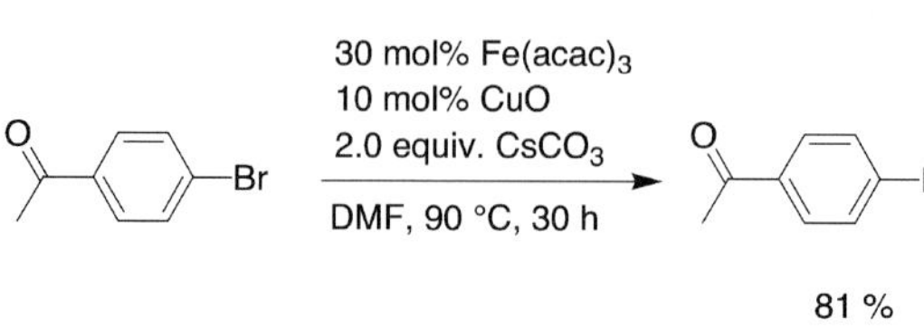

30 mol% $Fe(acac)_3$
10 mol% CuO
2.0 equiv. $CsCO_3$
DMF, 90 °C, 30 h
81 %

Scheme 7.6 Fe-catalyzed amination of aryl halides.

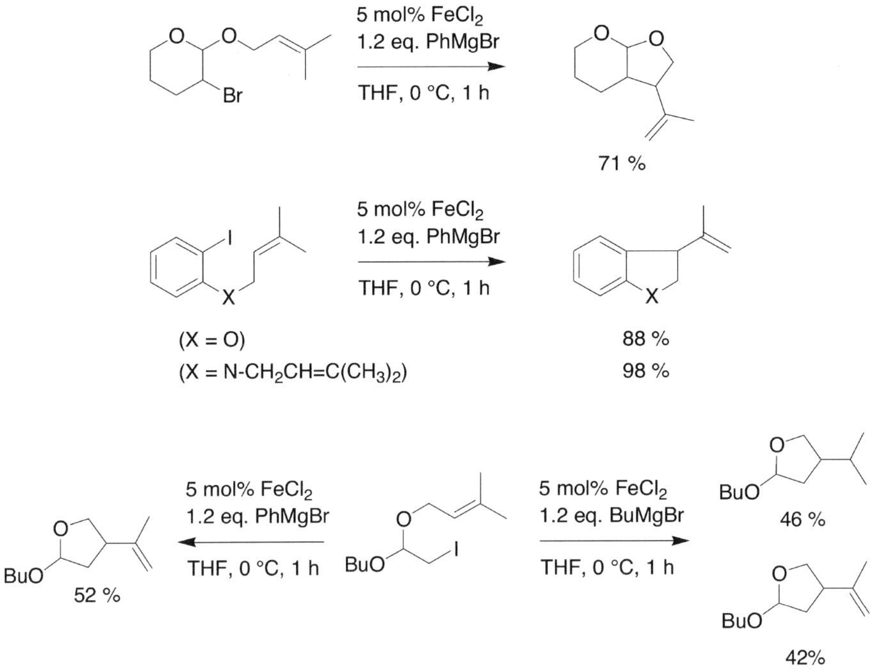

5 mol% $FeCl_2$
1.2 eq. PhMgBr
THF, 0 °C, 1 h
Br
71 %
(7.1)

5 mol% $FeCl_2$
1.2 eq. PhMgBr
THF, 0 °C, 1 h
(X = O) 88 %
(X = N-$CH_2CH=C(CH_3)_2$) 98 %
(7.2)

5 mol% $FeCl_2$
1.2 eq. PhMgBr
THF, 0 °C, 1 h
BuO
52 %
5 mol% $FeCl_2$
1.2 eq. BuMgBr
THF, 0 °C, 1 h
BuO 46 %
BuO 42%
(7.3)

Scheme 7.7 Heck-type reaction in the presence of Fe salts.

The replacement of a halogen by a low-valent ferrate catalyst is the key step in an intramolecular Heck-type reaction using both alkyl and aryl halides (Scheme 7.7) [11].

The catalytically active Fe complex is generated *in situ* upon mixing catalytic amounts of $FeCl_2$ with stoichiometric amounts of PhMgBr. A variety of alkyl and aryl halides were employed in this study [Equations (7.1) and (7.2), Scheme 7.7]. The corresponding cyclization products were obtained in good yields. Interestingly, it was shown that the regioselectivity of the β-hydride elimination depends strongly on the character of the Grignard reagent used [Equation (7.3), Scheme 7.7]. In the presence of PhMgCl lacking any β-hydrogen, the elimination occurred solely at the aliphatic moiety. Employing *n*-BuMgCl as the reductant, on the other hand, yielded a mixture of olefinic and saturated products due to a competing β-hydride elimination from the *n*-butyl substituent.

Aviv and Gross developed an interesting insertion reaction of diazo compounds into a secondary amine–hydrogen bond in the presence of Fe–corrole complexes (Scheme 7.8) [12]. Competition experiments performed in the presence of an amine and an alkene revealed the N–H-insertion reaction to be much faster than the cyclopropanation of the C=C bond. Apart from this chemoselectivity issue, the reactions are characterized by their very short reaction times: most insertion reactions were completed within 1 min at room temperature. Most recently, Woo's group reported on a similar process using commercially available iron tetraphenylporphyrin [Fe(TPP)] dichloride [13].

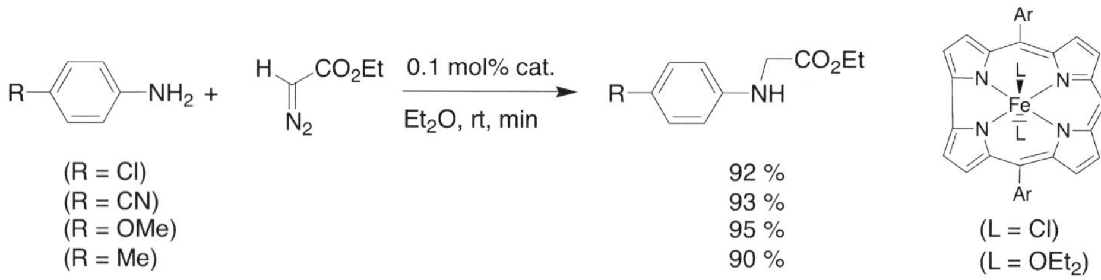

Scheme 7.8 Fe–corrole-catalyzed insertion of diazo compounds into N–H bonds.

7.2.2 Nucleophilic Substitution of Allylic and Propargylic C–X Bonds

7.2.2.1 Reactions Catalyzed by Lewis Acidic Fe Salts

Introduction The combination of a polarized C–X bond and an allylic π bond induces a fundamental change in substitution chemistry. Apart from reactions similar to those shown in Scheme 7.1, the partial positive charge on the allylic carbon atom allows for a further type of mechanism, i.e. conjugate substitution. On the one hand, the nucleophilic substitution might follow an S_N2-type mechanism (Scheme 7.1), in which the metal activates the leaving group. Depending on the nature of the leaving group and the Lewis acidic character of the catalyst, a concerted and hence stereo- and regioselective S_N2- or S_N2'-type reaction might take place. In the presence of strongly acidic catalysts and/or good leaving groups, allylic cations might be generated leading to a mixture of regio- and stereoisomeric products (Scheme 7.9).

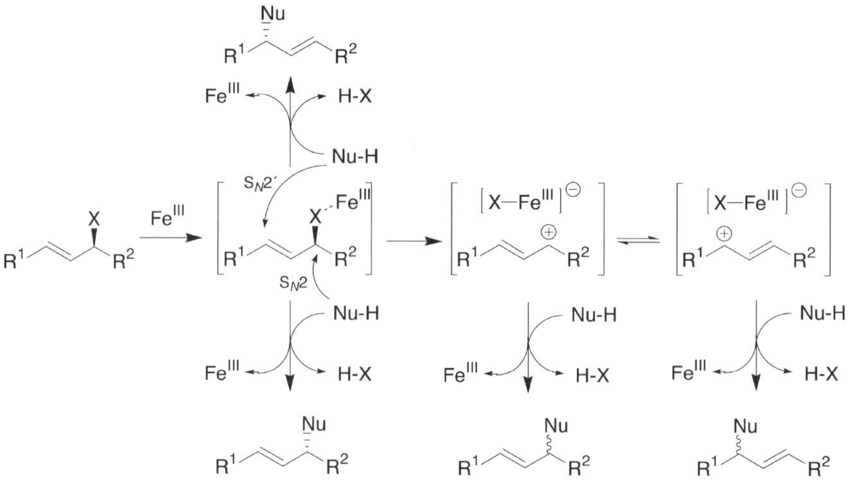

Scheme 7.9 Possible mechanisms in allylic substitution catalyzed by Lewis acidic Fe salts.

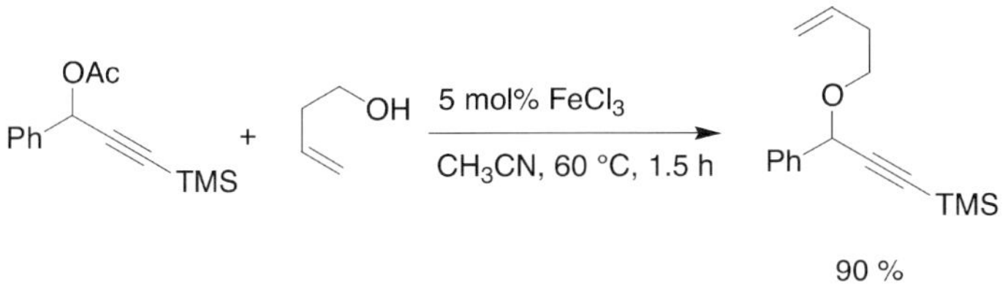

Scheme 7.10 Direct substitution of acetates by alcohols.

S_N2-type Reactions Other leaving groups are also prone to undergo direct S_N reactions in the presence of Fe salts. Zhan and Liu described the direct Fe-catalyzed etherification of propargylic acetates by *O*-nucleophiles to give rise to a variety of functionalized propargylic ethers (Scheme 7.10) [14].

Apart from halides, pseudohalides and acetates, $FeCl_3$ is able to activate hydroxyl groups in a similar manner. Hence various substitutions of hydroxyl groups have been developed, e.g. the condensation of alcohols or phenols with diphenylmethanol to give DPM-protected alcohols [Equation (7.4), Scheme 7.11] [15] or the direct coupling of allylic or benzylic alcohols with C–H-acidic compounds [Equation (7.5)] [16].

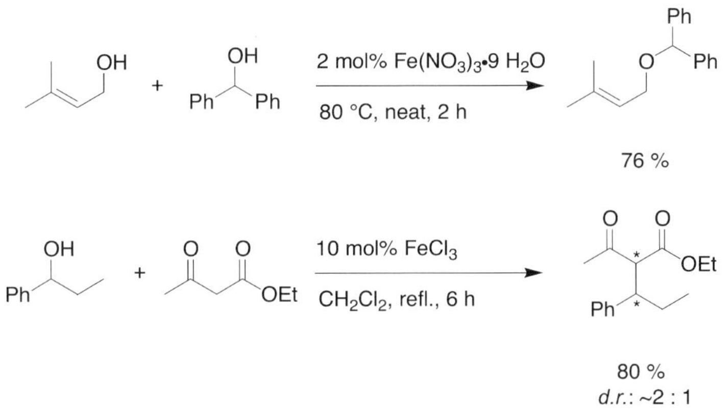

(7.4)

(7.5)

Scheme 7.11 Nucleophilic substitution of aliphatic alcohols.

The direct substitution of hydroxyl groups can also be extended towards propargylic alcohols. In the presence of $FeCl_3$, a plethora of *O*-, *N*- or *S*-nucleophiles are able to react with substituted propargylic alcohols [17]. Amongst the variety of nucleophiles, allylsilanes occupy an important position since this example resembles a cross-coupling reaction between an organometallic compound and an alcohol (Scheme 7.12) [17].

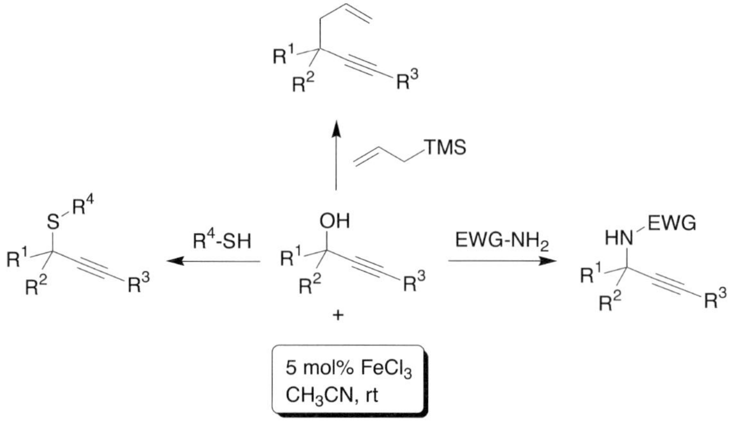

Scheme 7.12 Nucleophilic substitutions of propargylic alcohols.

Conjugate Nucleophilic Substitutions Although iron salts have found frequent use as Lewis acidic catalysts in direct S_N reactions, only one example of a catalytic activation of the allylic leaving group by an Fe salt and its subsequent conjugate substitution by an incoming nucleophile has been reported [18]. Using an immobilized Fe^{3+} source (Fe^{3+}–K10 montmorillonite), various allylic alcohols derived via a Baylis–Hillman reaction were coupled directly to aromatic compounds in a Friedel–Crafts-type reaction. Depending on the nature of the electron-withdrawing group in the 2-position, either *E*- or *Z*-configured trisubstituted alkenes were obtained in good to excellent yields (Scheme 7.13) [19].

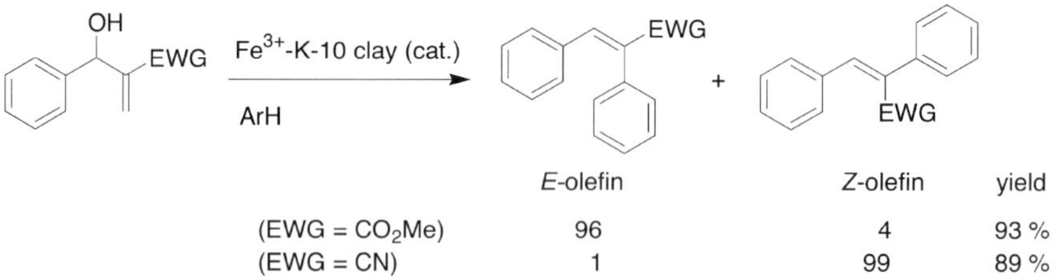

	E-olefin	*Z*-olefin	yield
(EWG = CO_2Me)	96	4	93 %
(EWG = CN)	1	99	89 %

Scheme 7.13 Lewis acid-catalyzed conjugate substitution of allylic alcohols.

An interesting Fe-catalyzed S_N2'-like carbene insertion reaction using diazo compounds and allyl sulfides (the Doyle–Kirmse reaction) was reported by Carter and Van Vranken in 2000 [20]. Various allyl thioethers were reacted with TMS-diazomethane in the presence of catalytic amounts of $Fe(dppe)Cl_2$ to furnish the desired insertion products with moderate levels of stereocontrol [Equation (7.6), Scheme 7.14]. The products obtained serve as versatile synthons in organic chemistry, e.g. reductive desulfurization furnishes lithiated compounds that can be used in Peterson-type olefinations to yield alkenes [Equation (7.7), Scheme 7.14] [21].

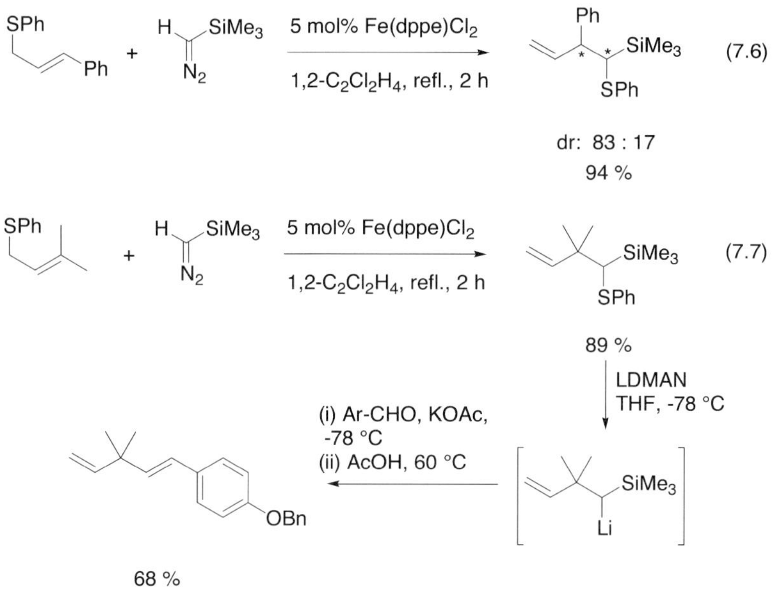

(7.6)

(7.7)

Scheme 7.14 Fe-catalyzed Doyle–Kirmse reaction. LDMAN = lithium 1-(dimethylamino)naphthalenide [22].

7.2.2.2 **Nucleophilic Substitutions Involving Ferrates**

Introduction Substitutions involving ferrates as the catalytically active species might also encounter regio- and stereoselectivity problems (Scheme 7.15). The nucleophilic substitution of the leaving group can occur either in an S_N2- [Equation (7.8),

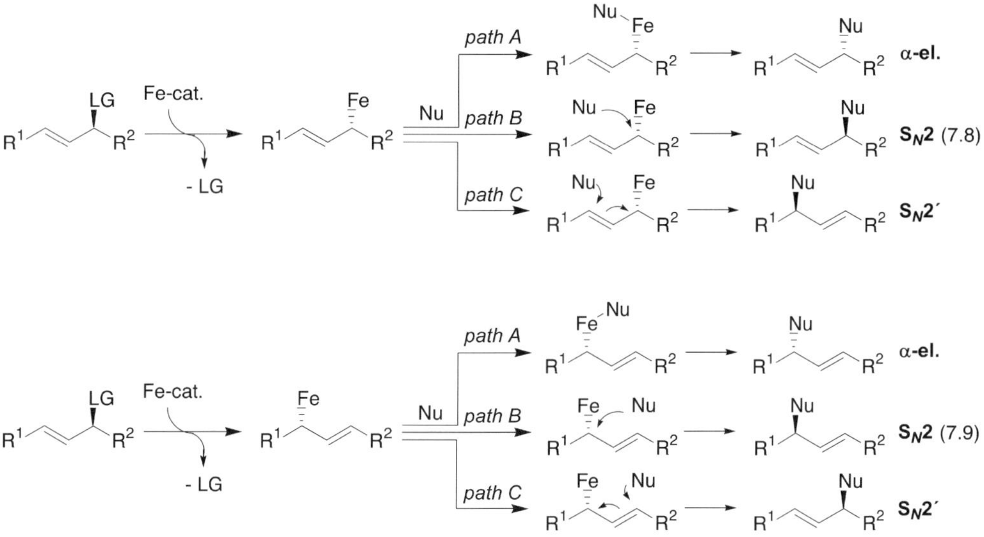

(7.8)

(7.9)

Scheme 7.15 Possible mechanisms in allyl Fe chemistry.

Scheme 7.15] or S_N2'-type mechanism [Equation (7.9)]. Depending on the nature of the nucleophile and catalyst employed, the subsequent nucleophilic substitution of the metal can follow either via α-elimination [*path A*, Equations (7.8) and (7.9), Scheme 7.15], via an S_N2 reaction (*path B*) or via an S_N2'-type reaction (*path C*). For reasons of clarity, only strictly concerted and stereospecific S_N2- or S_N2'-*anti*-type mechanistic scenarios are shown in Scheme 7.15. The situation might, however, be complicated if, e.g., the initial S_N2'-*anti* ionization event is competing with an S_N2'-*syn* reaction. Erosion in stereo- and regioselectivity can be the result of these competing reactions. Furthermore, fluxional intermediates such as π-allyl Fe complexes are not shown in Scheme 7.15 for reasons of clarity. These intermediates are known for a variety of late transition metal allyl complexes and will be referred to later. Moreover, apart from these ionic mechanisms, radicals might also be involved in the reaction. So far no distinct mechanistic study on allylic substitutions has been published.

S_N2-type Reactions In a series of reports, Yamamoto and coworkers summarized the results obtained in the reaction of primary allylic phosphates using a combination of a Grignard reagent and catalytic amounts of $Fe(acac)_3$ (i.e. the Kharasch reaction) [23]. The reaction proceeded at low temperatures with high selectivity in favor of the S_N substitution product independent of the double bond geometry (Scheme 7.16). This observation was taken as an experimental indication of a direct S_N-type reaction without intermediate formation of a π-allyl Fe complex [*path A* or *B*, Equation (7.8), Scheme 7.15]. Unfortunately, the stereoselective course has not been explored in detail. Thus, a differentiation between mechanistic pathways *A* and *B* (Scheme 7.15) is not possible. Interestingly, the use of $NiBr_2$ gave rise to the same *ipso*-substitution products whereas an S_N2' mechanism was observed with the use of CuCN in catalytic amounts.

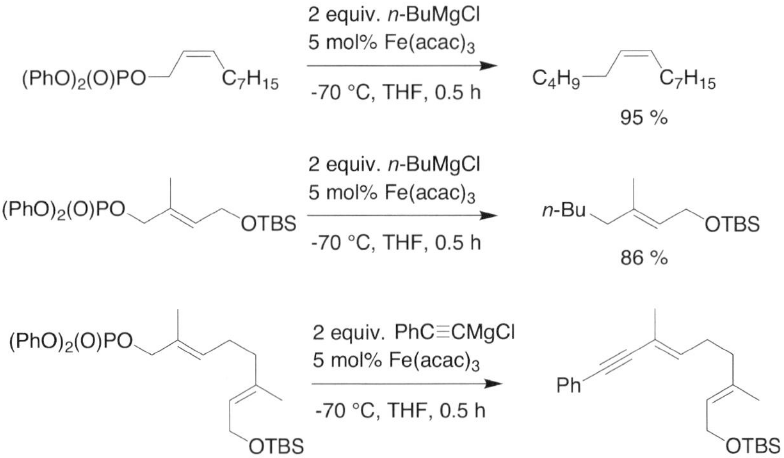

(PhO)₂(O)PO–CH₂CH=CH–C₇H₁₅ → (2 equiv. *n*-BuMgCl, 5 mol% Fe(acac)₃, -70 °C, THF, 0.5 h) → C₄H₉–CH₂CH=CH–C₇H₁₅, 95 % (7.10)

(PhO)₂(O)PO ... OTBS → (2 equiv. *n*-BuMgCl, 5 mol% Fe(acac)₃, -70 °C, THF, 0.5 h) → *n*-Bu ... OTBS, 86 % (7.11)

(PhO)₂(O)PO ... OTBS → (2 equiv. PhC≡CMgCl, 5 mol% Fe(acac)₃, -70 °C, THF, 0.5 h) → Ph ... OTBS (7.12)

Scheme 7.16 Fe-catalyzed substitution of primary allyl phosphates by Grignard reagents.

The analogous reaction between a propargylic chloride and a Grignard reagent in the presence of Fe salts, on the other hand, was shown to be highly dependent on the substitution pattern of the propargylic chloride. Whereas terminal acetylenes favor addition in an S_N' fashion, a TMS protecting group forces the reaction to follow a formal S_N-like mechanistic pathway (Scheme 7.17) [24].

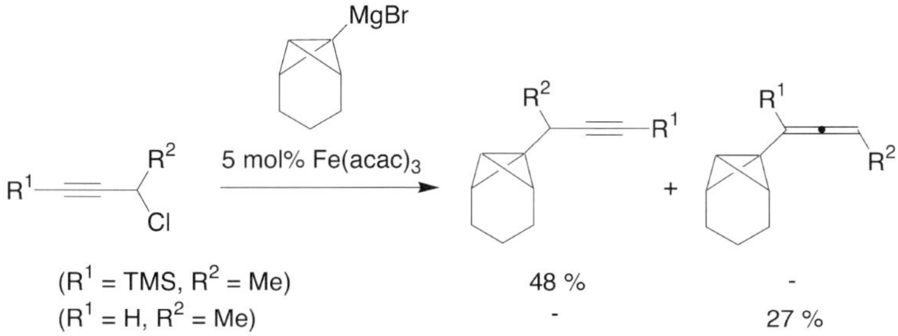

MgBr

5 mol% $Fe(acac)_3$

(R^1 = TMS, R^2 = Me)	48 %	-
(R^1 = H, R^2 = Me)	-	27 %

Scheme 7.17 Conjugate versus direct substitution of propargylic chlorides.

Conjugate Nucleophilic Substitutions An interesting application of Fe-catalyzed allylic substitutions has been reported by Nakamura and coworkers (Scheme 7.18) [25]. Treatment of a *meso*-oxanorbornene with a combination of $FeCl_3$ and RMgX in the presence of TMEDA yielded a ring-opened product, in which four new centers of chirality are formed with exclusive regio- and stereoselectivity. The reaction is thought to proceed via a carbometallation-reductive ring-opening–β-hydride elimination [according to *path A*, Equation (7.9), Scheme 7.15]. Interestingly, the ferrate catalyst attacks the double bond from the *exo* face pointing in the direction of a probable precoordination between the metal complex and the oxygen bridge. From a mechanistic point of view, the observed formation of byproducts arising from competing β-hydride elimination or even hydride reduction appears very interesting. It underlines the complexity of catalytic systems based on the use of Fe salts + Grignard reagent.

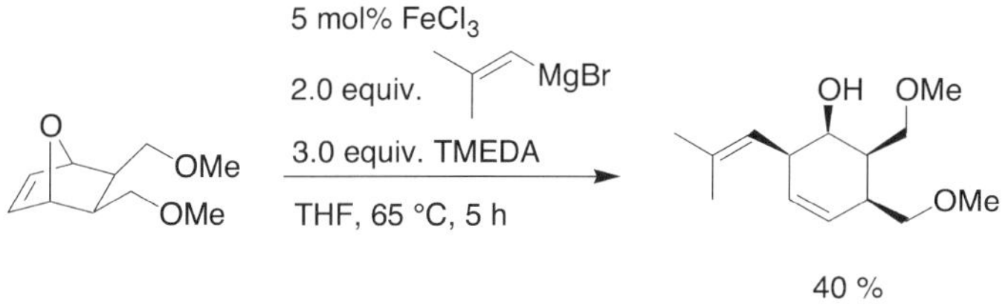

5 mol% $FeCl_3$
2.0 equiv. MgBr
3.0 equiv. TMEDA
THF, 65 °C, 5 h

40 %

Scheme 7.18 Fe-catalyzed conjugate addition–ring opening of oxanorbornenes.

An interesting S_N2'-*syn* addition of Grignard reagents to propargyl epoxides in the presence of $Fe(acac)_3$ was reported by Fürstner and Mendez (Scheme 7.19) [26]. Based on earlier results from Pasto and coworkers [27], the method was developed into a powerful tool for the synthesis of optically active allenol derivatives. The

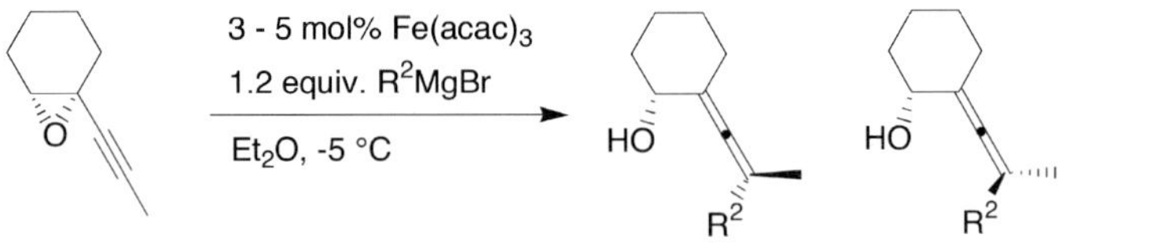

	syn	anti	yield
($R^2 = C_6H_{13}$)	86	14	93 %
(R^2 = *i*-Pr)	84	16	79 %
(R^2 = Ph)	66	34	98 %

Reagents: 3 - 5 mol% $Fe(acac)_3$, 1.2 equiv. R^2MgBr, Et_2O, -5 °C

Scheme 7.19 Fe-catalyzed conjugate addition to propargylic epoxides.

use of enantiomerically enriched oxiranes accessible via Shi epoxidation of the corresponding enynes allowed for control of the interesting center-to-axial chirality transfer. This method might be regarded as complementary to the well-elaborated S_N2'-*anti* addition using organocopper reagents. As for the carbometallation reported by Nakamura and coworkers (Scheme 7.18), a precoordination of the alkyl ferrate reagent to the oxygen of the epoxide was postulated to be the origin of the observed stereoselective course of this transformation [according to *path A*, Equation (7.9), Scheme 7.15].

Complementary to the conjugate substitution reaction in which the nucleophile is transferred directly from the tetraalkyl ferrate to the allylic ligand, preformed low-valent Fe complexes can form reactive allyl–iron complexes via an S_N2'-type mechanism [*path C*, Equations (7.8) and (7.9), Scheme 7.16]. These complexes react with incoming nucleophiles and electrophiles in a substitution reaction. Depending on the nature of the iron complex employed in the reaction, either σ- or π-allyl complexes are generated.

The development of the stoichiometric allyl–iron chemistry might serve as a textbook example of the development of pure academic research into a reliable synthetic methodology. This interesting subdivision of organoiron chemistry was opened up about 45 years ago when Emerson and Pettit reported on the isolation and characterization of a cationic π-allyl–$Fe(CO)_4$ complex [28]. Since these early discoveries, a plethora of π-allyl–Fe complexes have been isolated, most of which incorporate the metal in the oxidation state 0 and +II in the presence of a variety of ligands. The unique characteristic properties of π-allyl–Fe complexes (stability, purification by distillation, sublimation, column chromatography, etc.) attracted the attention of organic chemists soon after their discovery. However, whereas these compounds have found widespread use as stable planar chiral synthons for a variety of non-catalytic transformations [29], it was as early as 1979 that Roustan and coworkers reported on the use of catalytic amounts of an iron complex in order to perform an allylic substitution comparable to the well-established allyl–Pd complex chemistry [30]. The application of catalytic amounts of the Hieber complex $Na[Fe(CO)_3(NO)]$ [31] led to the substitution of an allylic chloride or acetate for a malonate [Equation (7.13), Scheme 7.20]. Most importantly, they observed a preference for the formation of the *ipso*-substitution product, i.e. the new C–Nu bond was formed preferentially at the C atom that was substituted by the leaving group in the

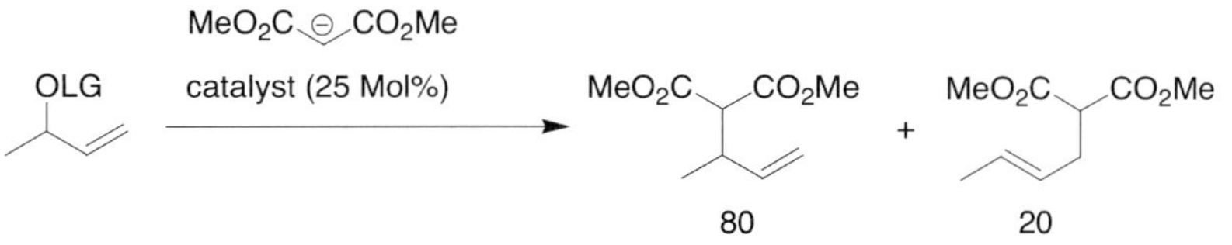

MeO$_2$C⊖CO$_2$Me
OLG — catalyst (25 Mol%) → MeO$_2$C CO$_2$Me (80) + MeO$_2$C CO$_2$Me (20)

		yield	
Roustan (1979):	Na[Fe(CO)$_3$(NO)], THF, refl.	85 %	(7.13)
Xu (1987):	[Bu$_4$N][Fe(CO)$_3$(NO)], CO$_{(g)}$, THF, refl.	35 %	(7.14)

Scheme 7.20 Fe-catalyzed allylic substitution.

starting material. In 1987, Xu and Zhou reported on an improved protocol of Roustan's original procedure in which they introduced the shelf-stable [Bu$_4$N][Fe(CO)$_3$(NO)] as a suitable catalyst for this type of reaction [Equation (7.14), Scheme 7.20) [32]. As observed by Roustan, the reaction proceeds with moderate to good regioselectivities and with formal retention of the configuration with a clear preference for the formation of the *ipso*-substitution product [33]. Further experiments indicated this allylic substitution to follow a σ-allyl mechanism, hence the observed regio- and stereoselectivities are a consequence of two subsequent S_N2'-*anti* reactions. The intermediate formation of a π-allyl–Fe complex was excluded.

Although this catalytic reaction appeared to be of synthetic interest, it has since then neither been applied in synthesis nor further developed. This might be attributed in part to problems with reproducibility and catalyst stability under the reaction conditions, although the Hieber complex was used in a stoichiometric manner for the preparation of a variety of π-allyl–Fe complexes. These latter compounds served as starting materials for a plethora of subsequent reactions [34]. The results obtained by Nakanishi and coworkers on the stability and reactivity of π-allyl–Fe–nitrosyl complexes proved such intermediates to be reactive towards a variety of nucleophiles; however, the Fe complexes formed upon nucleophilic substitution were catalytically inactive. Hence, in order to maintain the catalytic activity, the formation of intermediate π-allyl–Fe complexes had to be circumvented. About 3 years ago we started our research in this field and envisioned the use of a monodentate ligand to be a suitable way to stabilize the proposed catalytically active σ-allyl complex. The replacement of one CO by a non-volatile basic ligand was thought to prevent the formation of the catalytically inactive π-allyl–Fe complex (Scheme 7.21).

Indeed, the use of PPh$_3$ increased the catalyst's stability and reactivity, leading to a significantly improved protocol for the Fe-catalyzed allylic substitution of various allyl carbonates (Scheme 7.22) [35]. The substitution products were obtained in good to excellent yields with almost exclusive regioselectivity in favor of the *ipso*-substitution product. Furthermore, by using the leaving group as an *in situ* base no preformation of the nucleophile was necessary.

A variety of different allylic carbonates and pronucleophiles can be employed under the standard reaction conditions, giving rise to the allylated products in good to excellent yields and regioselectivities. Moreover, the reaction scope was extended to

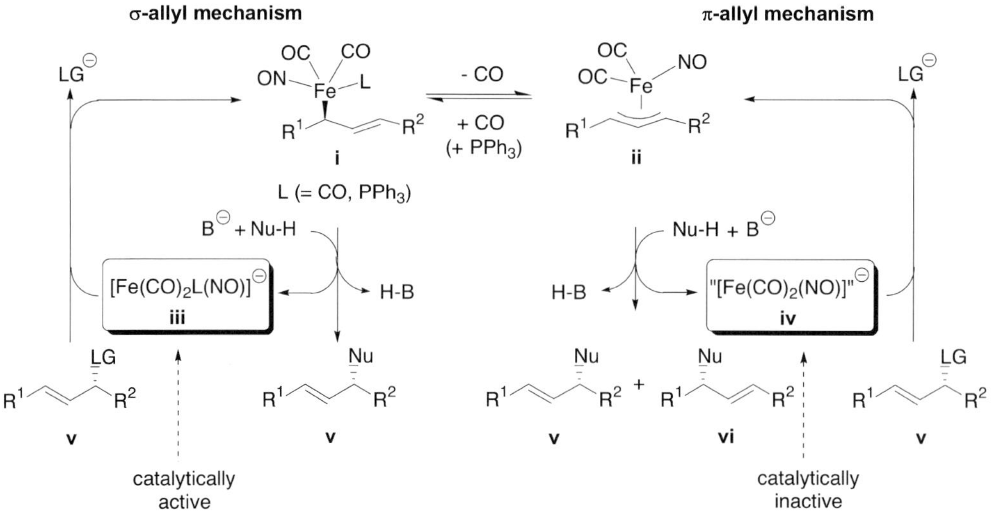

Scheme 7.21 σ- versus π-allyl mechanism in Fe-catalyzed allylic substitutions.

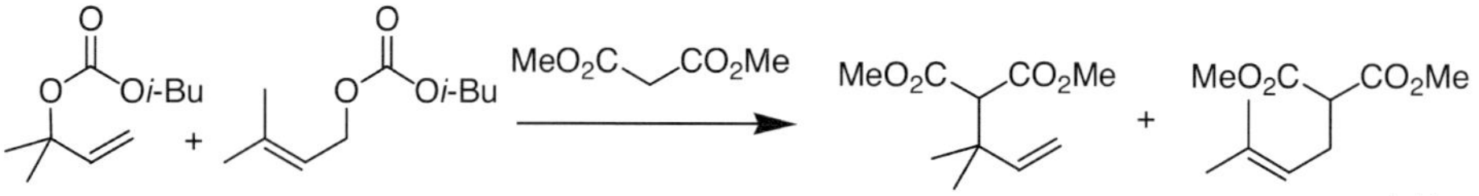

Scheme 7.22 Fe-catalyzed allylic alkylation. 2.5 mol% $[Bu_4N][Fe(CO)_3(NO)]$, 3 mol% PPh_3, DMF, 80 °C, 24 h.

the allylation of *N*-nucleophiles. Initial experiments indicated the inherent basicity of the substitution products to be problematic. However, in the presence of catalytic amounts of piperidine hydrochloride (pip·HCl) as a buffer, different aromatic amines were allylated with almost exclusive regioselectivities (Scheme 7.23) [36].

This protocol, which most likely involves the formation of a catalytically active σ-allyl–Fe species, was significantly improved by replacing the phosphane ligand with an *N*-heterocyclic carbene [37]. The addition of a *tert*-butyl-substituted NHC ligand allowed for full conversion in the exact stoichiometric reaction between allyl carbonate and pronucleophile. Good to excellent regioselectivities were obtained in the allylation of various *C*-nucleophiles. Most importantly, the π bond geometry in the reaction of isomeric (*E*)- and (*Z*)-carbonates remained intact, giving rise to the substitution products with full transfer of structural information from the carbonate to the allylation product (Scheme 7.24) [38].

By changing the ligand's topology, however, a significant change in the regioselective course of the reaction is observed. Whereas a *t*-Bu-substituted ligand allows for a regio- and stereoselective allylic substitution, an aryl-substituted ligand forces

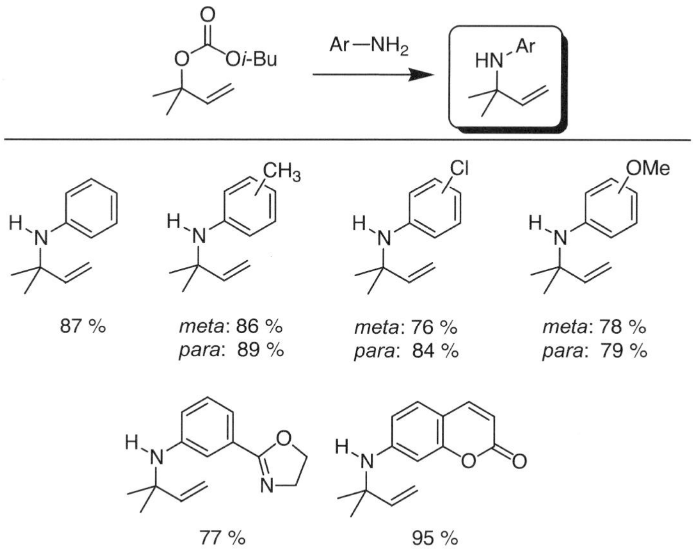

Ar—NH$_2$

87 %

meta: 86 %
para: 89 %

meta: 76 %
para: 84 %

meta: 78 %
para: 79 %

77 %

95 %

Scheme 7.23 Fe-catalyzed allylic amination. 5 mol% $[Bu_4N][Fe(CO)_3(NO)]$, 5 mol% PPh_3, 30 mol% pip·HCl, DMF, 80 °C.

the reaction to follow a π-allyl mechanism. This ligand dependant mechanistic dichotomy resembles a promising starting point for the development of an asymmetric Fe-catalyzed allylic substitution (Scheme 7.25). Moreover, the catalytic activity of an intermediate π-allyl–Fe complex was proven by employing catalytic amounts of a preformed π-allyl complex in the reaction. The regioselective course of the reaction

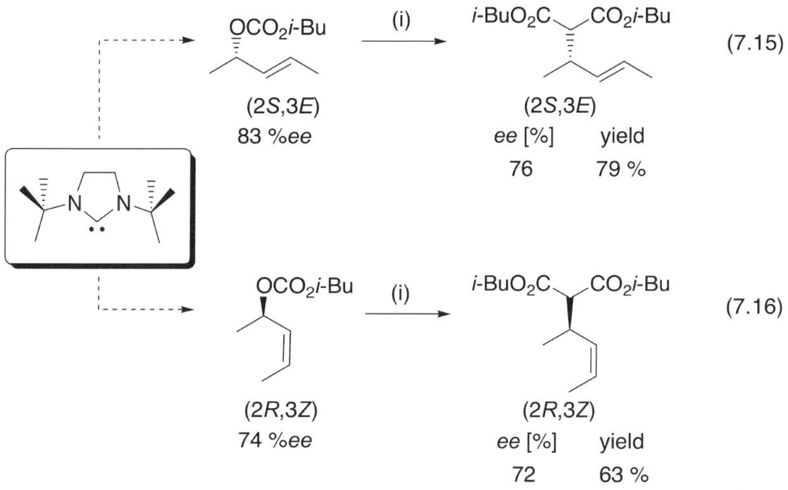

Scheme 7.24 Fe-catalyzed regio- and stereoselective allylic substitution in the presence of NHC ligand. (i) $[Bu_4N][Fe(CO)_3(NO)]$ (cat.), MTBE, 80 °C.

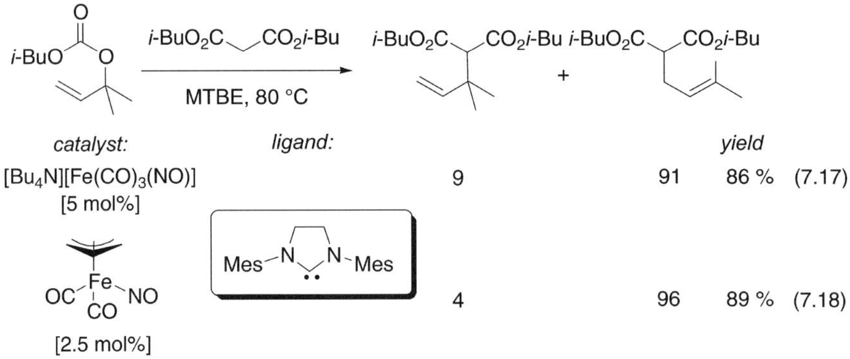

catalyst:	ligand:			yield	
$[Bu_4N][Fe(CO)_3(NO)]$ [5 mol%]		9	91	86 %	(7.17)
$(\pi\text{-allyl})Fe(CO)_2(NO)$ [2.5 mol%]	Mes–N(C:)N–Mes	4	96	89 %	(7.18)

Scheme 7.25 Fe-catalyzed allylic substitution via π-allyl mechanism.

using either catalyst was identical [Equations (7.17) and (7.18), Scheme 7.25), however, the π-allyl Fe-catalyst showed a remarkable increased reactivity and allowed for a reduction in catalyst loading and reaction time by a factor of two [Equation (2), Scheme 7.25].

Prior to these recent studies Nicholas and coworkers reported on the $Fe_2(CO)_9$-catalyzed reaction between allylic acetates and sodium dimethylmalonate (Scheme 7.26) [39]. Detailed mechanistic studies indicated that the Fe(0) complex acts as a precatalyst that is activated *in situ* upon reaction with the nucleophile. The

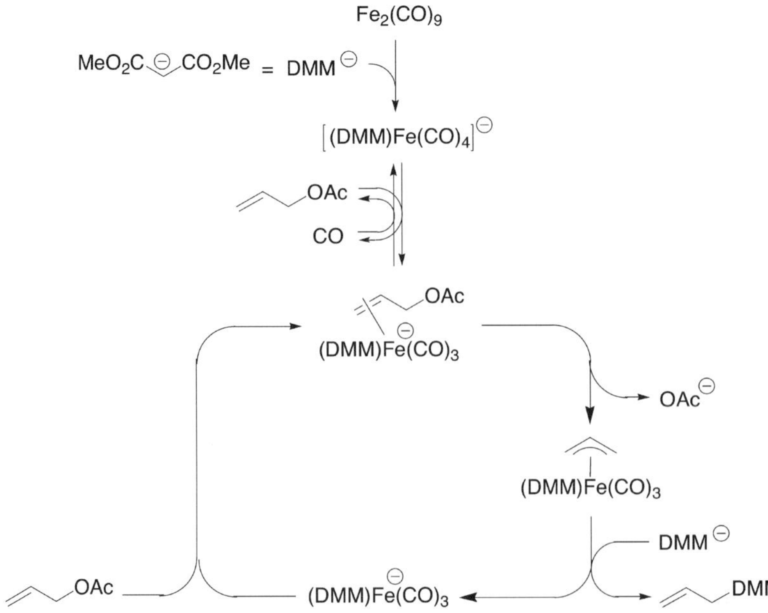

Scheme 7.26 Allylic alkylation using $Fe_2(CO)_9$ as the catalyst – *in situ* generation of a ferrate.

ferrate formed in the activation step undergoes a ligand exchange with the incoming allylic acetate. The π-allyl–Fe complex formed in this reaction is attacked by the nucleophile with formation of the product and regeneration of a catalytically active ferrate. Although this reaction represents one of the few examples of catalytically active π-allyl–Fe intermediates, no further investigation or synthetic application has been published since then.

7.3 Conclusion

The increasing demand both from society and from science for more general, efficient and affordable catalytic systems has spurred growing interest among chemists both in academia and in industry to develop new catalytic reactions based on ready available, environmentally benign and inexpensive catalysts. The nucleophilic substitutions represent an important class of reactions in organic synthesis. With regard to their importance, several highly efficient catalytic systems based on low-valent organoiron complexes have been developed within the past 20 years (Figure 7.2). All of these active catalysts are characterized by a nucleophilic metal center that allows efficient substitution of a leaving group. However, the exact mechanisms in the nucleophilic substitutions are currently under debate. More detailed studies on structure and activity relationships are needed in order to develop even more active and general catalytic systems.

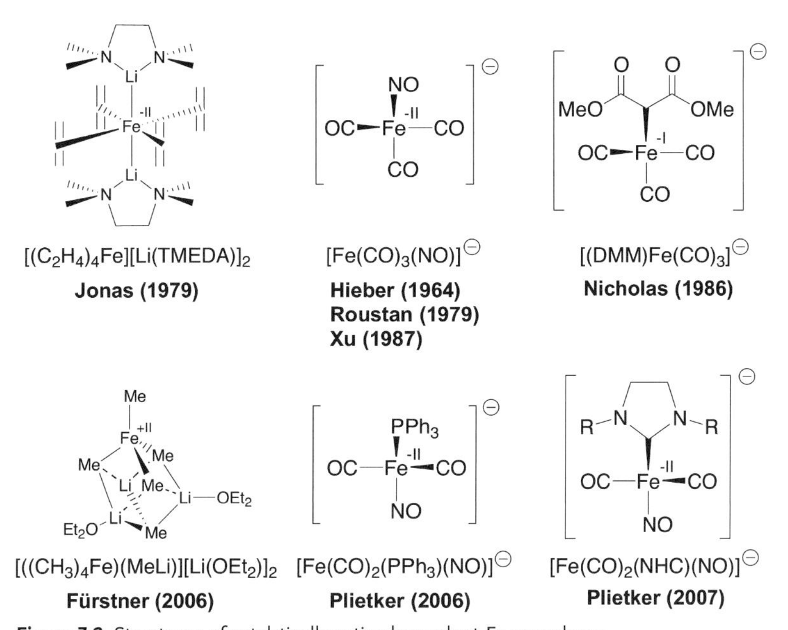

Figure 7.2 Structures of catalytically active low-valent Fe complexes.

References

1 J. A. Miller, M. J. Nunn, *J. Chem. Soc., Perkin Trans.* 1976, **1**, 416.
2 I. C. Barrett, J. D. Langille, M. A. Kerr, *J. Org. Chem.* 2000, **65**, 6268.
3 J. K. Kocchi, *Acc. Chem. Res.* 1974, **7**, 351
4 Th. Kauffmann, *Angew. Chem.* 1996, **108**, 401; *Angew. Chem. Int. Ed. Engl.* 1996, **35**, 386.
5 B. Bogdanovic, M. Schwickardi, *Angew. Chem.* 2000, **112**, 4788; *Angew. Chem. Int. Ed.* 2000, **39**, 4610.
6 T. Holm, *J. Am. Chem. Soc.* 1999, **121**, 515.
7 (a) K. Jonas, L. Schieferstein, C. Krüger, Y.-H. Tsay, *Angew. Chem.* 1979, **91**, 590; *Angew. Chem. Int. Ed. Engl.* 1979, **18**, 550; (b) A. Fürstner, H. Krause, C. W. Lehmann, *Angew. Chem.* 2006, **118**, 454; *Angew. Chem. Int. Ed.* 2006, **45**, 440.
8 Y. Moglie, F. Alonso, C. Vitale, M. Yus, G. Radivoy, *Appl. Catal. A* 2006, **313**, 94.
9 H. Guo, K. Kanno, T. Takahashi, *Chem. Lett.* 2004, **33**, 1356.
10 M. Taillefer, N. Xia, A. Ouali, *Angew. Chem.* 2007, **119**, 952; *Angew. Chem. Int. Ed.* 2007, **46**, 934.
11 Y. Hayashi, H. Shinokubo, K. Oshima, *Tetrahedron Lett.* 1998, **39**, 63.
12 I. Aviv, Z. Gross, *Synlett*, 2006, 951.
13 L. K. Baumann, H. M. Mbuvi, G. Du, L. K. Woo, *Organometallics*, 2007, **26**, 3995.
14 Z.-P. Zhan, H.-J. Liu, *Synlett*, 2006, 2278.
15 V. V. Namboodiri, R. S. Varma, *Tetrahedron Lett.* 2002, **43**, 4593.
16 U. Jana, S. Biswas, S. Maiti, *Tetrahedron Lett.* 2007, **48**, 4065.
17 Z.-P. Zhan, J. Yu, H. Liu, Y. Cui, R. Yang, W. Yang, J. Li, *J. Org. Chem.* 2006, **71**, 8298.
18 A related but non-catalytic Fe-catalyzed conjugate addition has been reported: P. R. Krishna, V. Kannan, G. V. M. Sharma, *Synth. Commun.* 2004, **34**, 55.
19 B. Das, A. Maijhi, J. Banerjee, N. Chowdhury, K. Venkateswarlu, *Chem. Lett.* 2005, **34**, 1492.
20 D. S. Carter, D. L. Van Vranken, *Org. Lett.* 2000, **2**, 1303.
21 J. B. Perales, N. F. Makino, D. L. Van Vranken, *J. Org. Chem.* 2002, **67**, 6711.
22 T. Cohen, M. Bhupathy, *Acc. Chem. Res.* 1989, **22**, 152.
23 (a) A. Yanagisawa, N. Nomura, H. Yamamoto, *Synlett* 1991, 513; (b) A. Yanagisawa, N. Nomura, H. Yamamoto, *Tetrahedron Lett.* 1994, **50**, 6017.
24 A. S. K. Hashmi, G. Szeimies, *Chem. Ber.* 1994, **127**, 1075.
25 (a) M. Nakamura, A. Hirai, E. Nakamura, *J. Am. Chem. Soc.* 2000, **122**, 978; (b) M. Nakamura, K. Matsuo, T. Inoue, E. Nakamura, *Org. Lett.* 2003, **5**, 1373.
26 A. Fürstner, M. Mendez, *Angew. Chem.* 2003, **115**, 5513; *Angew. Chem. Int. Ed.* 2003, **42**, 5355.
27 (a) D. J. Pasto, G. F. Hennion, R. H. Shults, A. Waterhouse, S.-K. Chou, *J. Org. Chem.* 1976, **41**, 3496; (b) D. J. Pasto, S.-K. Chou, A. Waterhouse, R. H. Shults, G. F. Hennion, *J. Org. Chem.* 1978, **43**, 1385.
28 G. F. Emerson, R. Pettit, *J. Am. Chem. Soc.* 1962, **84**, 4591.
29 (a) S. V. Ley, R. Liam, G. Meek, *Chem. Rev.* 1996, **96**, 423; (b) D. Enders, B. Jandeleit, S. von Berg, *Synthesis*, 1997, 421; (c) D. Enders, B. Jandeleit, S. von Berg, G. Raabe, J. Runsink, *Organometallics* 2001, **20**, 4312, and references therein.
30 (a) J.-L. Roustan, M. Abedini, H. H. Baer, *Tetrahedron Lett.* 1979, 3721; (b) J.-L. Roustan, M. Abedini, H. H. Baer, *J. Organomet. Chem.* 1989, **376**, C20.
31 (a) W. Hieber, H. Beutner, *Z. Naturforsch., Teil B* 1960, **15**, 323; (b) W. Hieber, H. Beutner, *Z. Anorg. Allg. Chem.* 1962, **320**, 101.
32 Y. Xu, B. Zhou, *J. Org. Chem.* 1987, **52**, 974.
33 B. Zhou, Y. Xu, *J. Org. Chem.* 1988, **53**, 4421.
34 (a) U. Eberhardt, G. Mattern, *Chem. Ber.* 1988, **121**, 1531; (b) S. G. Davies, A. J. Smallridge, *J. Organomet. Chem.* 1990, **386**, 195; (c) K. Itoh, S. Nakanishi, Y. Otsuji,

Bull. Chem. Soc. Jpn. 1991, **64**, 2965; (d) K. Itoh, S. Nakanishi, Y. Otsuji, *J. Organomet. Chem.* 1994, **473**, 215; (e) K. Itoh, Y. Otsuji, S. Nakanishi, *Tetrahedron Lett.* 1995, **36**, 5211; (f) S. Nakanishi, H. Yamaguchi, K. Okamoto, T. Takata, *Tetrahedron: Asymmetry,* 1996, **7**, 2219; (g) H. Yamaguchi, S. Nakanishi, K. Okamoto, T. Takata, *Synlett* 1997, 722; (h) S. Nakanishi, S. Memita, T. Takata, K. Itoh, *Bull. Chem. Soc. Jpn.* 1998, **71**, 403; (i) S. Nakanishi, K. Okamoto, H. Yamaguchi, T. Takata, *Synthesis* 1998, 1735.

35 B. Plietker, *Angew. Chem.* 2006, **118**, 1497; *Angew. Chem. Int. Ed.* 2006, **45**, 1469.

36 B. Plietker, *Angew. Chem.* 2006, **118**, 6200; *Angew. Chem. Int. Ed.* 2006, **45**, 6053.

37 F. Glorius, (ed.), *N-Heterocyclic Carbenes in Transition Metal Cataylsis* Springer, Hamburg, 2007.

38 B. Plietker, A. Dieskau, K. Möws, A. Jatsch, *Angew. Chem.* 2008, **120**, 204; *Angew. Chem.* 2008, **47**, 198.

39 (a) S. J. Ladoulis, K. M. Nicholas, *J. Organomet. Chem.* 1985, **285**, C13; (b) G. S. Silverman, S. Strickland, K. M. Nicholas, *Organometallics* 1986, **5**, 2117.

8
Addition and Conjugate Addition Reactions to Carbonyl Compounds

Jens Christoffers, Herbert Frey, and Anna Rosiak

8.1
Introduction

A solution of anhydrous iron(III) chloride and also its hexahydrate in organic solvents or water consists of several solvate or aqua complexes, e.g. $[FeCl_2(OH_2)_4]^+$, if water plays a role in the reaction mixture. Since Fe(III) is a d^5 ion, octahedral complexes count 17 valence electrons. Consequently, these kinetically labile species are in rapid equilibrium. The iron(III) ion is, of course, a strong and hard Lewis acid. In addition to this Lewis acidity, these complexes also behave as strong Brønsted acids in the presence of protic solvents such as alcohols or water. All three effects – kinetic lability and Brønsted and Lewis acidity – make $FeCl_3$ and $FeCl_3{\cdot}6H_2O$ to extraordinary catalysts for the activation and fast conversion of carbonyl compounds (structures **1** and **2**, Figure 8.1). Moreover, with certain chelating substrates, namely 1,3-dicarbonyl compounds, complexes such as in structure **3** are formed with a six-membered chelate ring, which shows high thermodynamic stability due to delocalization of π-electron density. In this species, the so-called 1,3-diketonato ligand is anionic, hence deprotonation of the respective 1,3-dicarbonyl compounds occurs under the reaction conditions. This chapter will, however, not only deal with simple iron(III) salts as catalysts, but also iron(II) compounds and more sophisticated tailored iron complexes will be included in this overview. The subject of iron-catalyzed reactions in organic synthesis has recently been reviewed twice [1].

This chapter will be divided into sections according to the electrophiles: aldehydes and ketones, imines and iminium salts, carboxylic acid derivatives and finally α,β-unsaturated carbonyl compounds, which undergo conjugate additions. Further subdivision will be made according to the nature of the nucleophile, i.e. *O*-, *N*-, *S*-, *P*- or *C*-nucleophiles. Finally, multicomponent heterocyclic syntheses will be mentioned, if they consist at least of one iron-catalyzed addition step to a carbonyl compound.

Iron Catalysis in Organic Chemistry. Edited by Bernd Plietker

ISBN: 978-3-527-31927-5

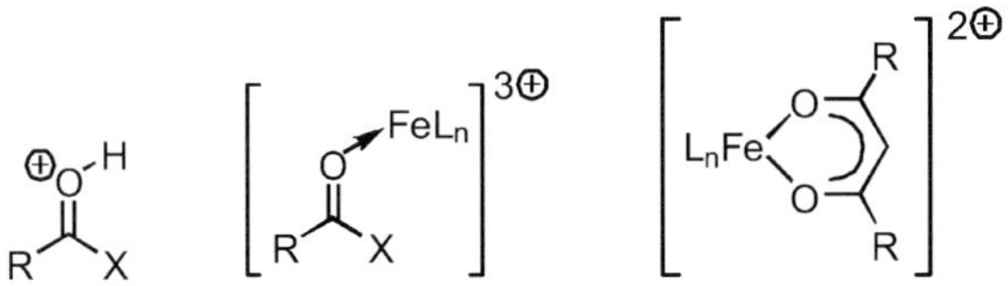

Figure 8.1 Activation of carbonyl compounds by iron(III) salts.

8.2 Additions to Aldehydes and Ketones

8.2.1 Oxygen Nucleophiles

Acetals are one of the most prominent protective groups for 1,2-diols and aldehydes or ketones [2]. Their formation or degradation proceeds under thermodynamic control, i.e. either removal or use of an excess of water or another condensation product. This equilibration is commonly catalyzed by various Brønsted or Lewis acids and it is therefore no wonder that anhydrous $FeCl_3$ has been reported for acetal formation [3]. For the methylenation of carbohydrates with formaldehyde, the use of heterobimetallic [Fe(III)–Sn(II)] catalysts was recommended [4]. For acetal cleavage, $FeCl_3 \cdot 6H_2O$ may be used [5]. An attractive variant here is the heterogenization of $FeCl_3$ on wet SiO_2, which simplifies the workup protocol [6]. In analogy with their *O*, *O*-congeners, the iron-catalyzed formation [7] and cleavage [8] of dithioacetals has also been reported.

The formation of 1,1-diacetates **5** from aldehydes **4** and Ac_2O dates back to early systematic studies by Knoevenagel [9]. A representative example is given with the preparation of compound **5a** in Scheme 8.1. Several groups have since then developed improved protocols for "acylal" formation [10]. It was realized that these *gem*-diacetates such as compound **5b** are perfect substrates for palladium-catalyzed

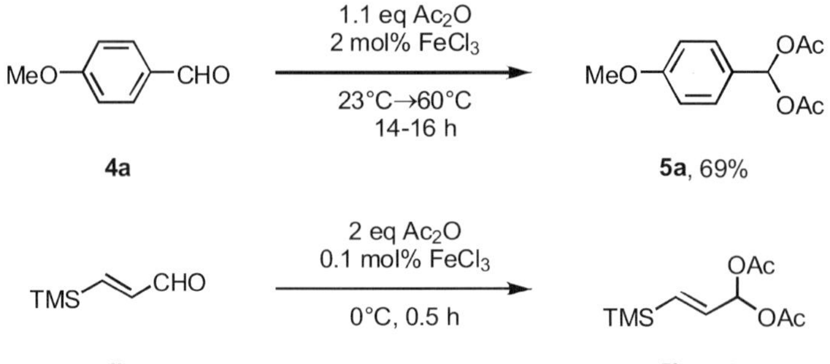

Scheme 8.1 $FeCl_3$-catalyzed acylal formation as reported in 1914 (product **5a**) and 2001 (product **5b**).

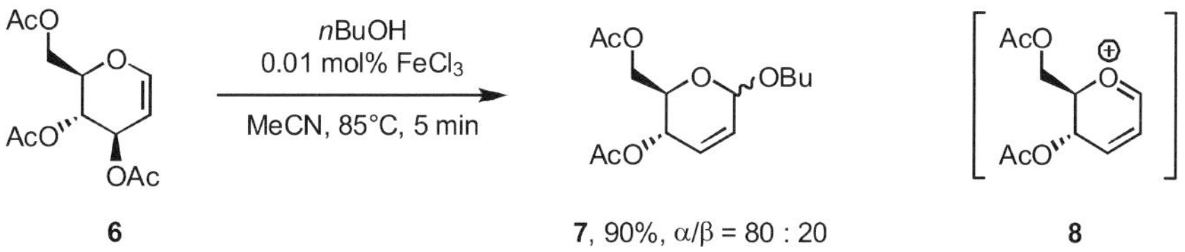

6 → (nBuOH, 0.01 mol% $FeCl_3$, MeCN, 85°C, 5 min) → 7, 90%, α/β = 80 : 20; [8]

Scheme 8.2 Iron-catalyzed Ferrier transformation of glucals to 2,3-unsaturated glycosides.

allylic substitution reactions if α,β-unsaturated aldehydes such as **4b** are converted (Scheme 8.1) [11].

Typical examples for acetals are *O*-glycosides. The anomeric carbon atom is always a stereogenic center. Its configuration (α or β) and its epimerization are important stereochemical issues in carbohydrate chemistry. Anomerization proceeds by elimination and addition of alkoxide or alcohol via an oxocarbenium ion. This equilibration is commonly catalyzed by Lewis or Brønsted acids and several reports used $FeCl_3$ in anhydrous CH_2Cl_2 for this purpose [12]. However, $FeCl_3$ was used not only for epimerization, but also for *O*-glycosidation [13], which can even be performed in an intramolecular fashion to give 1,6-anhydroglucopyranoses [14]. An interesting case of *O*-glycosidation is the so-called Ferrier rearrangement, being the Lewis acid-catalyzed transformation of a glucal to a 2,3-unsaturated pyranose [15]. This reaction can be catalyzed by iron salts [16]. The iron-catalyzed example depicted in Scheme 8.2 works with high efficiency. Triacetylglucal **6** converts with only 0.01 mol% $FeCl_3$ within 5 min to product **7** [17]. It is generally accepted that the Ferrier rearrangement proceeds via an oxoallyl cation intermediate such as species **8**.

8.2.2
Carbon Nucleophiles

As outlined in the preceding section, oxocarbenium ions can be generated from acetals by using $FeCl_3$ as a Lewis acid. In the presence of *C*-nucleophiles such as silanes, C—C bond formation can be achieved. Two representative examples of iron-catalyzed one-pot acetal formation and C—C coupling with allylsilane [18] or TMSCN [19] are given in Scheme 8.3. In both cases, the dibenzylacetal is formed in a first step and then converted with the *C*-nucleophilic reagent.

If alkenes are used instead of silanes, the intermediate oxocarbenium ion undergoes an "acetal"–ene reaction. An example is the reaction of acetal **11** with methylenecyclohexane to give the cyclohexenylmethyl-substituted product **12** (Scheme 8.4) [20]. Silenol ethers are electron-rich alkenes particularly suited for addition to cationic species. Pinacolone-derived enol ether **13**, for example, adds to an thioxocarbenium ion generated *in situ* from *S,S*-acetal **14** to give thioether **15** (Scheme 8.4) [21].

Barbier-type reactions of alkyl iodides with ketones can be performed with SmI_2 when Fe(III) compounds are used as catalysts [22]. Particularly efficient for the intramolecular reaction of ω-iodo ketones such as compound **16** turned out to be $Fe(dbm)_3$ (dbm = 1,3-diphenyl-1,3-propanedionato). The latter is an air-stable, THF-soluble, non-hygroscopic complex that can be very easily prepared from

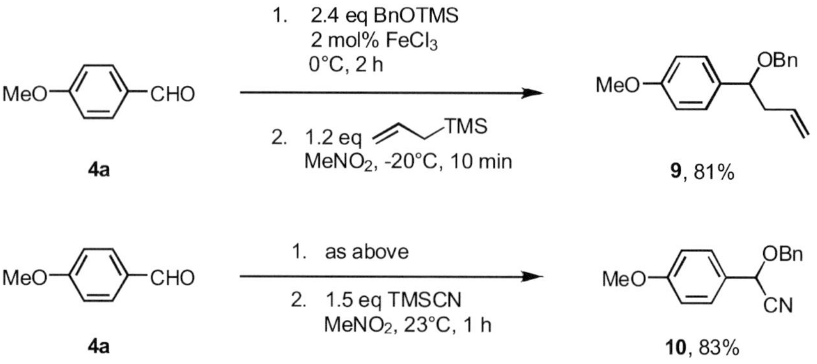

1. 2.4 eq BnOTMS
2 mol% $FeCl_3$
0°C, 2 h

MeO–C6H4–CHO → MeO–C6H4–CH(OBn)CH2CH=CH2

2. 1.2 eq allyl-TMS
$MeNO_2$, -20°C, 10 min

4a **9**, 81%

1. as above
2. 1.5 eq TMSCN
$MeNO_2$, 23°C, 1 h

MeO–C6H4–CHO → MeO–C6H4–CH(OBn)CN

4a **10**, 83%

Scheme 8.3 One-pot acetal formation, cleavage and C–C coupling.

dibenzoylmethane in gram quantities. An example of bicycloalkan-1-ol formation is given in Scheme 8.5 [23]. $FeCl_3$ undergoes addition to alkynes and the resulting iron–vinyl species can be converted with aldehydes. For example, coupling of 2 equiv. of alkyne **18a** with heptanal (**4c**) and stoichiometric amounts of $FeCl_3$ yields 1,5-dichloro-1,4-pentadiene derivative **19**, interestingly with the two C–C double bonds in opposite configuration (Scheme 8.5) [24a]. Conversion of butynol **18b** with isovaleraldehyde (**4d**) and stoichiometric amounts of $FeCl_3$ results in a Prins-type cyclization to give dihydropyran derivative **20** (Scheme 8.5) [24b].

The Lewis acid-catalyzed addition of nucleophilic allylsilanes to aldehydes (Sakurai reaction) can be catalyzed with $FeCl_3$ [25]. In contrast to silanes, allyl acetate is commonly considered as an electrophilic reagent. In an electrochemical process, which is catalyzed by iron(II), it can be transformed into a π-allyliron(II) species. The latter adds as an allyl anion equivalent to aldehydes to give homoallylic alcohols [26]. Instead of using electrochemistry, this process can also be performed

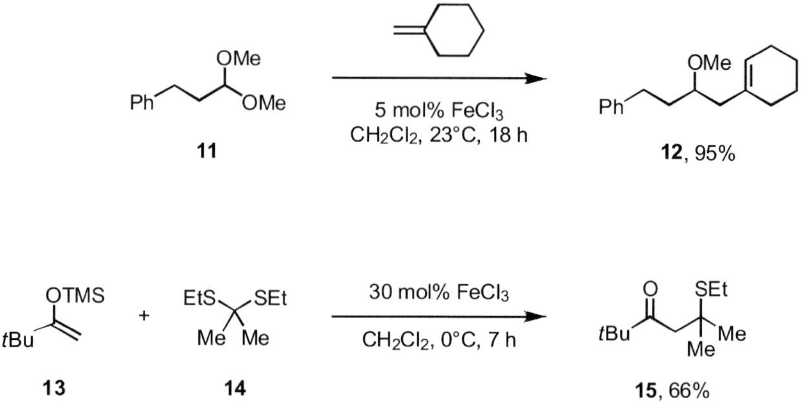

OMe, OMe, Ph → OMe, Ph

5 mol% $FeCl_3$
CH_2Cl_2, 23°C, 18 h

11 **12**, 95%

OTMS, tBu + EtS, SEt, Me, Me → O, SEt, tBu, Me, Me

30 mol% $FeCl_3$
CH_2Cl_2, 0°C, 7 h

13 **14** **15**, 66%

Scheme 8.4 Reaction of oxocarbenium ions and *S*-analogues with alkenes.

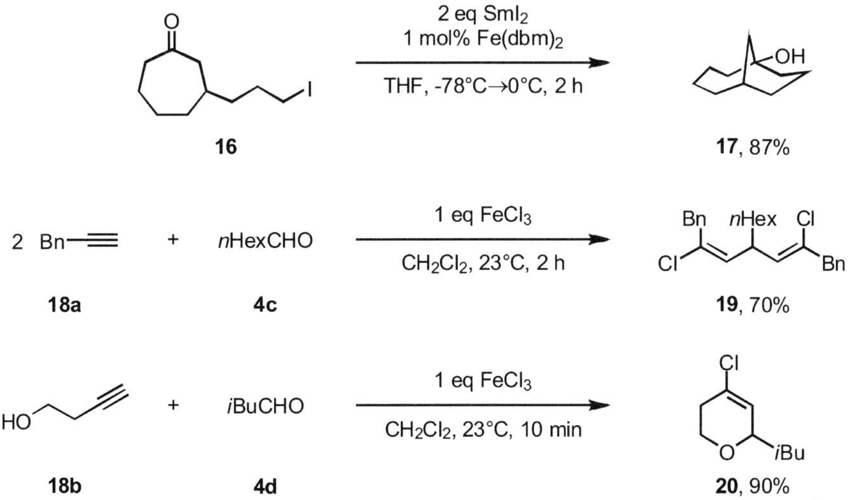

2 eq SmI_2
1 mol% $Fe(dbm)_2$
THF, -78°C→0°C, 2 h

16 → **17**, 87%

2 Bn–≡ + *n*HexCHO
1 eq $FeCl_3$
CH_2Cl_2, 23°C, 2 h

18a + **4c** → **19**, 70%

+ *i*BuCHO
1 eq $FeCl_3$
CH_2Cl_2, 23°C, 10 min

18b + **4d** → **20**, 90%

Scheme 8.5 Addition of iron alkyl- and vinyl-intermediates to ketones and aldehydes.

as a Reformatsky-type reaction with an excess of Mn as an appropriate reducing metal (Scheme 8.6) [27]. With Fe(II) only, the yields are low, however. They can be improved by using $ZnBr_2$ as a co-catalyst. Scheme 8.6 shows the formation of tertiary alcohol **22** from ketone **21** and allyl acetate with relatively large amounts of Zn, Fe and Mn compounds.

Reaction of aldehydes with ethyl diazoacetate normally results in the formation of β-oxo esters. When a cationic Fe(II) Lewis acid is used as the catalyst, an unexpected 1,2-aryl shift results in the formation of α-formyl arylacetic acid ester **23**, which is isolated as its enol tautomer (Scheme 8.6) [28]. The catalyst of this reaction,

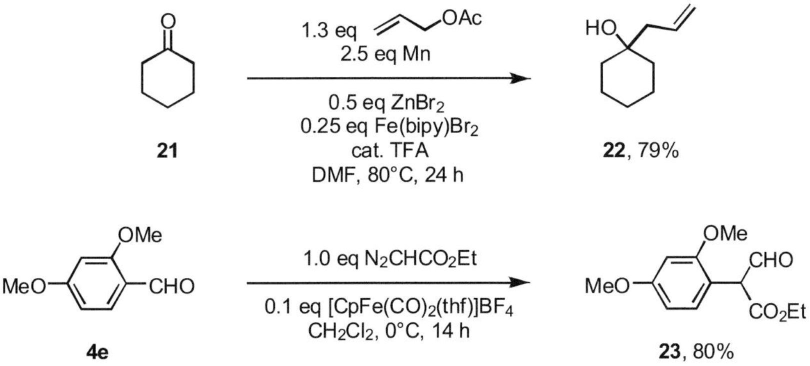

1.3 eq allyl-OAc
2.5 eq Mn
0.5 eq $ZnBr_2$
0.25 eq $Fe(bipy)Br_2$
cat. TFA
DMF, 80°C, 24 h

21 → **22**, 79%

1.0 eq N_2CHCO_2Et
0.1 eq $[CpFe(CO)_2(thf)]BF_4$
CH_2Cl_2, 0°C, 14 h

4e → **23**, 80%

Scheme 8.6 Umpolung of allyl acetate for the formation of homoallylic alcohols and reaction of aldehydes with ethyl diazoacetate.

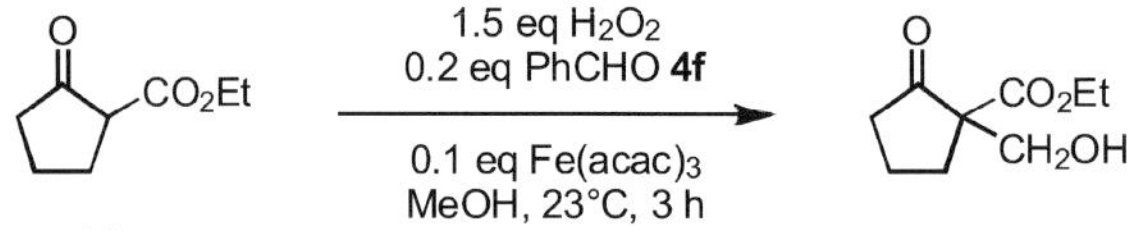

Scheme 8.7 Sequential oxidation of MeOH to $H_2C{=}O$ and aldol reaction with a β-oxo ester.

$[CpFe(CO)_2(thf)]^+$, is generated by protonation of the respective iron–methyl complex with HBF_4 in THF solution. Iron-catalyzed reactions of phenyldiazomethane with aldehydes resulted in mixtures of ketones and epoxides [29].

The classic C–C bond-forming processes of aldehydes and ketones are aldol reactions. In Scheme 8.7, an iron-catalyzed sequential methanol oxidation to formaldehyde and its aldol reaction with β-oxo ester **24a** is shown [30]. The oxidant is 30% aqueous H_2O_2. Curiously for an oxidation, the reaction has to be performed under an atmosphere of Ar in order to prevent α-hydroxylation of the β-oxo ester [31]. The role of benzaldehyde (**4f**) as substoichiometric additive is not completely clear.

Aldol reactions of aldehydes with cycloakanones were performed in ionic liquids and catalyzed by $FeCl_3{\cdot}6H_2O$ [32]. Mukaiyama aldol reactions of silylenol ethers with aldehydes can be carried out in aqueous media; however, among several Lewis acidic catalysts investigated, iron compounds were not the optimal ones [33]. If silyl ketene acetals are applied as carbon nucleophiles in Mukaiyama aldol reactions, cationic Fe(II) complexes give good results. As catalysts, $CpFe(CO)_2Cl$ [34] and $[CpFe(dppe)(acetone)]BF_4$ [35] [dppe = 1,2-bis(diphenylphosphano)ethane] were applied (Scheme 8.8). No diastereomeric ratio was reported for product **26a**.

Asymmetric Mukaiyama aldol reactions in aqueous media [EtOH–H_2O (9 : 1)] were reported with $FeCl_2$ and PYBOX ligands **27a** [36] and **27b** [37]. The latter provides product **28** with higher yield and diastereo- and enantioselectivity (Scheme 8.9). The *ee* values given are for the *syn*-diastereoisomer. Whereas ligand **27a** is a derivative of L-serine, compound **27b** has four stereogenic centers, since it was prepared from

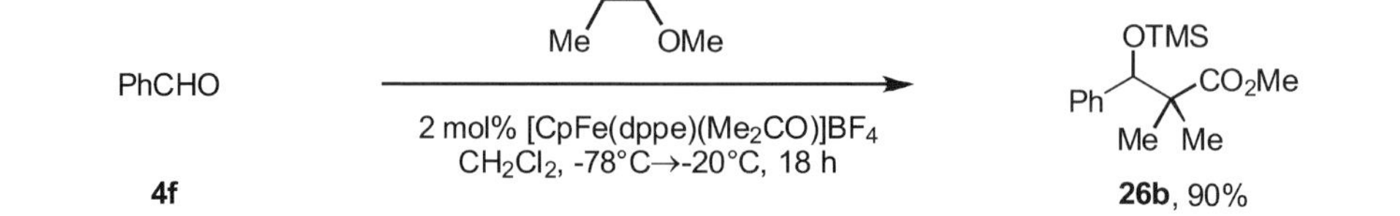

Scheme 8.8 Mukaiyama aldol reactions of silyl ketene acetals with aldehydes.

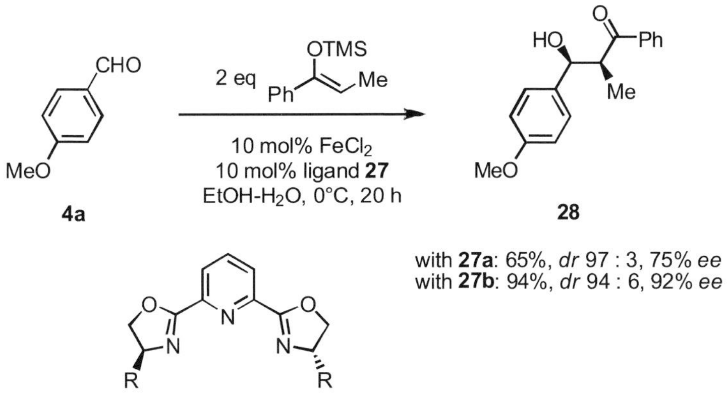

with **27a**: 65%, *dr* 97 : 3, 75% *ee*
with **27b**: 94%, *dr* 94 : 6, 92% *ee*

27a: R = CH_2OH
27b: R = (*R*)-CH(Me)OSiPh_2*t*Bu

Scheme 8.9 Asymmetric Mukaiyama aldol reactions with Fe(II)–PYBOX catalysts.

L-threonine. With ligand **27b**, the selectivities can even be improved when Zn $(OTf)_2$ is used instead of $FeCl_2$.

8.3
Additions to Imines and Iminium Ions

Doubly acceptor-activated imines with an intramolecular alkene moiety such as **29** can cyclize according to an electrophilic mechanism to give pyrrolidine, piperidine or azepine derivatives. This reaction is induced by stoichiometric amounts of Lewis acids, preferably trialkylsilyl triflates [38]. In certain cases $FeCl_3$ can also be used, for example for the preparation of azepinolactone **30** from imine **29** (Scheme 8.10) [39]. Catalytic amounts of $FeCl_3$ are required for the addition of diethyl phosphite to an

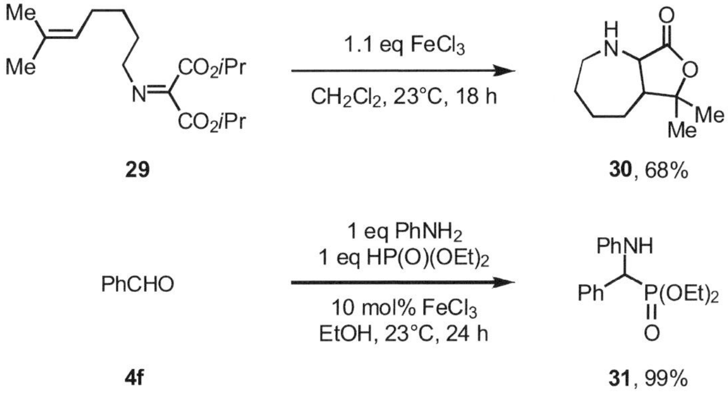

Scheme 8.10 Preparation of azepinolactone **30** and an α-aminophosphonate **31**.

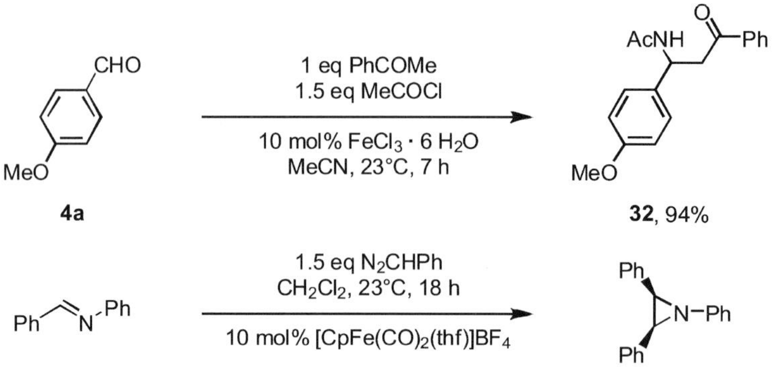

CHO

MeO

4a

1 eq PhCOMe
1.5 eq MeCOCl

10 mol% $FeCl_3 \cdot 6\ H_2O$
MeCN, 23°C, 7 h

AcNH O Ph

MeO

32, 94%

Ph N Ph

33

1.5 eq N_2CHPh
CH_2Cl_2, 23°C, 18 h

10 mol% $[CpFe(CO)_2(thf)]BF_4$

Ph N–Ph Ph

34, 95%

Scheme 8.11 Preparation of an β-amino ketone and an aziridine.

iminium ion generated *in situ* from benzaldehyde (**4f**) and aniline. The α-aminophosphonate **31** (an *N,P*-acetal) was isolated in quantitative yield (Scheme 8.10) [40].

An iron-catalyzed multicomponent reaction of aldehyde **4a**, acetophenone, acetyl chloride and acetonitrile, which was used as the solvent, gave β-amino ketones such as **32** (Scheme 8.11) [41]. It was assumed that the sequence starts with an aldol reaction of aldehyde and ketone and then proceeds further with a displacement of a β-acetoxy group by the nucleophilic nitrile-nitrogen.

The reaction of imines such as **33** with ethyl diazoacetates yields complex product mixtures consisting of aziridines and β-enamino esters. When phenyldiazomethane is used as the nucleophilic component in this iron-catalyzed reaction, aziridines such as **34** are obtained in high yield and as single diastereoisomers (Scheme 8.11) [42]. The catalyst is the same Fe(II)-complex that was applied for the preparation of α-formyl ester **23** (cf. Scheme 8.6).

8.4 Additions to Carboxylic Acids and Their Derivatives

8.4.1 Oxygen Nucleophiles

Esterification is an important functional group interconversion on both the laboratory and bulk scales [43]. Since it is commonly catalyzed by Brønsted or Lewis acids, it is not surprising that the use of iron salts has been reported in that field [44]. A method using Fe(III) as a catalyst for consecutive synthetic steps can be regarded as particularly efficient. A classical example is the cleavage of an ether and subsequent esterification, which can be regarded as a one-pot protective group interconversion. In one of the first reports on $FeCl_3$ as a catalyst for this transformation, methyl, *n*-butyl, benzyl and silyl ethers have been investigated in addition to *tert*-butyl

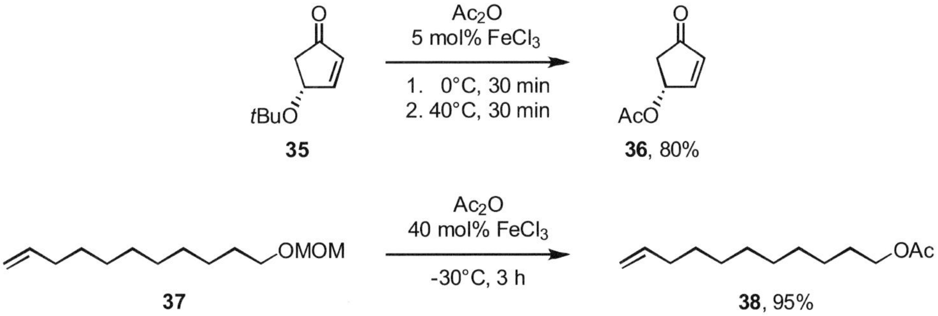

Scheme 8.12 Iron-catalyzed protective group interconversions.

ethers [45]. The latter, however, turned out to be the most promising substrates for this reaction, which has since then found numerous applications in organic synthesis [46]. Speculations on the mechanism of the ether cleavage (S_N1 on *t*Bu, S_N1 or S_N2 on the alkyl residue) were answered by the experiment illustrated in Scheme 8.12: cleavage on the cyclopentanone derivative **35** proceeded with full stereospecificity and retention of configuration [47].

Since it is well known that acetals are readily cleaved by iron(III) salts (see Sections 8.2.1 and 8.2.2), it is unsurprising that corresponding protective group interconversions (acetal → acetate) have also been reported [48]. In Scheme 8.12, an example of the transformation ROMOM → ROAc is given [49]. Both methods, ether and acetal cleavage, are best performed when Ac_2O is used as the solvent. One advantage of these iron(III) salts compared with Brønsted acids as catalysts was reported to be that C–C double bond isomerization or migration is prevented.

8.4.2
Carbon Nucleophiles

In analogy with iron-catalyzed Barbier-type reactions with SmI_2 (cf. Scheme 8.5), intramolecular nucleophilic acyl substitutions (S_Nt) can be used to prepare cyclic ketones from esters [50]. An illustrative example is shown in Scheme 8.13 [51]. Again, tris(1,3-diphenyl-1,3-propanedionato)iron(III) [$Fe(dbm)_3$] is used as the catalyst. Compound **40** is obtained as one racemic diastereoisomer.

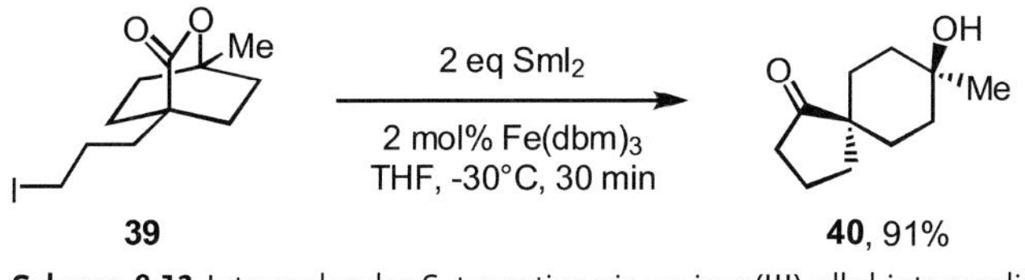

Scheme 8.13 Intramolecular S_Nt reaction via an iron(III) alkyl intermediate.

8.5
Conjugate Addition to α,β-Unsaturated Carbonyl Compounds

8.5.1
Carbon Nucleophiles

8.5.1.1 Michael Reactions

The Michael reaction is the conjugate addition of a soft enolate, commonly derived from a β-dicarbonyl compound **24**, to an acceptor-activated alkene such as enone **41a**, resulting in a 1,5-dioxo constituted product **42** (Scheme 8.14) [52]. Traditionally, these reactions are catalyzed by Brønsted bases such as tertiary amines and alkali metal alkoxides and hydroxides. However, the strongly basic conditions are often a limiting factor since they can cause undesirable side- and subsequent reactions, such as aldol cyclizations and retro-Claisen-type decompositions. To address this issue, acid- [53] and metal-catalyzed [54] Michael reactions have been developed in order to carry out the reactions under milder conditions.

The first investigations on iron-catalyzed Michael reactions utilized $Fe(acac)_3$ as catalyst. However, this metal complex is itself catalytically almost inactive. Yields of only up to 63% could be achieved, if $BF_3{\cdot}OEt_2$ is used as a co-catalyst [55]. Polystyrene-bound $Fe(acac)_3$ catalysts were also reported to give yields up to 63% [56]. $FeCl_3$ was used as a co-catalyst for clay-supported Ni(II). Yields achieved with this heterogeneous system ranged from 40 to 98% [57]. The double Michael addition of acrylonitrile to ethyl cyanoacetate is smoothly catalyzed by a complex generated from $[Fe(N_2)(depe)_2]$ [depe = 1,2-bis(diethylphosphano)ethane]. At 23 °C and after 36 h, an 88% yield is obtained with 1 mol% of this Fe(0) catalyst [58].

A breakthrough in the field of iron-catalyzed Michael reactions was achieved in 1997 with the use of $FeCl_3{\cdot}6H_2O$ [59]. In Scheme 8.15, the reaction of β-oxo ester **24a** with MVK (**41a**) is given as an example, which was scaled up to 50 g of product **42a** [60]. The efficiency of this iron-catalyzed process is remarkable: Inert conditions are not required, since oxygen and moisture are tolerated. As long as the starting materials and the product are liquid at ambient temperature, no solvents are necessary. A few milligrams of $FeCl_3{\cdot}6H_2O$ are simply added to a stoichiometric mixture of the starting materials.

With this method, Brønsted basic conditions are avoided, resulting in excellent chemoselectivities. Because of the quantitative conversions, workup and purification are extraordinary simple: separation of the product from the catalyst is achieved by

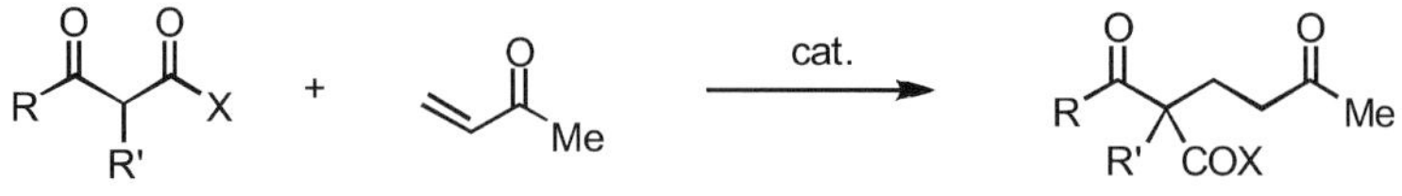

R(C=O)CH(R')(C=O)X (**24**, Michael donor) + MVK (**41a**, Michael acceptor) —cat.→ **42** (R(C=O)C(R')(COX)CH₂CH₂(C=O)Me)

Scheme 8.14 Michael reaction of β-diketones (X = alkyl) and β-oxo esters (X = alkoxy) with methyl vinyl ketone (**41a**, MVK).

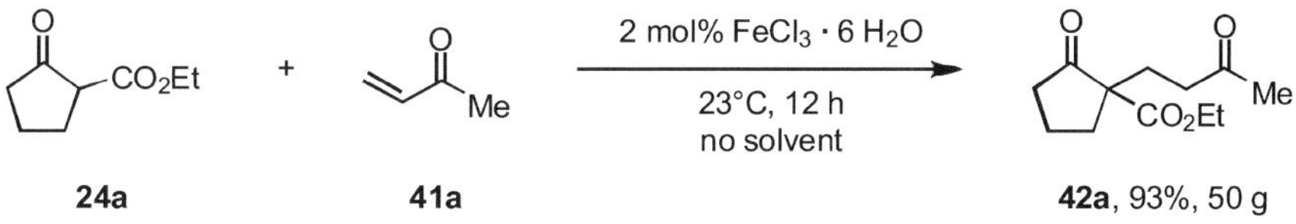

CO_2Et + Me → (2 mol% $FeCl_3 \cdot 6\,H_2O$; 23°C, 12 h; no solvent) CO_2Et, Me

24a **41a** **42a**, 93%, 50 g

Scheme 8.15 Catalysis of the Michael reaction by $FeCl_3{\cdot}6H_2O$.

either distillation or filtration through silica. Apart from these practical aspects, the catalyst $FeCl_3{\cdot}6H_2O$ is in terms of ecological and economic considerations the transition metal compound of first choice. Figure 8.2 gives an impression of the scope of this method [61]. Best results are usually obtained with cyclic β-oxo esters; with 1 mol% $FeCl_3{\cdot}6H_2O$ yields generally exceed 90% (**42a–c**). Acyclic β-oxo esters and β-diketones require 5 mol% of the catalyst (products **42d–g**). The conversion of substituted enones such as chalcone requires elevated temperatures, e.g. 50 °C in case of products **42c** and **e**. Succinyl succinates can be converted in double Michael reactions to yield products such as **42h** as *cis*-diastereomers [62]. Furthermore, piperidine derivatives [63] and lactones [64] can be applied as Michael donors. Malonates as donors do not react under the conditions of iron catalysis, since they form no chelate complexes **3** (cf. Figure 8.1) with Fe(III). Similarly, β-cyano esters and β-cyano ketones are not converted with iron, but ruthenium catalysts with optimal results [65]. With regard to the acceptor, reactions of cyclic enones are only observed in rare cases and with low yields [66].

Medium ring size formation is often a challenging task in organic synthesis. With $FeCl_3{\cdot}6H_2O$ as the catalyst, compound **24b**, with both a Michael donor and acceptor moiety, undergoes intramolecular reaction to furnish the annulated product **42i** with two seven-membered rings as a single diastereomer [67]. An attempt at macrocyclization by iron-catalyzed Michael reaction was not fruitful [68] (Scheme 8.17).

Quantitative conversion is one of the essential preconditions to achieve a significant molecular weight in stepwise polymerization process. Consequently, an iron-catalyzed Michael reaction would be a suitable elementary step for a polyaddition. Bis-donor **24c** and bis-acceptor **41b**, readily accessible from common starting materials [69], were converted with $FeCl_3{\cdot}6H_2O$ to yield a poly-addition product

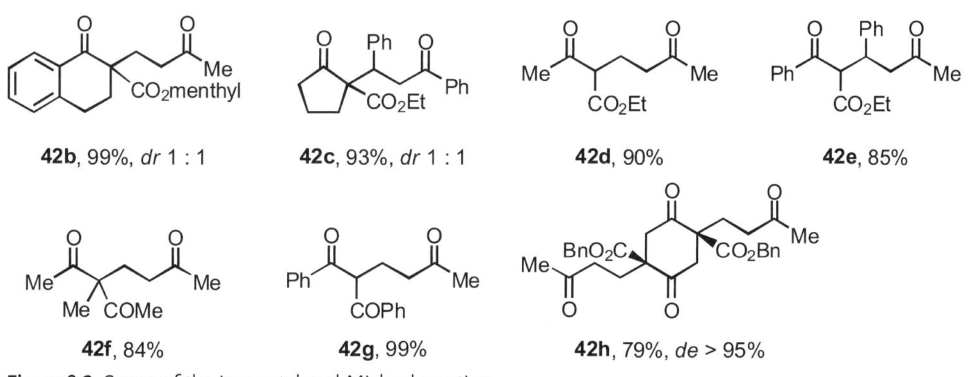

Figure 8.2 Scope of the iron-catalyzed Michael reaction.

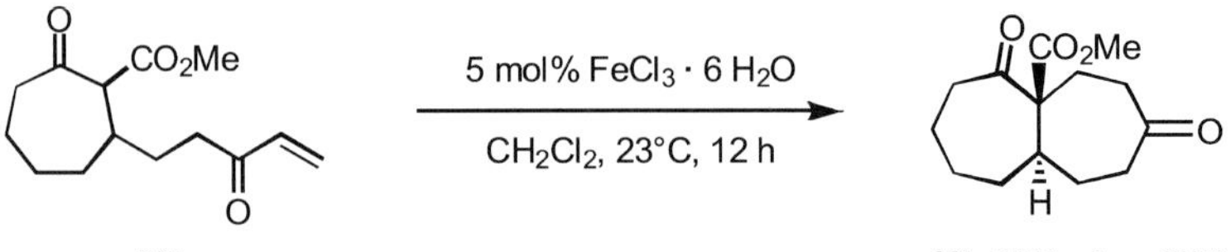

5 mol% $FeCl_3 \cdot 6\,H_2O$, CH_2Cl_2, 23°C, 12 h

24b → 42i, 80%, de > 95%

Scheme 8.16 An intramolecular Michael reaction catalyzed by $FeCl_3{\cdot}6H_2O$.

42j. The material is an analogue of ethylene–carbon monoxide copolymers [70] bearing an interesting constitution with carbonyl moieties all along the carbon backbone. According to molecular mass analysis performed by GPC, the oligomeric product **42j** contained an average of 24 monomeric units ($n \approx 12$ in Scheme 8.17) [69]. Although macromolecules such as **42j** will never be able to compete from an economic point of view with so far known polyaddition products, it may be of academic interest to compare their properties with those of other structures.

With respect to the mechanism of the iron catalysis, the activity of $FeCl_3{\cdot}6H_2O$ is closely related to its ability to give dionato chelate complexes **3** with β-dicarbonyl compounds. Without prior deprotonation – even in Brønsted acidic media – these deeply colored iron complexes are instantly formed. With this property, Fe(III) is unique among all other transition metals, which require a stoichiometric amount of base for dionato complex formation. Known for over 100 years, the significant color of the complexes has been utilized for the detection of β-oxo esters and β-diketones.

The chelate ligand in dionato complex **3** is planar and it is particularly stabilized by π-delocalization. In addition to this thermodynamic stability, the iron center has 17 valence electrons in an octahedron, hence its coordination sphere is kinetically labile. By ligand exchange, the acceptor **41a** is coordinated at a vacant site to form species **44** (Scheme 8.18). The function of the center metal is not only to hold the acceptor in proximity to the donor. Additionally, the acceptor is activated by Lewis acidity of the center metal. Subsequently, the nucleophilic carbon atom of the dionato ligand is

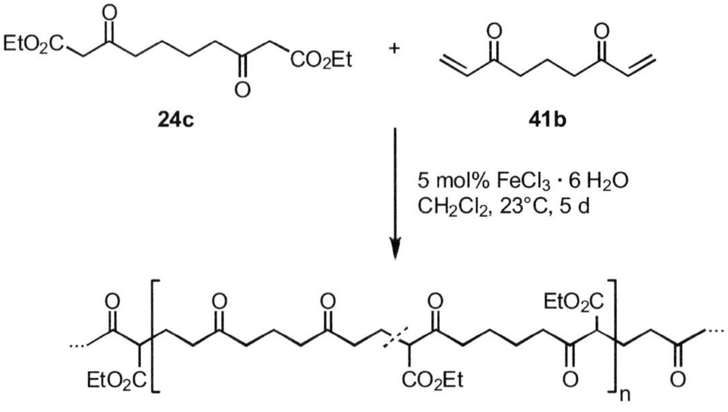

24c + 41b

5 mol% $FeCl_3 \cdot 6\,H_2O$, CH_2Cl_2, 23°C, 5 d

42j, $M_w = 6100$, $M_w/M_n = 1.7$, n ca. 12

Scheme 8.17 A poly-Michael reaction catalyzed by $FeCl_3{\cdot}6H_2O$.

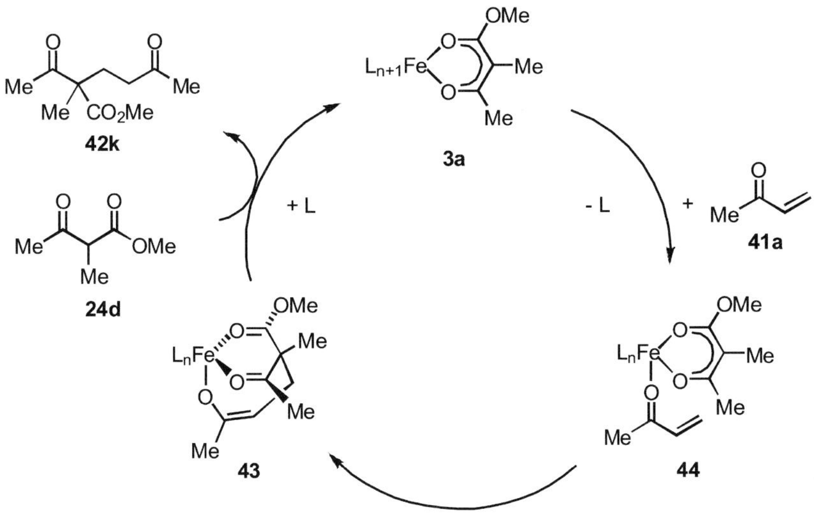

Scheme 8.18 Mechanism of the iron-catalyzed Michael reaction.

alkylated by the acceptor to form the bicyclic intermediate **43** with a coordinating enolate side-chain. From this species **43** the product **42k** is liberated readily and complex **3a** is regenerated by ligand exchange, since π-delocalization is obviously impossible in structure **43**.

The idea of an one-center template mechanism was initially supported by first-order kinetics in iron. Moreover, intermediates **3a**, **43** and **44** ands also their transition states in the catalytic cycle (Scheme 8.18) were proved by computational studies [71]. Moreover, mass spectrometric (ESI) [72] and spectroscopic (EXAFS and Raman) studies indicated complex **45** with two equatorial β-diketonate ligands to be the catalytically active species in solution (Scheme 8.19) [73]. Actually, 4 equiv. of $FeCl_3 \cdot 6H_2O$ are needed to generate 1 equiv. of complex **45** under reaction conditions;

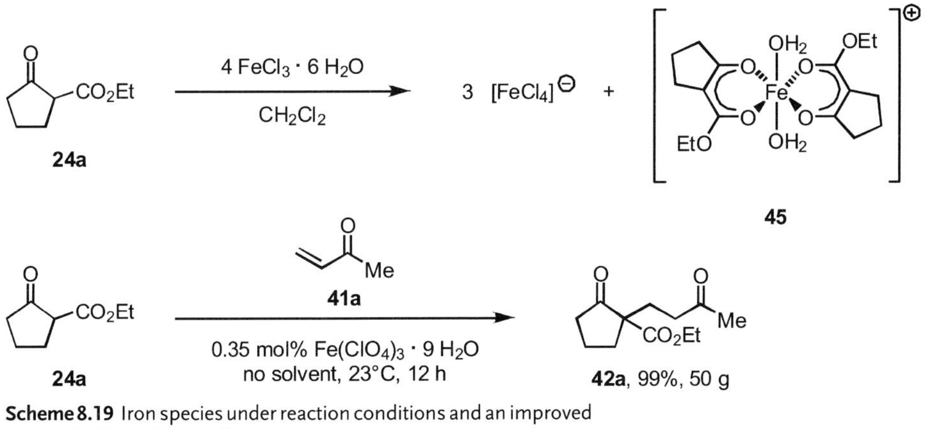

Scheme 8.19 Iron species under reaction conditions and an improved protocol with a chloride-free catalyst.

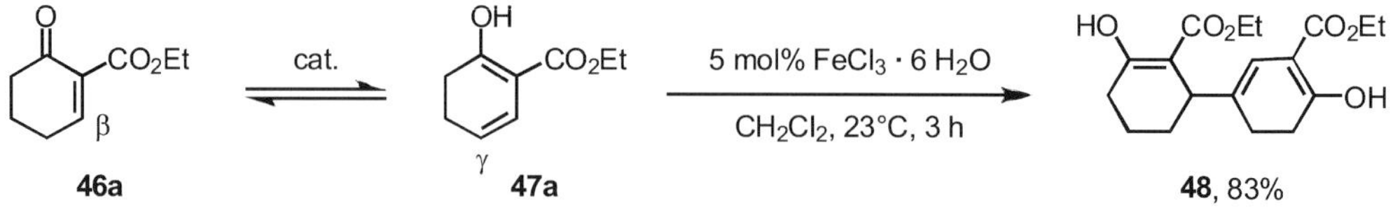

46a ⇌ (cat.) 47a → (5 mol% $FeCl_3 \cdot 6\,H_2O$, CH_2Cl_2, 23°C, 3 h) 48, 83%

Scheme 8.20 Enone–dienol tautomerism and iron-catalyzed dimerization.

3 equiv. of $[FeCl_4]^-$ are formed as the dominating iron-complex in solution. This species actually behaves like a thermodynamic iron sink under reaction conditions. It removes three quarters of the Fe(III), and one quarter remains catalytically active. It is not surprising that chloride-free pre-catalysts such as $Fe(ClO_4)_3{\cdot}9H_2O$ result in improved efficiency. With only 0.35 mol% of this catalyst, a 99% yield of product **42a** on a 50 g scale is obtained under solvent-free reaction conditions (Scheme 8.19).

8.5.1.2 **Vinylogous Michael Reactions**

Cycloalkenones with an additional 2-acceptor substituent such as ester **46a** (Scheme 8.20) show an interesting enone–dienol tautomerism. The equilibrium of **46a** and **47a** shown in Scheme 8.20 is the vinylogous case of keto–enol tautomerism and it is catalyzed by either Brønsted acid or base. In contrast to the keto–enol case, remarkable kinetic stability of both tautomers **46a** and **47a** under neutral conditions can be recognized. They can, for example, be separated by chromatography on SiO_2. With a catalytic amount of $FeCl_3{\cdot}6H_2O$, the mixture of enone **46a** and dienol **47a** is not stable, but converts within 3 h into a unique product **48** [74]. Closer inspection of the constitution of this dimer **48** makes it clear that it is formally a product of a Michael reaction of the acceptor **46a** in the β-position with the donor **47a** in the vinylogous γ-position. A vinylogous Michael reaction of such a kind is a very rare, unprecedented principle in organic synthesis.

The iron catalysis of vinylogous Michael reactions is not only restricted to dimerizations. The γ-donor **46b** can be converted with MVK (**41a**) to give the 1,7-dioxo-constituted product **49** when the catalyst is Fe(III) (Scheme 8.21) [75]. If NaOMe in MeOH is applied as the catalyst, reaction of the dienolate of donor **46b** in the α-position with acceptor **41a** proceeds via a "normal" Michael reaction and 1,5-dioxo-constituted product **50** is obtained.

Benzoquinone derivatives applied as acceptors in the vinylogous Michael reaction lead to products with two six-membered rings connected by a new C–C single bond. This principle could be further developed into a new method for the synthesis of highly functionalized biaryl compounds. The iron-catalyzed conversion of the vinylogous donor **46c** (Scheme 8.22), for example, with 1,4-naphthoquinone **41c** gives an intermediate **51**, which, however, turned out not to be stable. First, the quinoid diketo moiety tautomerizes to the more favorable hydroquinone system. Subsequently, oxa-Michael addition of a hydroxy group to the enone function – in equilibrium with the dienol – occurs, closing a dehydrofuran ring. Reversion of the conjugate addition is prevented by oxidation to the annulated furan *in situ* and the naphthodihydrobenzo-

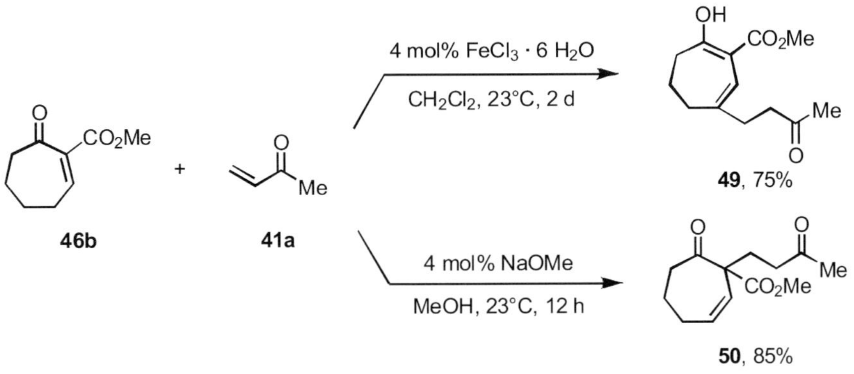

Scheme 8.21 Base-catalyzed Michael and iron-catalyzed vinylogous Michael reactions.

furan derivative **53** was isolated as a stable compound [76]. Further oxidation to the fully aromatic system **52** with MnO_2 required prior acetylation of the phenolic hydroxy groups.

Vinylogous donor **46d** (Scheme 8.23) was treated with an excess of 1,2-naphthoquinone (**41d**) to give a cross-coupled product, which was already oxidized under the reaction conditions by air to the aromatic system **54**. Reductive acetylation furnished highly substituted biaryl product **55** [77]. The latter example proves that the method clearly deserves further development towards a new approach to highly substituted biaryl products. It is operationally simple, compared with other cross-coupling strategies, since neither inert or anhydrous conditions nor protective groups are required and the catalyst $FeCl_3{\cdot}6H_2O$ is a cheap and non-toxic reagent (Schemes 8.22 and 8.23).

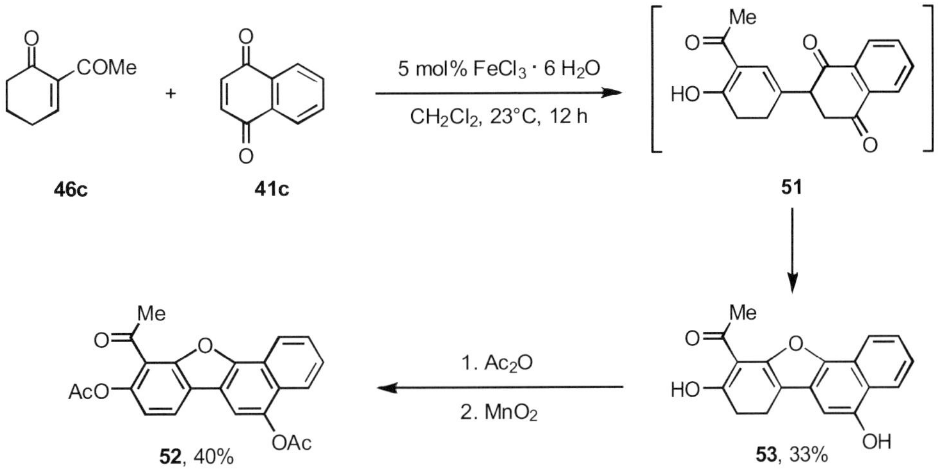

Scheme 8.22 Synthesis of a naphthobenzofuran derivative by vinylogous Michael reaction.

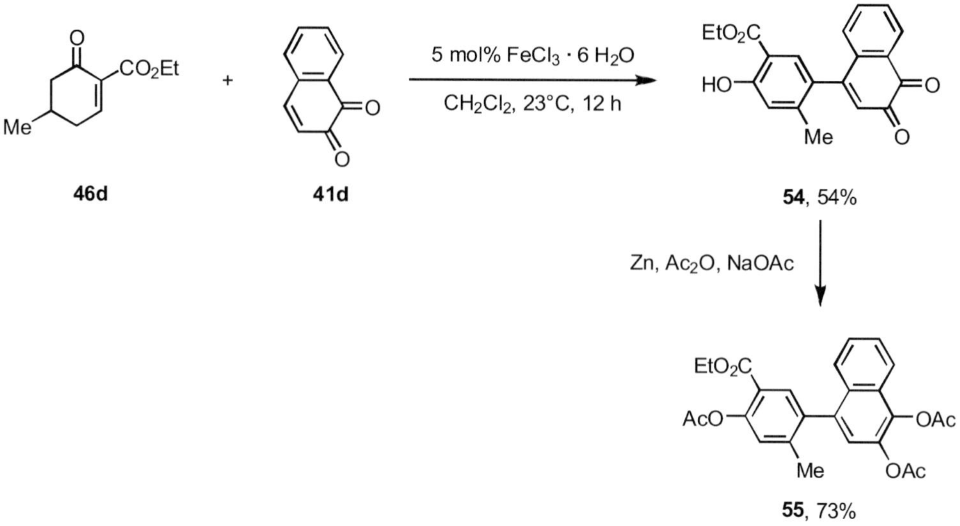

5 mol% $FeCl_3 \cdot 6\,H_2O$
CH_2Cl_2, 23°C, 12 h
46d + 41d → 54, 54%
Zn, Ac_2O, NaOAc
55, 73%

Scheme 8.23 Biaryl synthesis by vinylogous Michael reaction.

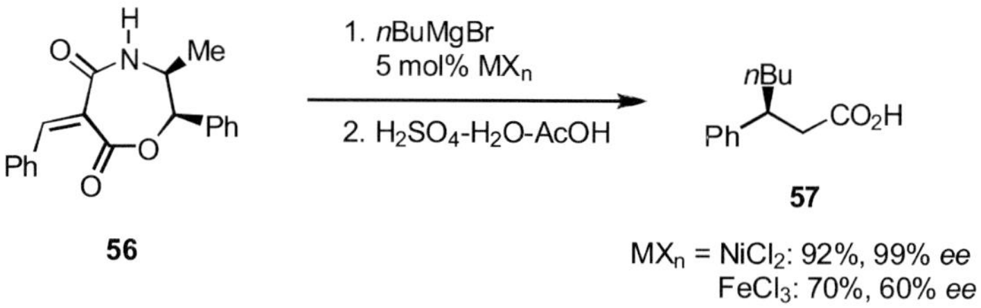

1. *n*BuMgBr 5 mol% MX_n
2. H_2SO_4-H_2O-AcOH
56 → 57
MX_n = $NiCl_2$: 92%, 99% *ee*
$FeCl_3$: 70%, 60% *ee*

Scheme 8.24 Asymmetric conjugate addition with ephedrine as chiral auxiliary.

8.5.1.3 Asymmetric Michael Reactions

The first asymmetric iron-catalyzed conjugate addition was reported in 1977. Benzylidene malonate **56** with an ephedrine moiety as chiral auxiliary was converted with Grignard reagents such as *n*BuMgBr in the presence of catalytic amounts of various metal salts. The optically active phenylpropionic acid **57** was obtained with

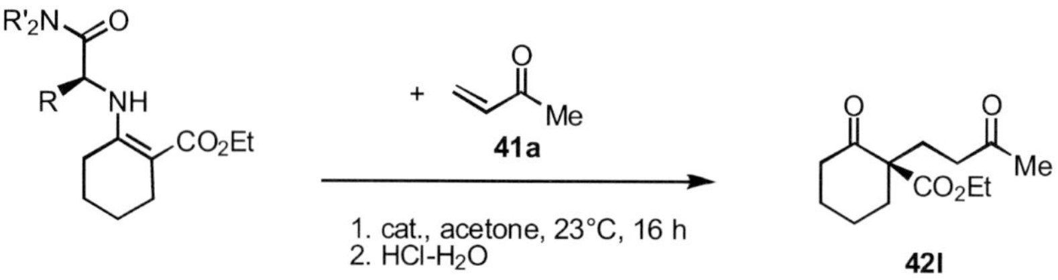

+ 41a
1. cat., acetone, 23°C, 16 h
2. HCl-H_2O
→ 42l
58a: R = *i*Bu, R' = Me, cat. = 5 mol% $FeCl_3$: 77% *ee*
58b: R = *i*Pr, R' = Et, cat. = 2.5 mol% $Cu(OAc)_2 \cdot H_2O$: 86%, 98% *ee*

Scheme 8.25 Application of α-amino acid amides as chiral auxiliaries for the asymmetric Michael reaction.

$FeCl_3$ after saponification of the oxazepine-moiety followed by decarboxylation, in 70% yield and with 60% *ee* [78]. Superior results were obtained, however, by using $NiCl_2$ as catalyst (92% yield, 99% *ee*).

Quaternary stereocenters can be obtained with high selectivity with α-amino acid amides as chiral auxiliaries, which were first converted with β-oxo esters to give enamines such as compounds **58**. According to a combinatorial strategy, various enamino esters **58** were screened in Michael additions with MVK (**41a**) and several metal salts as catalysts. With $FeCl_3$, however, the maximum stereoselectivity achieved was only 77% *ee* (with enamine **58a** derived from L-isoleucine dimethylamide). $Cu(OAc)_2 \cdot H_2O$ turned out be the optimal catalyst for this transformation. With L-valine diethylamide as chiral auxiliary in compound **58b**, reaction proceeds with 86% yield and 98% *ee* after aqueous workup [79]. Importantly, this valuable method for the construction of quaternary stereocenters [80] under ambient conditions seems to be generally applicable to a number of Michael donors [81]. In all cases, the auxiliary can be quantitatively recovered after workup.

Asymmetric catalysis with chiral ligands [82] is commonly considered to be advantageous instead of using chiral auxiliaries. Catalytic asymmetric Michael reactions are known [83], but not with iron as the catalytically active metal. Only two reports on iron catalyzed catalytic asymmetric Michael reaction with dipeptides [84] or diamino thioethers [85] exist, but the enantioselectivities were disappointing (18% *ee* and 10% *ee*, respectively).

Several Lewis acids were investigated in enantioselective radical additions to cinnamoyl oxazolidone **59** using chiral bisoxazoline ligand. With $Fe(NTf_2)_2$ and *t*Bu-BOX ligands, up to 80% *ee* were achieved (Scheme 8.26) [86]. Selectivities up to 98% were obtained when $Mg(NTf_2)_2$ was used as the precatalyst.

8.5.1.4 Michael Reactions in Ionic Liquids and Heterogeneous Catalysis

Ionic liquids are attracting increasing interest as environmentally benign solvents, because they possess a number of interesting properties. Among these, one of the most important is their virtually non-existent vapor pressure, which makes them easily confinable and also allows easy recyclability of catalytic systems after distillation of volatile products.

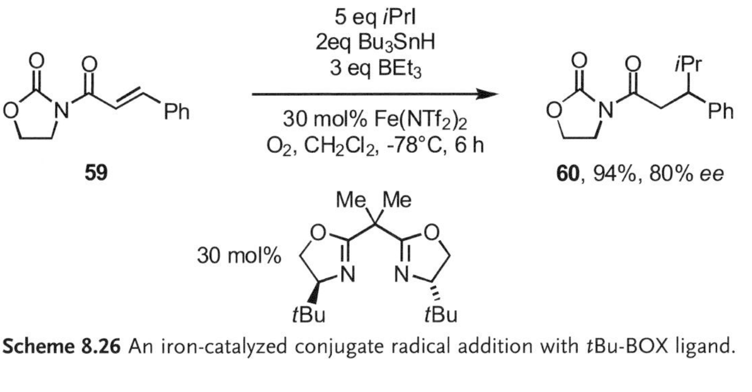

59 → **60**, 94%, 80% *ee*

5 eq *i*PrI, 2eq Bu_3SnH, 3 eq BEt_3; 30 mol% $Fe(NTf_2)_2$, O_2, CH_2Cl_2, -78°C, 6 h; 30 mol% *t*Bu-BOX ligand

Scheme 8.26 An iron-catalyzed conjugate radical addition with *t*Bu-BOX ligand.

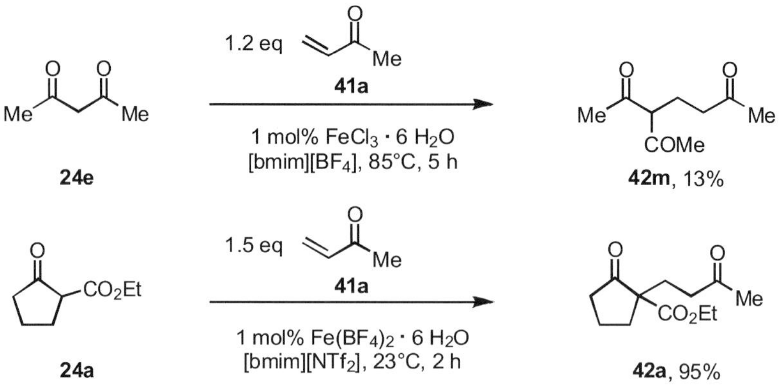

1.2 eq **41a**

1 mol% $FeCl_3 \cdot 6\ H_2O$
[bmim][BF_4], 85°C, 5 h

24e → **42m**, 13%

1.5 eq **41a**

1 mol% $Fe(BF_4)_2 \cdot 6\ H_2O$
[bmim][NTf_2], 23°C, 2 h

24a → **42a**, 95%

Scheme 8.27 Michael reactions in ionic liquids.

Several metal catalysts have been investigated for Michael reaction of acacH (**24e**) with MVK (**41a**) in [bmim][BF_4] (bmim = butylmethylimidazolium) as solvent. $Ni(acac)_2$ turned out to be the optimal catalyst (94% yield). With $FeCl_3 \cdot 6H_2O$ the yield of product **42m** was limited (Scheme 8.27) [87]. It is presumed that the formation of tetrachloroferrates is a rate-limiting factor, as was observed earlier in common organic solvents or under solvent-free conditions (cf. Scheme 8.19) [88].

A chloride-free catalyst, $Fe(BF_4)_2 \cdot 6H_2O$, was used for the reaction of β-oxo ester **24a** with MVK (**41a**) in the ionic liquid [bmim][NTf_2] [89]. Product **42a** was obtained in about the same yield (95%) as for the solvent-free protocol with $Fe(ClO_4)_3 \cdot 9H_2O$ (99% yield) (Scheme 8.27). Both protocols with ionic liquids are, however, operationally less simple than the solvent-free methods reported before, because of the use significant amounts of Et_2O for workup and purification of the products.

Indole (**61**) shows significant nucleophilic reactivity in its 3-position. If the carbonyl compound is activated by a Lewis acid, indole (**61**) adds to aldehydes to furnish bis(indolyl)methanes such as **62** or they undergo conjugate addition to α,β-unsaturated ketones with formation of β-indolyl ketones such as **63**. Both transformations can be catalyzed with iron salts in ionic liquids (Scheme 8.28). Bis(indolyl) methane **62** is obtained in 96% yield with $FeCl_3 \cdot 6H_2O$ as the catalyst in [omim][PF_6] (omim = methyloctylimidazolium) [90]. The conjugate addition yielding 92% of ketone **63** was catalyzed by $Fe(BF_4)_2 \cdot 6H_2O$ in [bmim][NTf_2] [91]. Both protocols utilized larger amounts of CH_2Cl_2 for workup and purification of the products.

Fe(III)-exchanged fluorotetrasilicic mica acts as a highly effective and reusable catalyst for solvent-free Michael reactions of β-oxo esters with MVK (**41a**). The immobilized catalyst shows higher activity than homogenous Fe(III) catalysts. Product **42a** (cf. Scheme 8.27), for example, is formed in 99% yield even if the catalyst is reused four times [92].

A rather unusual Fe(III) species for catalysis is $[Cp_2Fe]^+$, ferrocenium. A polymer-bound ferrocenium catalyst was obtained by oxidizing a poly(vinylferrocene-*block*-isoprene)copolymer with AgOTf. The activity of this catalyst was tested with the reaction of β-oxo ester **24a** and MVK (**41a**) (cf. Scheme 8.27) [93].

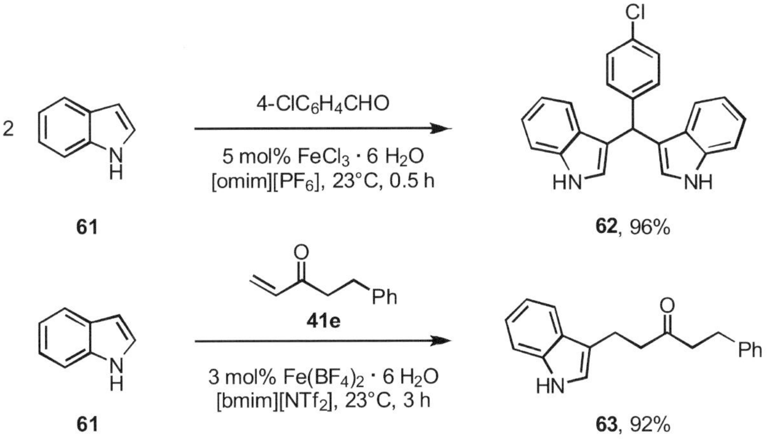

2 **61** → 4-ClC$_6$H$_4$CHO, 5 mol% FeCl$_3$ · 6 H$_2$O, [omim][PF$_6$], 23°C, 0.5 h → **62**, 96%

61 → **41e**, 3 mol% Fe(BF$_4$)$_2$ · 6 H$_2$O, [bmim][NTf$_2$], 23°C, 3 h → **63**, 92%

Scheme 8.28 Iron-catalyzed reactions of indoles.

8.5.2
Nitrogen Nucleophiles

The aza-Michael reaction yields, complementary to the Mannich reaction, β-amino carbonyl compounds. If acrylates are applied as Michael acceptors, β-alanine derivatives such as **64** and **65** are obtained. The aza-Michael reaction can be catalyzed by Brønsted acids or different metal ions. Good results are also obtained with FeCl$_3$, as shown in Scheme 8.29. The addition of HNEt$_2$ to ethyl acrylate (**41f**), for example, requires 10 mol% of the catalyst and a reaction time of almost 2 days [94]. The addition of piperidine to α-amino acrylate **41g** is much faster and yields α,β-diaminocarboxylic acid derivative **65** [95].

Conjugate additions of carbamates to α,β-unsaturated enones require – apart from metal halide – TMSCl as a stoichiometric additive [96]. The addition of ethyl carbamate to cyclohexenone (**41h**) requires only 50 mol% TMSCl, which was an exceptionally low amount compared with other Michael acceptors. With 10 mol% of the catalyst, the yield of 3-aminocyclohexenone derivative **66** was good (93%) [97]. Aza-Michael reactions also proceed in aqueous media with good results if Co(II),

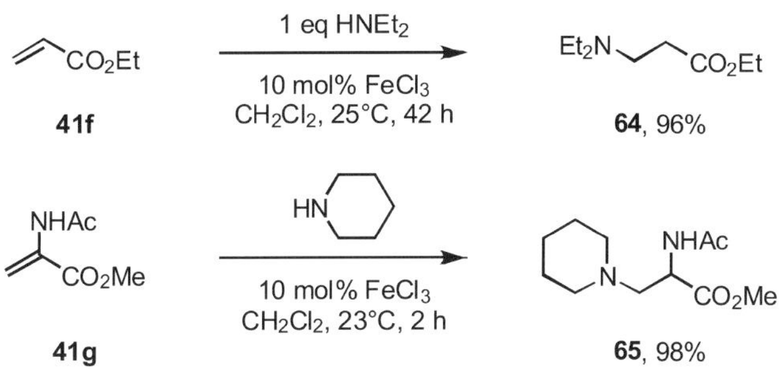

41f (CO$_2$Et) → 1 eq HNEt$_2$, 10 mol% FeCl$_3$, CH$_2$Cl$_2$, 25°C, 42 h → Et$_2$N–CH$_2$CH$_2$–CO$_2$Et **64**, 96%

41g (NHAc, CO$_2$Me) → piperidine (HN), 10 mol% FeCl$_3$, CH$_2$Cl$_2$, 23°C, 2 h → **65**, 98% (NHAc, CO$_2$Me)

Scheme 8.29 Iron-catalyzed synthesis of β-amino- and α,β-diaminocarboxylic acid derivatives.

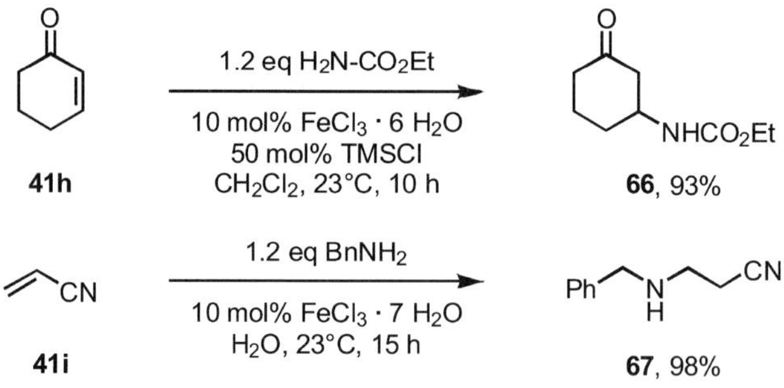

41h — 1.2 eq $H_2N\text{-}CO_2Et$; 10 mol% $FeCl_3 \cdot 6\ H_2O$, 50 mol% TMSCl, CH_2Cl_2, 23°C, 10 h → $NHCO_2Et$ 66, 93%

41i (CN) — 1.2 eq $BnNH_2$; 10 mol% $FeCl_3 \cdot 7\ H_2O$, H_2O, 23°C, 15 h → Ph N H CN 67, 98%

Scheme 8.30 Aza-Michael reactions of carbamates in aqueous media.

Cu(II) or Fe(III) salts are applied as catalysts. In Scheme 8.30, the reaction of nitrile **41i** with benzylamine gives an almost quantitative yield if $FeCl_3{\cdot}7H_2O$ is applied as the catalyst, which is, however, in contrast to the hexahydrate, not commercially available [98]. Water is normally considered as an environmentally benign solvent. The authors curiously needed CH_2Cl_2 and other organic solvents for workup and purification of the product **67** (Scheme 8.30).

8.6 Synthesis of Heterocycles

8.6.1 Pyridine and Quinoline Derivatives

According to the classical Hantzsch synthesis of pyridine derivatives, an α,β-unsaturated carbonyl compound is first formed by Knoevenagel condensation of an aldehyde with a β-dicarbonyl compound. The next step is a Michael reaction with another equivalent of the β-dicarbonyl compound (or its enamine) to form a 1,5-diketone, which finally undergoes a cyclocondensation with ammonia to give a 1,4-dihydropyridine with specific symmetry in its substitution pattern.

An iron-catalyzed reaction of an α,β-unsaturated oxime such as **68** with a β-oxo ester also gave pyridine derivatives such as nicotinic acid **69** [99]. Under the reaction conditions (150–160 °C, without solvent) first Michael adducts such as intermediate **70** are presumably formed, which further condense via intermediate **71**. This method is not restricted to a centric symmetry in the substitution pattern, which is an advantage compared with the Hantzsch synthesis. Moreover, the method starts with hydroxylamine being two oxidation stages above ammonia; therefore, no oxidation in the final stage from dihydro- to pyridine is necessary (Scheme 8.31).

Three equivalents of $FeCl_3$ are required for the reaction of chalcone **41j** with ethyl cyanoacetate to give α-pyridone derivative **72** (Scheme 8.32) [100]. The reaction is carried out at 140 °C under strongly acidic conditions ($FeCl_3$ dissolved in propionic acid). It proceeds presumably by an initial Michael addition yielding intermediate **73**. Excess of iron(III) is required, because this is the oxidizing reagent for the introduction of the second C—C double bond in intermediate **74**.

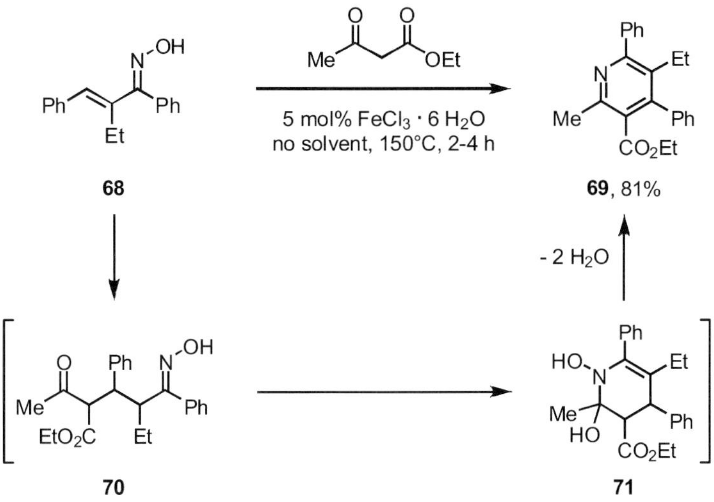

Scheme 8.31 Iron-catalyzed synthesis of pyridine derivatives.

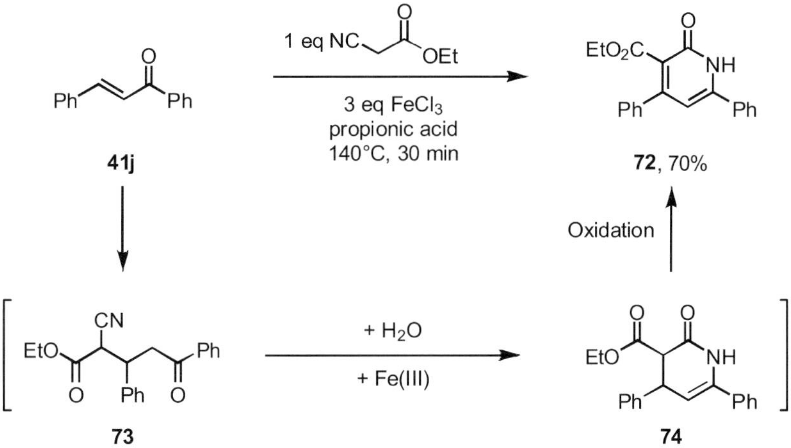

Scheme 8.32 Iron-mediated synthesis of 2-pyridone derivatives.

Friedländer quinoline synthesis has been accomplished in ionic liquids with substoichiometric amounts of $FeCl_3 \cdot 6H_2O$ (Scheme 8.33) [101]. The substitution pattern of products **77** is, however, relatively limited to 2,4-diaryl-functionalized quinolines, as depicted in Scheme 8.33.

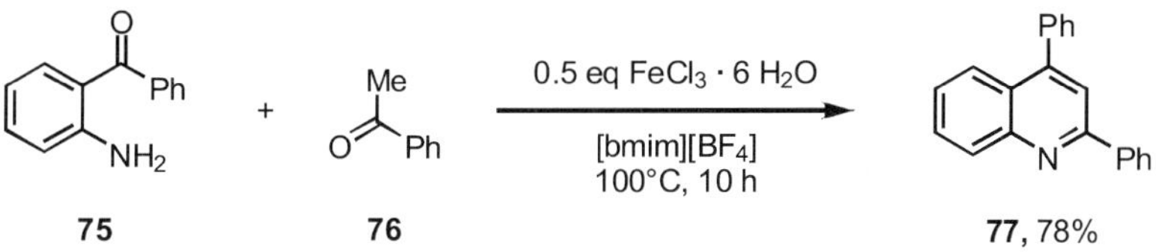

Scheme 8.33 Iron-mediated Friedländer quinoline synthesis.

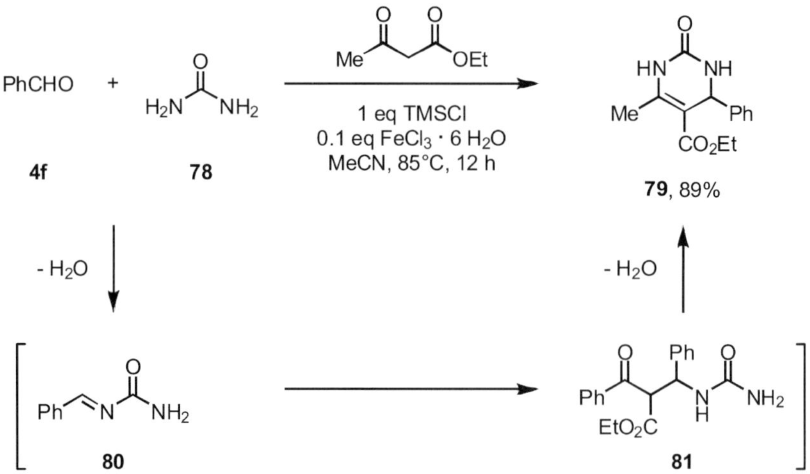

Scheme 8.34 Iron-mediated Bignelli reaction.

8.6.2
Pyrimidine and Pyrazine Derivatives

According to a protocol first reported in 1893 by Bignelli, 3,4-dihydropyridines, e.g. **79** (Scheme 8.34), are obtained by acid-catalyzed reaction of aldehydes **4**, urea (**78**) and β-oxo esters [102]. In recent years, protocols for Bignelli reactions have been continuously improved by using various Lewis or Brønsted acids, e.g. iron salts [103]. The latest state of the art is the use of a combination of 0.1 equiv. of $FeCl_3{\cdot}6H_2O$ and 1 equiv. of TMSCl, which gives **79** in 89% yield [104]. The stepwise mechanism of this two-fold condensation involves, as indicated in Scheme 8.34, a conjugate addition of the β-oxo ester to an acylimine **80**, being an aza analogue of chalcone.

Aminoacetonitrile (**83**) can be condensed with α-ketoximes such as **82** using a stoichiometric amount of $FeCl_3$ (Scheme 8.35). The reaction presumably proceeds with initial imine formation (intermediate **85**) followed by tautomerization to give ketenimine **86** and ring closure to 2-aminopyrazine *N*-oxide **87**, which can be isolated. In a one-pot protocol this product **87** can be further reduced with $Pd/C/H_2$ to give aminopyrazine **84** (80% over two steps) [105].

8.6.3
Benzo- and Dibenzopyrans

Iron(III) catalyzes effectively the unusual cyclocondensation of 2-hydroxybenzaldehyde (**4h**) with 2,2-dimethoxypropane (**88**) (Scheme 8.36) [106]. The product, dihydrobenzopyran derivative **89**, is obtained as a single diastereoisomer. It can be assumed that this reaction proceeds via an enol ether as intermediate. Alternatively, the same product is obtained from a reaction mixture containing aldehyde **4h**, acetone and trimethyl orthoformate.

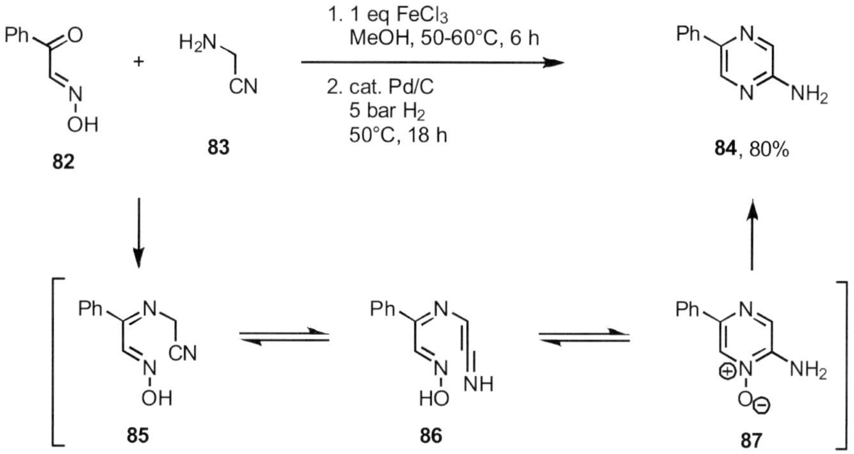

Scheme 8.35 An iron-mediated α-aminopyrazine synthesis.

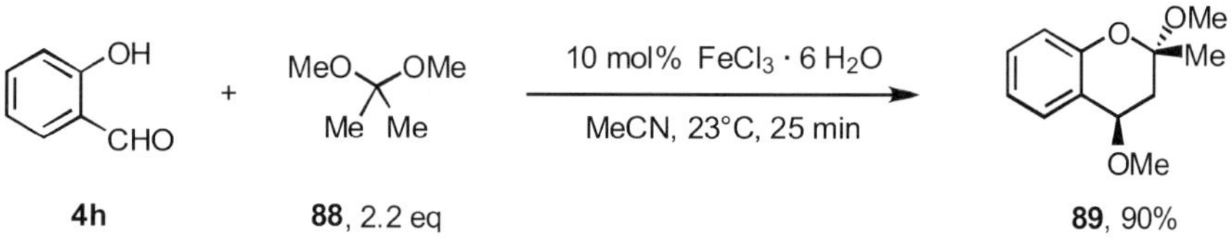

Scheme 8.36 Iron catalyzed benzopyran formation.

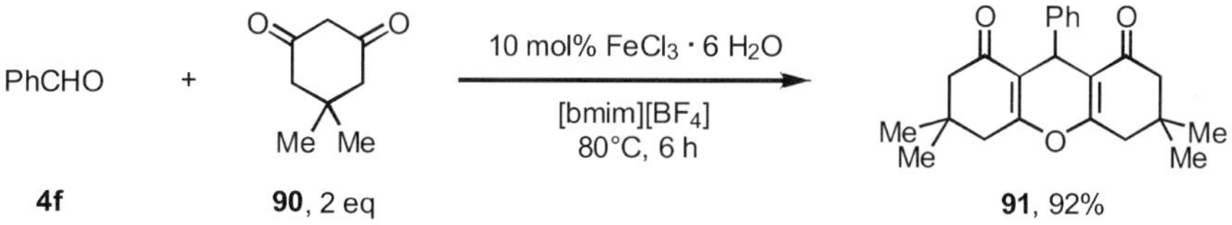

Scheme 8.37 Iron-catalyzed dibenzopyran synthesis.

Benzaldehyde (**4f**) can be condensed with 2 equiv. of dimedone (**90**) and a substoichiometric amount of $FeCl_3{\cdot}6H_2O$ to give xanthenedione derivative **91** in good yield (Scheme 8.37) [107]. The reaction is carried out in an ionic liquid as solvent. Only minor amounts of EtOH and H_2O are required for workup and purification of the product.

References

1 (a) D. D. Díaz, P. O. Miranda, J. I. Padrón, V. S. Martín, *Curr. Org. Chem.* 2006, **10**, 457–476; (b) C. Bolm, J. Legros, J. Le Paih, L. Zani, *Chem. Rev.* 2004, **104**, 6217–6254.

2 T. W. Greene, P. G. M. Wuts, *Protective Groups in Organic Synthesis*, 2nd edn., Wiley, New York, 1991.

3 (a) K. Iwanami, T. Oriyama, *Chem. Lett.* 2004, **33**, 1324–1325; (b) P. P. Singh,

M. M. Gharia, F. Dasgupta, H. C. Srivastava, *Tetrahedron Lett.* 1977, 439–440.

4 W. Danikiewicz, M. Olejnik, J. Wójcik, S. K. Tyrlik, B. Nalewajko, *J. Mol. Catal. A* 1997, **123**, 25–33.

5 S. E. Sen, S. L. Roach, J. K. Boggs, G. J. Ewing, J. Magrath, *J. Org. Chem.* 1997, **62**, 6684–6686.

6 (a) K. Katoh, M. Kirihara, Y. Nagata, Y. Kobayashi, K. Arai, J. Minami, S. Terashima, *Tetrahedron* 1994, **50**, 6239–6258; (b) A. Fadel, R. Yefsah, J. Salaün, *Synthesis* 1987, 37–40; (c) I. W. J. Still, Y. Shi, *Tetrahedron Lett.* 1987, **28**, 2489–2490; (d) K. S. Kim, Y. H. Song, B. H. Lee, C. S. Hahn, *J. Org. Chem.* 1986, **51**, 404–407.

7 H. K. Patney, *Tetrahedron Lett.* 1991, **32**, 2259–2260.

8 A. Kamal, E. Laxman, P. S. M. M. Reddy, *Synlett* 2000, 1476–1478.

9 E. Knoevenagel, *Liebigs Ann. Chem.* 1914, **402**, 111–148.

10 (a) C. Wang, M. Li, *Synth. Commun.* 2002, **32**, 3469–3473; (b) T.-S. Li, Z.-H. Zhang, Y.-J. Gao, *Synth. Commun.* 1998, **28**, 4665–4671; (c) K. S. Kochhar, B. S. Bal, R. P. Deshpande, S. N. Rajadhyaksha, H. W. Pinnick, *J. Org. Chem.* 1983, **48**, 1765–1767.

11 B. M. Trost, C. B. Lee, *J. Am. Chem. Soc.* 2001, **123**, 3671–3686.

12 (a) A. Roën, J. I. Padrón, J. T. Vázquez, *J. Org. Chem.* 2003, **68**, 4615–4630; (b) R. Bukowski, L. M. Morris, R. J. Woods, T. Weimar, *Eur. J. Org. Chem.* 2001, 2697–2705; (c) N. Ikemoto, O. K. Kim, L.-C. Lo, V. Satyanarayana, M. Chang, K. Nakanishi, *Tetrahedron Lett.* 1992, **33**, 4295–4298.

13 S. K. Chatterjee, P. Nuhn, *Chem. Commun.* 1998, 1729–1730.

14 P. O. Miranda, I. Brouard, J. I. Padrón, J. Bermejo, *Tetrahedron Lett.* 2003, **44**, 3931–3934.

15 (a) R. J. Ferrier, *J. Chem. Soc., Perkin Trans.* 1979, **1**, 1455–1458; (b) R. J. Ferrier, N. Prasad, *J. Chem. Soc. C* 1969, 570–575; (c) R. J. Ferrier, N. Prasad, G. H. Sankey, *J. Chem. Soc. C* 1968, 974–977.

16 (a) R. D. Tilve, M. V. Alexander, A. C. Khandekar, S. D. Samant, V. R. Kanetkar, *J. Mol. Catal. A.* 2004, **223**, 237–240; (b) K. Krohn, U. Flörke, D. Gehle, *J. Carbohydr. Chem.* 2002, **21**, 431–443.

17 C. Masson, J. Soto, M. Bessodes, *Synlett* 2000, 1281–1282.

18 T. Watahiki, Y. Akabane, S. Mori, T. Oriyama, *Org. Lett.* 2003, **5**, 3045–3048.

19 K. Iwanami, T. Oriyama, *Chem. Lett.* 2004, **33**, 1324–1325.

20 A. Ladépêche, E. Tam, J.-E. Ancel, L. Ghosez, *Synthesis* 2004, 1375–1380.

21 M. T. Reetz, S. Hüttenhain, P. Walz, U. Löwe, *Tetrahedron Lett.* 1979, 4971–4974.

22 (a) G. A. Molander, J. B. Etter, *J. Org. Chem.* 1986, **51**, 1778–1786; (b) G. A. Molander, J. B. Etter, *Tetrahedron Lett.* 1984, **25**, 3281–3284.

23 G. A. Molander, J. A. McKie, *J. Org. Chem.* 1991, **56**, 4112–4120.

24 (a) P. O. Miranda, D. D. Díaz, J. I. Padrón, M. A. Ramírez, V. S. Martín, *J. Org. Chem.* 2005, **70**, 57–62; (b) P. O. Miranda, D. D. Díaz, J. I. Padrón, J. Bermejo, V. S. Martín, *Org. Lett.* 2003, **5**, 1979–1982.

25 (a) T. Watahiki, T. Oriyama, *Tetrahedron Lett.* 2002, **43**, 8959–8962; (b) S. E. Denmark, E. J. Weber, T. M. Wilson, T. M. Willson, *Tetrahedron* 1989, **45**, 1053–1065.

26 M. Durandetti, C. Meignein, J. Périchon, *J. Org. Chem.* 2003, **68**, 3121–3124.

27 M. Durandetti, J. Périchon, *Tetrahedron Lett.* 2006, **47**, 6255–6258.

28 S. J. Mahmood, M. M. Hossain, *J. Org. Chem.* 1998, **63**, 3333–3336.

29 S. J. Mahmood, A. K. Saha, M. M. Hossain, *Tetrahedron* 1998, **54**, 349–358.

30 V. Lecomte, C. Bolm, *Adv. Synth. Catal.* 2005, **347**, 1666–1672.

31 J. Christoffers, A. Baro, T. Werner, *Adv. Synth. Catal.* 2004, **346**, 143–151.

32 X. Zhang, X. Fan, H. Niu, J. Wang, *Green Chem.* 2003, **5**, 267–269.

33 (a) N. Aoyama, K. Manabe, S. Kobayashi, *Chem. Lett.* 2004, **33**, 312–313; (b) S. Kobayashi, S. Nagayama, T. Busujima, *J. Am. Chem. Soc.* 1998, **120**, 8287–8288.

34 L. Colombo, F. Ulgheri, L. Prati, *Tetrahedron Lett.* 1989, **30**, 6435–6436.

35 T. Bach, D. N. A. Fox, M. T. Reetz, *J. Chem. Soc., Chem. Commun.* 1992, 1634–1636.

36 J. Jankowska, J. Paradowska, J. Mlynarski, *Tetrahedron Lett.* 2006, **47**, 5281–5284.

37 J. Jankowska, J. Paradowska, B. Rakiel, J. Mlynarski, *J. Org. Chem.* 2007, **72**, 2228–2231.

38 L. F. Tietze, M. Bratz, *Chem. Ber.* 1989, **122**, 997–1002.

39 L. F. Tietze, M. Bratz, *Liebigs Ann. Chem.* 1989, 559–564.

40 J. Wu, W. Sun, W.-Z. Wang, H.-G. Xia, *Chin. J. Chem.* 2006, **24**, 1054–1057.

41 A. T. Khan, T. Parvin, L. H. Choudary, *Tetrahedron* 2007, **63**, 5593–5601.

42 M. F. Mayer, M. M. Hossain, *J. Org. Chem.* 1998, **63**, 6839–6844.

43 J. Otera, *Esterification*, Wiley-VCH, Weinheim, 2003.

44 T. Hiyama, H. Oishi, H. Saimoto, *Tetrahedron Lett.* 1985, **26**, 2459–2462.

45 B. Ganem, V. R. Small, Jr., *J. Org. Chem.* 1974, **39**, 3728–3730.

46 (a) A. Alexakis, J. M. Duffault, *Tetrahedron Lett.* 1988, **29**, 6243–6246; (b) A. Alexakis, M. Gardette, S. Colin, *Tetrahedron Lett.* 1988, **29**, 2951–2954; (c) R. K. Haynes, D. E. Lambert, P. A. Schober, S. G. Turner, *Aust. J. Chem.* 1987, **40**, 1211–1222.

47 B. M. Eschler, R. K. Haynes, M. D. Ironside, S. Kremmydas, D. D. Ridley, T. W. Hambley, *J. Org. Chem.* 1991, **56**, 4760–4766.

48 (a) G. V. M. Sharma, A. K. Mahalingam, M. Nagarajan, A. Ilangovan, P. Radhakrishna, *Synlett* 1999, 1200–1202; (b) R. A. Holton, R. R. Juo, H. B. Kim, A. D. Williams, S. Harusawa, R. E. Lowenthal, S. Yogai, *J. Am. Chem. Soc.* 1988, **110**, 6558–6560.

49 M. P. Bosch, I. Petschen, A. Guerrero, *Synthesis* 2000, 300–304.

50 G. A. Molander, S. R. Shakya, *J. Org. Chem.* 1994, **59**, 3445–3452.

51 G. A. Molander, J. A. McKie, *J. Org. Chem.* 1993, **58**, 7216–7227.

52 J. Christoffers, in I. Horvath (ed.), *Encyclopedia of Catalysis*, Vol. 5, Wiley, New York, 2003, pp. 99–118.

53 H. Kotsuki, K. Arimura, T. Ohishi, R. Maruzasa, *J. Org. Chem.* 1999, **64**, 3770–3773.

54 (a) J. Comelles, M. Moreno-Mañas, A. Vallribera, *Arkivoc* 2005, 207–238; (b) J. Christoffers, *Eur. J. Org. Chem.* 1998, 1259–1266.

55 (a) P. Kočovský, D. Dvořák, *Collect. Czech. Chem. Commun.* 1988, **53**, 2667–2674; (b) P. Kočovský, D. Dvořák, *Tetrahedron Lett.* 1986, **27**, 5015–5018.

56 C. P. Fei, T. H. Chan, *Synthesis* 1982, 467–468.

57 P. Laszlo, M.-T. Montaufier, S. L. Randriamahefa, *Tetrahedron Lett.* 1990, **31**, 4867–4870.

58 M. Hirano, S. Kiyota, M. Imoto, S. Komiya, *Chem. Commun.* 2000, 1679–1680.

59 (a) J. Christoffers, *Synlett* 2001, 723–732; (b) J. Christoffers, *Chem. Commun.* 1997, 943–944; (c) J. Christoffers, *J. Chem. Soc., Perkin Trans. 1* 1997, 3141–3149.

60 J. Christoffers, *Org. Synth.* 2002, **78**, 249–253.

61 (a) A. Hinchcliffe, C. Hughes, D. A. Pears, M. R. Pitts, *Org. Proc. Res. Dev.* 2007, **11**, 477–481; (b) M. E. Jung, S.-J. Min, K. N. Houk, D. Ess, *J. Org. Chem.* 2004, **69**, 9085–9089; (c) J. Christoffers, H. Oertling, N. Önal, *J. Prakt. Chem.* 2000, **342**, 546–553.

62 J. Christoffers, Y. Zhang, W. Frey, P. Fischer, *Synlett* 2006, 624–626.

63 J. Christoffers, H. Scharl, *Eur. J. Org. Chem.* 2002, 1505–1508.

64 J. Christoffers, H. Oertling, W. Frey, *Eur. J. Org. Chem.* 2003, 1665–1671.

65 (a) S.-I. Murahashi, H. Takaya, *Acc. Chem. Res.* 2000, **33**, 225–233; (b) T. Naota, H. Taki, M. Mizuno, S.-I. Murahashi, *J. Am. Chem. Soc.* 1989, **111**, 5954–5955.

66 (a) A. Mekonnen, R. Carlson, *Eur. J. Org. Chem.* 2006, 2005–2013; (b) Y. Mori, K. Kakumoto, K. Manabe, S. Kobayashi,

Tetrahedron Lett. 2000, **41**, 3107–3111; (c) G. Bartoli, M. Bosco, M. C. Bellucci, E. Marcantoni, L. Sambri, E. Torregiani, *Eur. J. Org. Chem.* 1999, 617–620.

67 J. Christoffers, *Tetrahedron Lett.* 1998, **39**, 7083–7084.

68 J. Christoffers, H. Oertling, *Tetrahedron* 2000, **56**, 1339–1344.

69 J. Christoffers, H. Oertling, M. Leitner, *Synlett* 2000, 349–350.

70 E. Drent, P. H. M. Budzelaar, *Chem. Rev.* 1996, **96**, 663–681.

71 S. Pelzer, T. Kauf, C. van Wüllen, J. Christoffers, *J. Organomet. Chem.* 2003, **684**, 308–314.

72 (a) P. Gruene, C. Trage, D. Schröder, H. Schwarz, *Eur. J. Inorg. Chem.* 2006, 4546–4552; (b) C. Trage, D. Schröder, H. Schwarz, *Chem. Eur. J.* 2005, **11**, 619–627.

73 M. Bauer, T. Kauf, J. Christoffers, H. Bertagnolli, *Phys. Chem. Chem. Phys.* 2005, **7**, 2664–2670.

74 J. Christoffers, *J. Org. Chem.* 1998, **63**, 4539–4540.

75 J. Christoffers, *Eur. J. Org. Chem.* 1998, 759–761.

76 J. Christoffers, A. Mann, *Eur. J. Org. Chem.* 1999, 2511–2514.

77 J. Christoffers, A. Mann, *Eur. J. Org. Chem.* 2000, 1977–1982.

78 (a) T. Mukaiyama, T. Takeda, K. Fujimoto, *Bull. Chem. Soc. Jpn.* 1978, **51**, 3368–3372; (b) T. Mukaiyama, T. Takeda, M. Osaki, *Chem. Lett.* 1977, 1165–1168.

79 J. Christoffers, A. Mann, *Angew. Chem.* 2000, **112**, 2871–2874; *Angew. Chem. Int. Ed.* 2000, **39**, 2752–2754.

80 A. Baro, J. Christoffers, in J. Christoffers, A. Baro (eds.), *Quaternary Stereocenters: Challenges and Solutions for Organic Synthesi*, Wiley-VCH, Weinheim, 2005, pp. 83–115.

81 J. Christoffers, *Chem. Eur. J.* 2003, **9**, 4862–4867.

82 S. T. Handy, *Curr. Org. Chem.* 2000, **4**, 363–395.

83 J. Christoffers, A. Baro, *Angew. Chem.* 2003, **115**, 1726–1728; *Angew. Chem. Int. Ed.* 2003, **42**, 1688–1690.

84 J. Christoffers, *J. Prakt. Chem.* 1999, **341**, 495–498.

85 J. Christoffers, A. Mann, *Eur. J. Org. Chem.* 1999, 1475–1479.

86 M. P. Sibi, G. Petrovic, *Tetrahedron: Asymmetry* 2003, **14**, 2879–2882.

87 M. M. Dell'Anna, V. Gallo, P. Mastrorilli, C. F. Nobile, G. Ramanazzi, G. P. Suranna, *Chem. Commun.* 2002, 434–435.

88 V. Gallo, P. Mastrorilli, C. F. Nobile, G. Ramanazzi, G. P. Suranna, *J. Chem. Soc., Dalton Trans.* 2002, 4339–4342.

89 H. Uehara, S. Nomura, S. Hayase, M. Kawatsura, T. Itoh, *Electrochemistry* 2006, **74**, 635–638.

90 S.-J. Ji, M.-F. Zhou, D.-G. Gu, Z.-Q. Jiang, T.-P. Loh, *Eur. J. Org. Chem.* 2004, 1584–1587.

91 T. Itoh, H. Uehara, K. Ogiso, S. Nomura, S. Hayase, M. Kawatsura, *Chem. Lett.* 2007, **36**, 50–51.

92 (a) K.-i. Shimizu, M. Miyagi, T. Kan-no, T. Hatamachi, T. Kodama, Y. Kitayama, *J. Catal.* 2005, **229**, 470–479; (b) K.-i. Shimizu, M. Miyagi, T. Kan-no, T. Kodama, Y. Kitayama, *Tetrahedron Lett.* 2003, **44**, 7421–7424.

93 D. A. Durkee, H. B. Eitouni, E. D. Gomez, M. W. Ellsworth, A. T. Bell, N. P. Balsara, *Adv. Mater.* 2005, **17**, 2003–2006.

94 J. Cabral, P. Laszlo, L. Mahẽ, M.-T. Montaufier, S. L. Randriamahefa, *Tetrahedron Lett.* 1989, **30**, 3969–3972.

95 M. Pérez, R. Pleixats, *Tetrahedron* 1995, **51**, 8355–8362.

96 L.-W. Xu, C.-G. Xia, *Synthesis* 2004, 2191–2195.

97 L.-W. Xu, C.-G. Xia, X.-X. Hu, *Chem. Commun.* 2003, 2570–2571.

98 L.-W. Xu, L. Li, C.-G. Xia, *Helv. Chim. Acta* 2004, **87**, 1522–1526.

99 A. M. Chibiryaev, N. De Kimpe, A. V. Tkachev, *Tetrahedron Lett.* 2000, **41**, 8011–8013.

100 S. Wang, T. Tan, J. Li, H. Hu, *Synlett* 2005, 2658–2660.

101 J. Wang, X. Fan, X. Zhang, L. Han, *Can. J. Chem.* 2004, **82**, 1192–1196.

102 C. O. Kappe, *Tetrahedron* 1993, **49**, 6937–6963.

103 (a) L.-W. Xu, Z.-T. Wang, C.-G. Xia, L. Li, P.-Q. Zhao, *Helv. Chim. Acta* 2004, **87**, 2608–2612; (b) S. Martínez, M. Meseguer, L. Casas, E. Rodríguez, E. Molins, M. Moreno-Mañas, A. Roig, R. M. Sebastián, A. Vallribera, *Tetrahedron* 2003, **59**, 1553–1556; (c) J. Lu, Y. Bai, *Synthesis* 2002, 466–470; (d) J. Lu, H. Ma, *Synlett* 2000, 63–64.

104 Z.-T. Wang, L.-W. Xu, C.-G. Xia, H.-Q. Wang, *Tetrahedron Lett.* 2004, **45**, 7951–7953.

105 T. Itoh, K. Maeda, T. Wada, K. Tomimoto, T. Mase, *Tetrahedron Lett.* 2002, **43**, 9287–9290.

106 J. S. Yadav, B. V. S. Reddy, V. Geetha, *Synth. Commun.* 2002, **32**, 763–769.

107 X. S. Fan, Y. Z. Li, X. Y. Zhang, X. Y. Hu, J. J. Wang, *Chin. Chem. Lett.* 2005, **16**, 897–899.

9
Iron-catalyzed Cycloadditions and Ring Expansion Reactions

Gerhard Hilt and Judith Janikowski

9.1
Introduction

This chapter will focus on preparative applications of iron catalysts in the synthesis of cyclic ring systems by means of intra- and intermolecular cycloadditions, Alder–ene reactions and ring expansion reactions. Previous reviews concerning iron-catalyzed chemistry have either focused their attention on different aspects involving iron-containing compounds as active catalysts [1] or have concentrated on certain reactions [2] and on the synthesis of specific substance classes [3]. A more recent general review concerning all aspects of modern applications in iron-catalyzed reactions has been summarized by Bolm *et al.* [4].

9.2
Cycloisomerization and Alder–Ene Reaction

The Alder–ene reaction is an atom-economic reaction which forms a new carbon carbon–bond from two double bond systems (alkenes, carbonyl groups, etc.) with double bond migration [5]. This reaction follows the Woodward–Hoffmann rules if the reaction is performed under thermal conditions. However, when transition metal catalysts are involved, thermally forbidden Alder–ene reactions can also be realized (Scheme 9.1). Examples of such processes are the formal [4 + 4]-Alder–ene reaction catalyzed by low-valent iron catalysts.

Intramolecular examples of iron-catalyzed formal Alder–ene reactions, which are also denoted cycloisomerization reactions, were described in the late 1980s by the groups of Tietze and Takacs in reactions directed towards cyclopentane [6, 7], cyclohexane [8], piperidine [9] and tetrahydropyran derivatives [10].

The reaction described by Tietze [8] is of a more traditional approach utilizing $FeCl_3$ as a Lewis acid catalyst for the activation of a carbonyl group in a [4 + 2]-Alder–ene reaction (Scheme 9.2). The cyclization of **1** to **2** proceeded smoothly at

Iron Catalysis in Organic Chemistry. Edited by Bernd Plietker

ISBN: 978-3-527-31927-5

Scheme 9.1 Types of Alder–ene reactions.

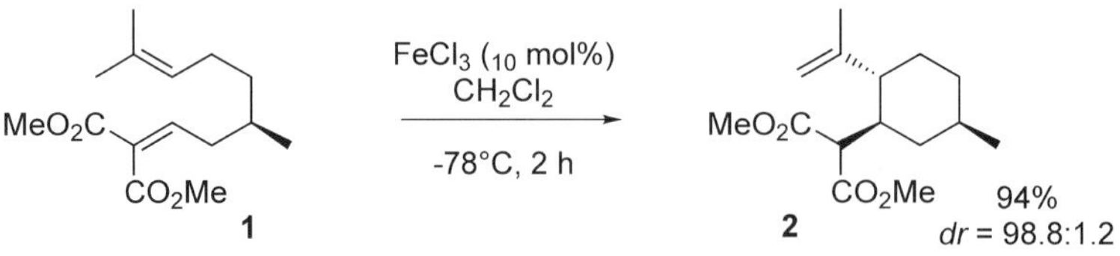

Scheme 9.2 Intramolecular Alder–ene reaction.

low temperatures with high diastereoselectivity regarding the relative stereochemistry of the methyl and the malonate groups. For the diastereoselectivity of the malonate and the isopropenyl groups, an exclusive *trans* configuration is observed.

On the other hand, Takacs and coworkers added organometallic reducing agents to the reaction mixture and promoted the formation of low-valent iron(0) bipyridine complexes. The mechanism of the low-valent iron-catalyzed Alder–ene reaction involves coordination of the two starting materials within the ligand sphere of the iron, which makes the Woodward–Hoffmann rules for such reactions obsolete [11]. Thereby, the scope of the reactions was broadened so that alkenes and 1,3-dienes could also be used as educts in a formal [4 + 4]-cycloisomerization (Scheme 9.3) [12]. Intriguingly, the diastereoselectivity of the cyclopentane products can be influenced by either the application of the 2*Z*-isomer **3** or the 2*E*-isomer **4**. Especially the *E*-isomers **4** gave almost exclusive *cis* selectivity [13].

However, when triene esters were applied in the iron-catalyzed [4 + 4]-Alder–ene reaction (Scheme 9.4), the 2*E*/2*Z* geometry of the conjugated double bond was irrelevant for the *cis*/*trans* ratio of the product, indicating an iron-mediated isomerization over the course of the reaction [14].

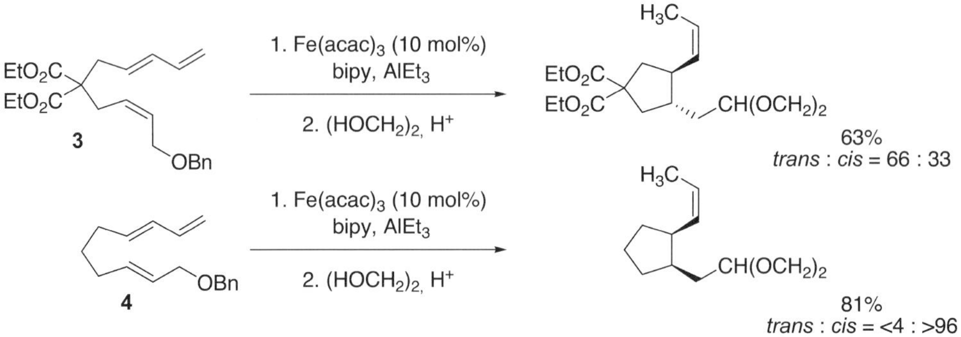

Scheme 9.3 Alder–ene reaction with low-valent iron catalysts.

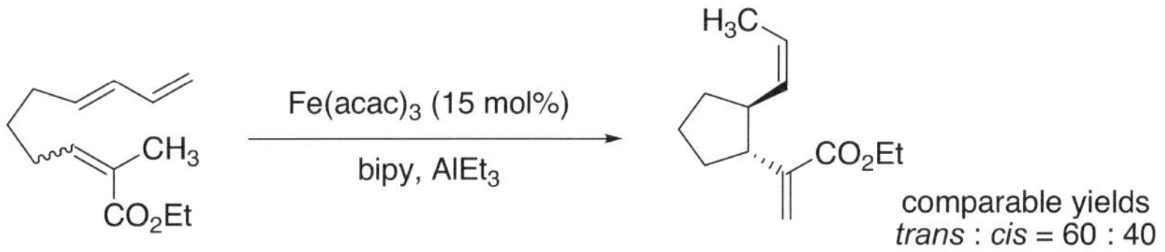

Fe(acac)$_3$ (15 mol%)
bipy, $AlEt_3$
comparable yields
trans : *cis* = 60 : 40

Scheme 9.4 Diastereoselectivity of the Alder–ene reaction.

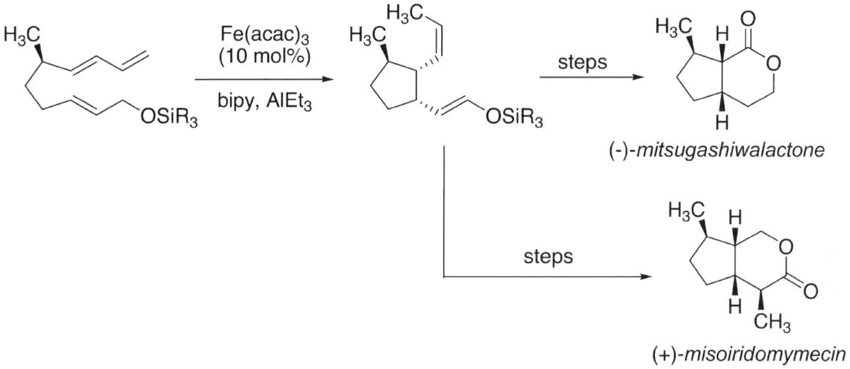

Fe(acac)$_3$ (10 mol%)
bipy, $AlEt_3$
steps
(-)-*mitsugashiwalactone*
steps
(+)-*misoiridomymecin*

Scheme 9.5 The Alder–ene reaction as a key step in synthesis.

The application of this chemistry to the synthesis of terpene natural products such as (−)-mitsugashiwalactone and (+)-isoiridomyrmecin [15] (Scheme 9.5) involved low-valent iron-catalyzed Alder–ene reactions in key steps in these synthesis.

In the same manner, the natural products (−)-protometinol analogue [16] (Scheme 9.6) and (−)-gibboside [17] were synthesized utilizing an intramolecular Alder–ene reaction in the key step. However, the synthesis of the cyclization

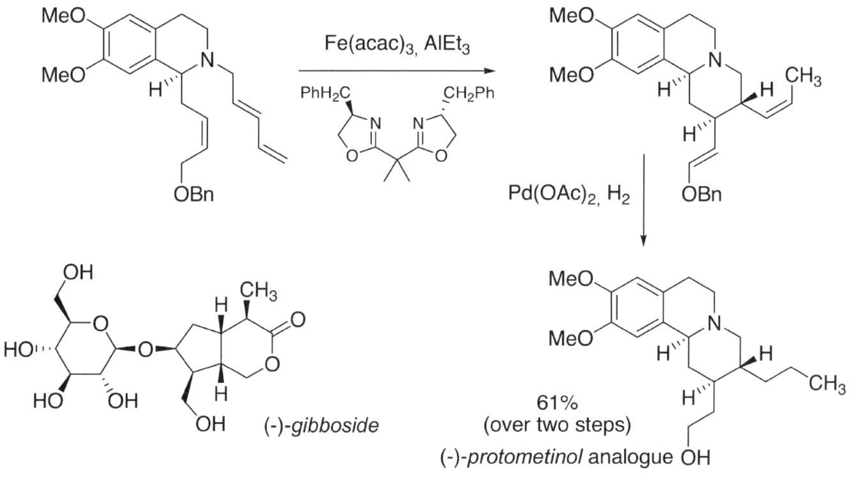

Fe(acac)$_3$, $AlEt_3$
Pd(OAc)$_2$, H_2
(-)-*gibboside*
61%
(over two steps)
(-)-*protometinol* analogue

Scheme 9.6 Synthesis of natural products and analogues.

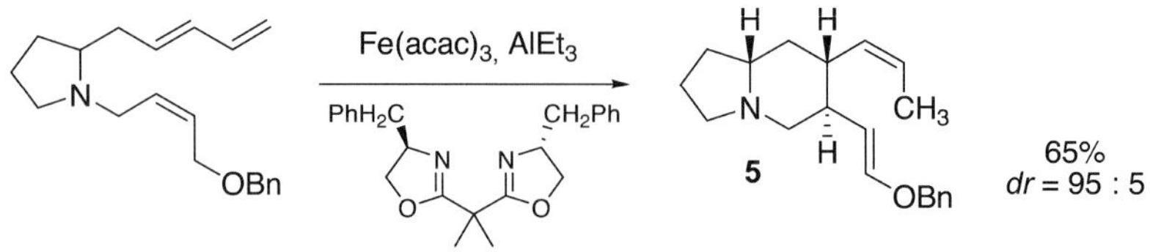

$Fe(acac)_3$, $AlEt_3$

PhH_2C CH_2Ph

OBn

5

65%
dr = 95 : 5

Scheme 9.7 Iron-catalyzed approach to heterocyclic compounds.

precursors was accomplished following straightforward but essentially traditional linear reaction sequences.

Heterocyclic products such as indolizidines and quinolizidine (Scheme 9.7) could also be obtained, as was shown by Takacs *et al.* for the synthesis of **5** in a stereoselective fashion applying a chiral bisoxazoline ligand [18]. The product was obtained in 65% yield with excellent diastereoselectivity and the enol ether was formed exclusively as the *E*-isomer.

The simple diastereoselectivity of carbocyclic and heterocyclic derivatives has been thoroughly investigated and discussed [19]. The cyclization is dependent on the nature of the heteroatoms in the ring, the pre-existing ring size and the nature of the ligand. As shown in Scheme 9.7, bisoxazoline ligands are superior with respect to efficiency and stereoselectivities to the bipyridine ligands.

The intramolecular iron-catalyzed Alder–ene reaction of enynes in the carbocyclization reaction was recently reported by Fürstner *et al.* (Scheme 9.8) [20]. A low-valent cyclopentadienyliron catalyst, specifically the $[CpFe(C_2H_4)_2][Li(tmeda)]$ complex, is a reactive catalyst for enyne cycloisomerization reactions. Bicyclic products, also incorporating large ring systems, are thereby accessible, and the Thorpe–Ingold effect seems to be helpful for these types of reactions.

An interesting application was recently reported by Pearson and Wang [21] applying an iron–diene complex in a [6 + 2]-Alder–ene reaction (Scheme 9.9). In this example, the iron acts not only as a coordinating and activating group for the

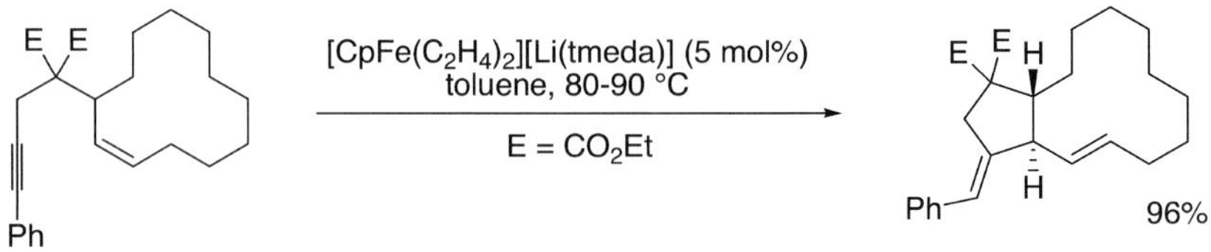

$[CpFe(C_2H_4)_2][Li(tmeda)]$ (5 mol%)
toluene, 80-90 °C

E = CO_2Et

96%

Scheme 9.8 Iron-catalyzed enyne cycloisomerization.

Bu_2O, CO
142 °C
75%

6 + 7

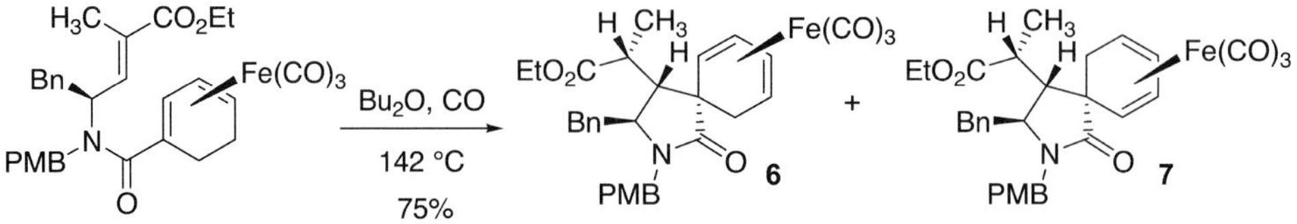

Scheme 9.9 Synthesis of spiro compounds by Alder–ene reaction.

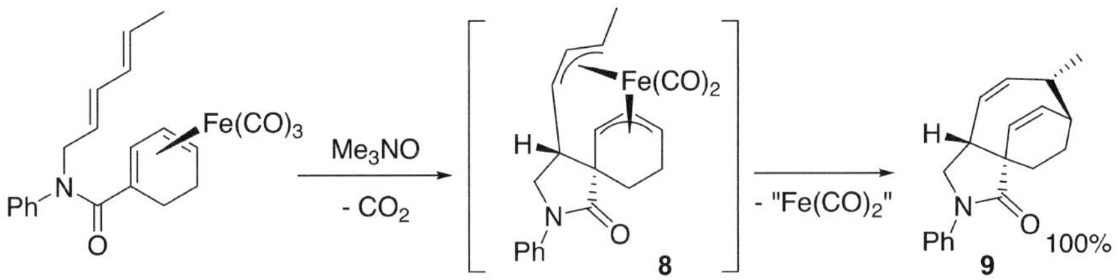

Scheme 9.10 Formal iron-catalyzed [4 + 4]-cycloaddition to a polycyclic product.

1,3-diene subunit but also as an intramolecular promoter enabling the Alder–ene reaction to proceed.

The two products **6** and **7** are formed in a 1:1 mixture; however, the stereochemistry of the side-chain is controlled effectively, so that after decomplexation and further manipulations both products can be applied in the synthesis of deoxycytochalasin. When pendant dienes were used instead of the allyl amide side-chain after the initial step of the formal [6 + 2]-Alder–ene reaction (Scheme 9.10), the intermediate **8** undergoes another reductive cyclization to the final polycyclic product **9** in quantitative yield as a single diastereomer, as reported by Pearson and Wang [22]. The overall process can be seen as a formal [4 + 4]-cycloaddition reaction of the cyclohexadiene with the pendant diene.

9.3
[2 + 1]-Cycloadditions

A carbene or nitrene transfer reaction to a carbon–carbon or carbon–heteroatom double bond system leads to the formation of three-membered rings, such as a cyclopropane, an aziridine or an epoxide. These processes can be catalyzed by applying iron catalysts and the different cyclic systems are discussed here.

9.3.1
Iron-catalyzed Aziridine Formation

The formation of aziridines can be realized by a nitrene transfer reaction to a carbon–carbon double bond catalyzed by an iron(III)–porphyrin complex, as reported by Zhang's group, applying bromamine-T as a suitable nitrene source [23]. The aziridination was successful for aromatic, aliphatic and cyclic alkenes (Scheme 9.11). Accordingly, the reaction is not limited to terminal alkenes and internal alkenes are also acceptable substrates. When internal alkenes were used, the *E*-configured alkenes gave better results than the *Z*-configured alkenes, while the stereochemistry of the starting material was not retained in the products. The *Z*-isomer suffers considerable isomerization to the *trans* product during the course of the nitrene transfer. Isomerization to the *cis* product was also observed for the *E*-isomer.

Chiral modifications of the catalyst system are possible, but the synthesis is tedious and the chiral inductions of such complexes have not prevailed so far.

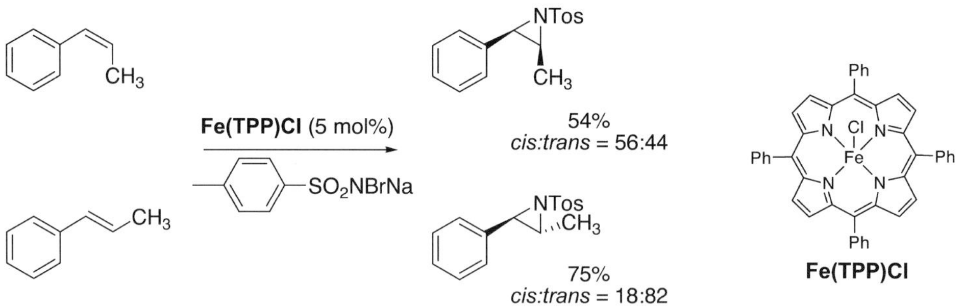

Scheme 9.11 Diastereoselectivity in styrene aziridinations.

The transfer of a nitrene from PhI=NTos as a commercial source can also be catalyzed by a bimetallic iron complex which exhibits moderate activity, as reported by Avenier and Latour (Scheme 9.12) [24]. The iron complex **10** is better suited for aliphatic substrates where better yields were reported compared with copper catalyst systems applied to the same reactions.

The nitrene transfer from PhI=NTos to alkenes catalyzed by the $CpFe(CO)_2^+$ fragment gave better results (85% for styrene) [25], but the characteristics of the chemistry of the cationic intermediates as postulated by the reaction mechanism are closely connected to the alternative formation of aziridines by a carbene transfer to imines.

The aziridination of imines catalyzed by the $CpFe(CO)_2^+$ fragment by carbene transfer is a most remarkable reaction [26]. The imine is consumed first to form both the *cis* and *trans* products. The catalyst then coordinates predominantly to the *trans* product and leads to its decomposition, leaving the *cis* product untouched [27].

This diastereoselective decomposition can be rationalized by the steric hindrance in the *cis* products as the coordination of the $CpFe(CO)_2^+$ fragment is prohibited by the relative orientation of the substituents (Scheme 9.13). Consequently, the rate of decomposition is enhanced when the substituents stabilize cationic ring-opened intermediates, such as in the methoxy derivative **11** (0% yield). On the other hand, the rate of decomposition of the *trans* products is reduced, based on the destabilization of cationic intermediates when an electron-withdrawing substituent, such as in the nitro derivative **13**, is used (78%, *cis* : *trans* = 4 : 1).

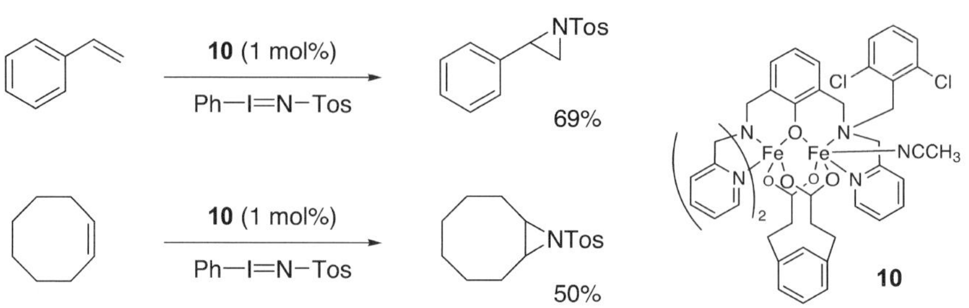

Scheme 9.12 Aziridination by a bimetallic complex.

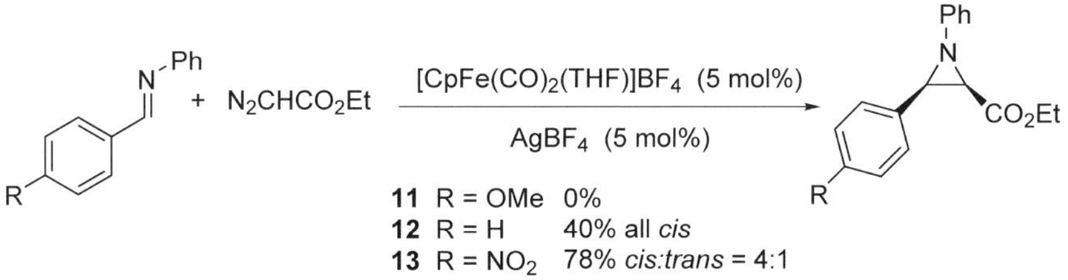

+ N_2CHCO_2Et → $[CpFe(CO)_2(THF)]BF_4$ (5 mol%), $AgBF_4$ (5 mol%)

11 R = OMe 0%
12 R = H 40% all *cis*
13 R = NO_2 78% *cis:trans* = 4:1

Scheme 9.13 Iron-catalyzed carbene transfer reactions.

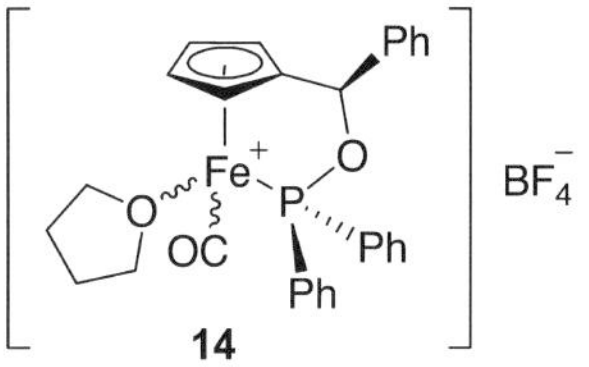

14 BF_4^-

Figure 9.1 Chiral modified iron catalyst.

A rather low level of asymmetric induction (5% *ee*) was reported for the generation of the product **12** (R = H in Scheme 9.13) in a comparable yield of 42% by Mayer and Hossain, applying the chiral modified CpFe complex **14** (Figure 9.1) [28].

The iron-catalyzed process can also be performed with iron–PYBOX systems, as reported by Redlich and Hossain [29]. The PYBOX ligands are powerful tools in organic synthesis. However, for the iron-catalyzed synthesis of aziridines (Scheme 9.14), the results are not as convincing as for similar copper–PYBOX systems. Not only are the yields moderate (up to 54%), but also the enantiomeric excesses (up to 49%) are not in a synthetically useful range, which precludes their use in elaborate applications.

9.3.2
Iron-catalyzed Epoxide Formation

The formation of epoxides is a well-investigated synthetic problem and two approaches, either from a double bond system by transfer of oxygen starting from an alkene, or carbene transfer to a carbonyl group, have attracted much interest. The use of the $CpFe(CO)_2^+$ fragment was also investigated by Hossain and coworkers with a view to its use for the synthesis of epoxides (Scheme 9.15) [30, 31]. However, the $CpFe(CO)_2^+$ fragment not only catalyzes the carbene transfer, but also acts as

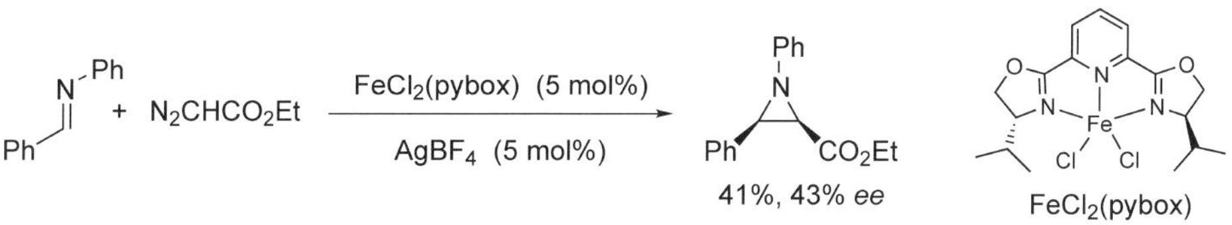

+ N_2CHCO_2Et → $FeCl_2$(pybox) (5 mol%), $AgBF_4$ (5 mol%)

41%, 43% *ee*

$FeCl_2$(pybox)

Scheme 9.14 Asymmetric iron-catalyzed aziridine synthesis.

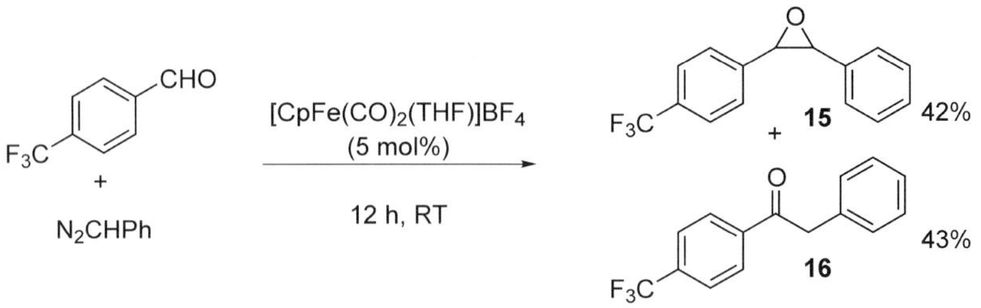

Scheme 9.15 Iron-catalyzed epoxidation and ring-opening reaction.

a Lewis-acid, promoting the rearrangement of the epoxide to a ketone. Therefore, the total conversion of the initial carbonyl starting material is fairly good. The yields of the individual products, the desired epoxide **15** and the ketone **16**, are moderate. However, longer reaction times lead to the rearranged ketone **16** in up to 95% yield [32].

9.3.3
Iron-catalyzed Cyclopropane Formation

The cyclopropanation of alkenes can be achieved by carbene transfer, usually from a diazo compound to an alkene, and can be catalyzed by iron complexes generally bearing different nitrogen- and phosphorus-containing ligands. An early report by Roger and Lapinte proposed an electron transfer-induced carbene transfer reaction by preformed cationic $Cp^*(CO)_2Fe{=}CH_2$ or $Cp^*(dppe)Fe{=}CH_2$ complexes [33]. Although this approach is very interesting from a mechanistic point of view, the regeneration of the methylene subunit in **18** could not be realized from a simple carbene precursor, so that stoichiometric amounts of the reagent $Cp^*(CO)_2Fe{=}CH_2{}^+$ had to be used for the regioselective cyclopropanation of alkenes and monoprotected dienes such as the isoprene complex **17** in Scheme 9.16 [34].

In a long-term research project, Hossain and coworkers investigated the usefulness of the $CpFe(CO)_2{}^+$ fragment [35–38] in the cyclopropanation reaction of alkenes by a carbene transfer utilizing diazo esters as the carbene source (Scheme 9.17). The cyclopropanation products of styrene derivatives could be obtained in good yields of up to 80% and excellent *cis* selectivity by using an excess of the alkene, whereas the cyclopropanation of aliphatic alkenes was less effective, yielding the desired cyclopropane derivative in up to 51% yield.

Nevertheless, Hossain and coworkers also investigated the carbene transfer reaction using a chiral bimetallic iron–carbene complex, which also exhibits excellent

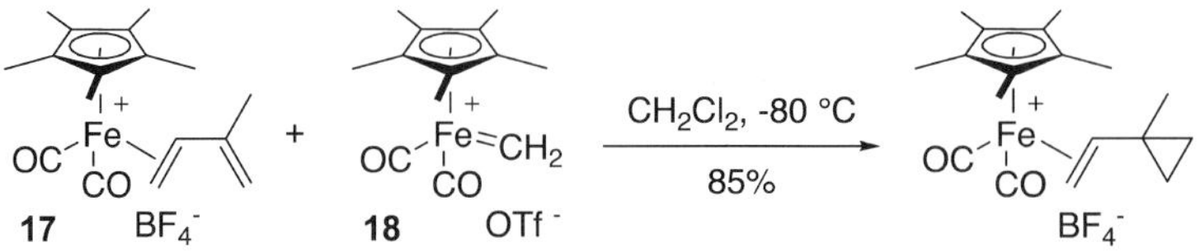

Scheme 9.16 Cyclopropanation by a carbene transfer from a $CpFe(CO)_2$ fragment.

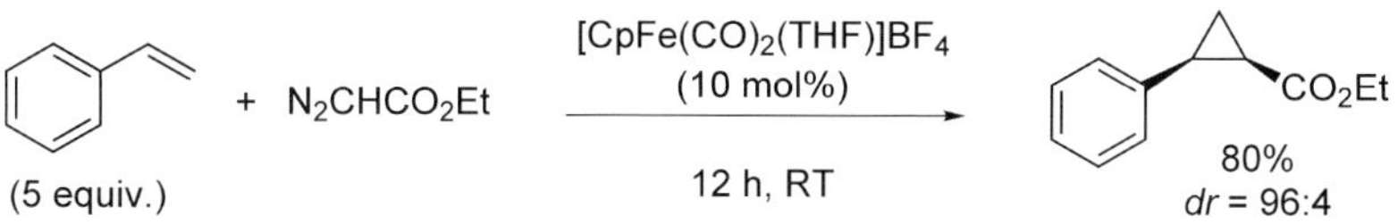

Scheme 9.17 Diastereoselective iron-catalyzed carbene transfer reaction.

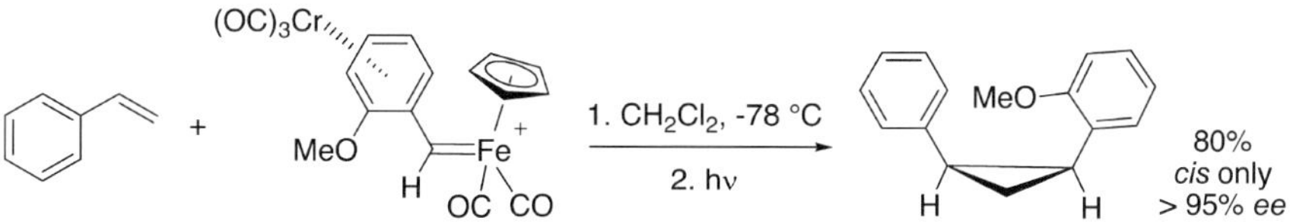

Scheme 9.18 Diastereo- and enantioselective cyclopropanation.

cis selectivities and excellent stereoselectivity (after photochemical decomplexation) when the carbene ligand contained the stereo-information in the form of a planar chiral chromium tricarbonyl compound (Scheme 9.18) [39].

Another class of iron complexes capable of carbene transfer to alkenes which utilize the planar ligand motif that is found in porphyrin **19** [40–42] and also in the salen-type ligands **22** [43, 44] (Figure 9.2). An unusual non-planar motif was realized by Braunstein *et al.* [45] in the bimetallic complex **21**, where the copper ion bridges the two oxazoline ligands in addition to functioning as a ligand by forming a metal–metal bond.

The application of such complexes in the cyclopropanation of alkenes was investigated, mostly in benchmark reactions between styrene and diazo esters or aromatic diazomethane derivatives (Scheme 9.19) to form the *cis*- and *trans*-cyclopropene derivatives **23** and **24**.

The porphyrin ligands **19** ($R = 4\text{-MeC}_6H_4$) gave good yields of cyclopropanation products **25**/**26** (79%; *cis* : *trans* = 1 : 14) whereas the amount of stilbenes formed by

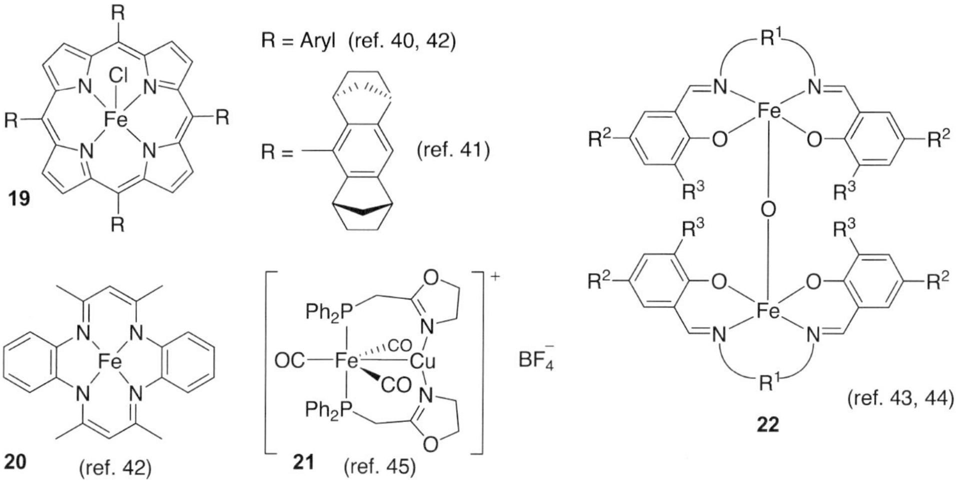

Figure 9.2 Catalysts for the cyclopropanation of alkenes.

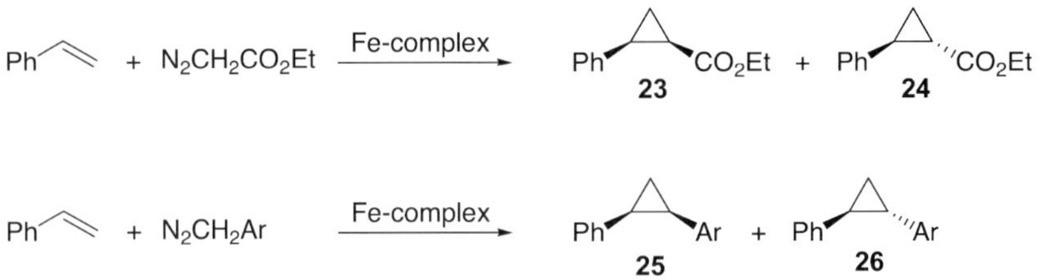

Scheme 9.19 Benchmark reactions in carbene transfer reactions to alkenes.

dimerization of the diazo compound was relatively low (21%). On the other hand, complex **20** led predominantly to dimerization of the diazo derivative (84%) and only a rather small yield of the cyclopropane (16%; *cis* : *trans* = 1 : 1.9) was obtained. The chiral porphyrin derivatives **19** gave the cyclopropyl ester in up to 71% yield, good *cis*: *trans* ratios (up to 1 : 23) and acceptable to good enantioselectivties for the *trans* product (up to 81%), whereas the enantioselectivities for the *cis* product were poor (<6%). The iron complexes containing the Jacobsen ligand **22** ($R^1 = CH_2CH_2$, R^2, $R^3 = t$Bu) gave the cyclopropane ester **23**/**24** in up to 84% yield and a moderate *cis* : *trans* ratio (around 1 : 2), whereas with sterically more hindered diazo esters *cis*:*trans* ratios of up to 1 : 4 could be reached. The bimetallic complex **21** formed cyclopropyl ester **23**/**24** in excellent yield (91%); however, the *cis* : *trans* ratio was only moderate (around 1 : 2).

9.4
[2 + 2]-Cycloaddition

The generation of four-membered ring systems can be accomplished by a cycloaddition process under photochemical conditions or with special substrates under thermal conditions. Iron–vinylidene complexes belong to such a class of special substrates where a thermal [2 + 2]-cycloaddition is possible. If imines are used, a hetero-[2 + 2]-cycloaddition with an iron–vinylidene complex leads to an iron–carbene complex attached to an azetidine ring system, as reported by Barrett and coworkers (Scheme 9.20) [46, 47]. The oxidation of these iron–carbene complexes leads to β-lactams **27**. Interestingly, the application of 2-thiazolines generates penam

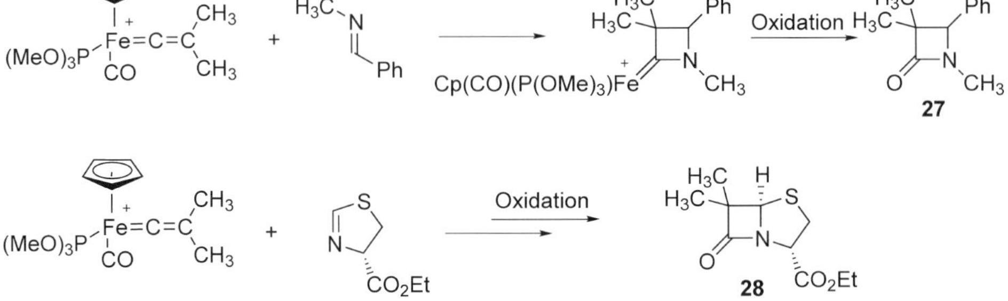

Scheme 9.20 Iron-mediated synthesis of azetidine derivatives.

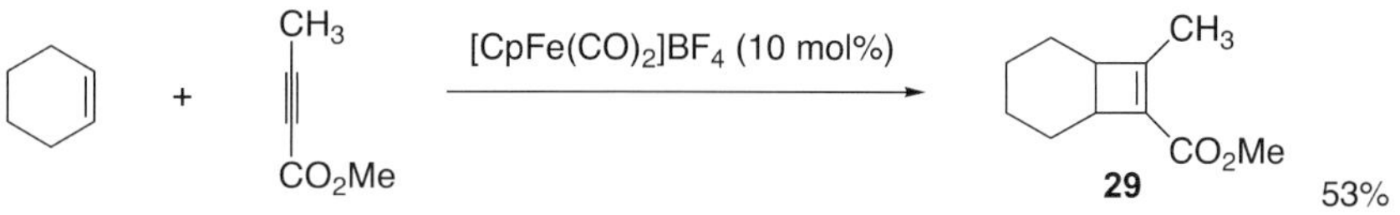

Scheme 9.21 Iron-catalyzed [2 + 2]-cycloaddition to cyclobutenes.

derivatives as products. However, the isolation of the intermediates and products of type **28** is described as being problematic so that reproducible yields were not reported.

The intermolecular [2 + 2]-cycloaddition of alkenes and alkynes utilizing an iron complex as a catalyst was reported by Rosenblum and Scheck [48]. The application of the $[CpFe(CO)_2]BF_4$ complex (Scheme 9.21) gave the desired cyclobutene derivatives **29** in up to 53% yield.

However, judging from the observation of open-chain side-products, the reaction is not concerted and the carbon monoxide ligands can also lead to insertion reactions so that iron–acyl complexes are also generated as products (in up to 66% yield).

The reaction of iron–carbonyl complexes with alkynes led to cyclobutenediones, which is formally a [2 + 1 + 1]-cycloaddition process for the formation of a cyclobutene derivative (Scheme 9.22) [49]. Nevertheless, in this reaction the liberation of the ligand is initiated by addition of stoichiometric amounts of copper(II) salts and the use of various alkynes leads to interesting products such as **30** in good yields.

Recently, Chirik's group reported an iron-catalyzed [2 + 2]-cycloaddition process with α,ω-dienes (Scheme 9.23) [50]. The tridentate pyridine–diimine complex **31** gave excellent conversions with a short reaction time (TOF > 240 h^{-1}) and a broad substrate scope is accepted by the catalyst. Esters, amides, amines and even 1,6-heptadiene can be used as substrates without requiring the Thorpe–Ingold effect.

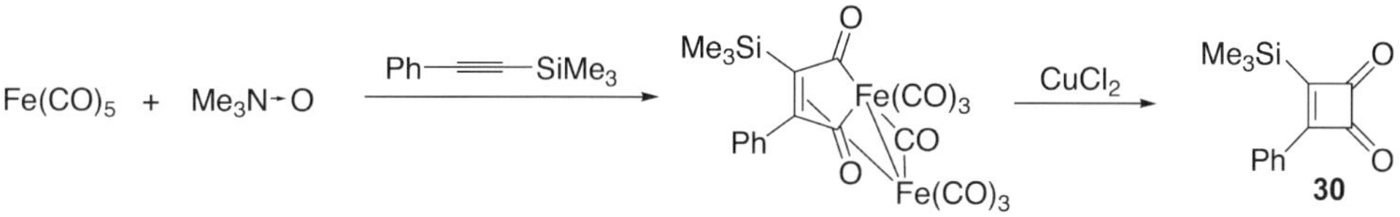

Scheme 9.22 Iron-mediated synthesis of cyclobutenediones.

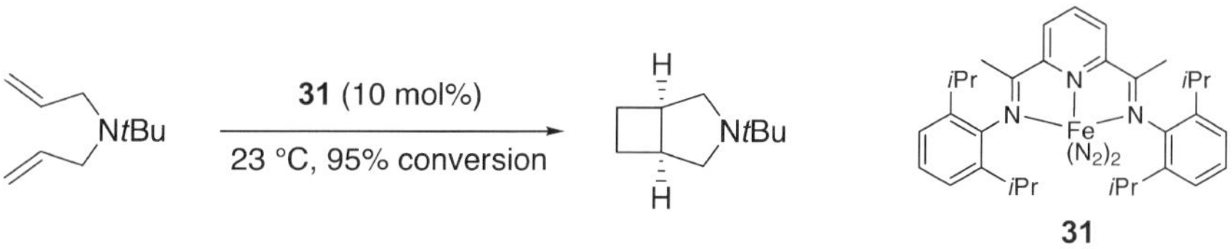

Scheme 9.23 Intramolecular iron-catalyzed [2 + 2]-cycloaddition of α,ω-dienes.

9.5
[4 + 1]-Cycloadditions

The synthesis of five-membered ring systems can be achieved by a formal [4 + 1]- or a [2 + 2 + 1]-cycloaddition process essentially depending on the point of view with which one chooses to consider the reaction. Here we discuss the progress which has been made with iron complexes as catalysts for such transformations.

The work reported by Eaton and coworkers can be summarized as reactions where an allene system in conjugation with a further unsaturated functionality reacts with carbon monoxide in the presence of an iron–carbonyl complex such as $Fe(CO)_5$ under photochemical and thermal conditions when $Fe_2(CO)_9$ is used. When diallenes are used ($X = R_2C{=}C$, Scheme 9.24), five-membered carbocyclic products are obtained [51, 52], whereas when allenyl ketones (X = O) are applied, five-membered lactones are generated [53, 54]. The use of allenylimines (X = NR) leads to five-membered lactams under these conditions [55].

The reaction with unsymmetrical allene subunits result in the predominant formation of the *E*-configured exocyclic double bond, typically with a ratio of 80 : 20 for the lactones and 90 : 10 for the lactams. An identical trend is observed for the diallene system, where both exocyclic double bonds exhibit predominantly the *E*-configuration in the product.

A similar type of transformation is observed in a three-component reaction when ketimines, alkenes and carbon monoxide are reacted in the presence of $Fe_2(CO)_9$, as reported by Anders' group. In these cases the cyclization process leads to unusual polycyclic spiro-heterocycles such as **32** shown in Scheme 9.25 [56, 57].

The reaction seems to be limited to this very special functional group of the ketimines whereas the terminal alkene component can be varied somewhat. The scope of the reaction is rather limited based on the number of isomers generated from symmetrical (two isomers) and unsymmetrical alkenes (up to six isomers) and, as a consequence, the usefulness of this reaction for the synthesis of more complex molecules is uncertain at the moment. A theoretical investigation has been undertaken to elucidate the reaction pathway and the role of the ketimine subunit. It could be shown that the ketimine functionality is important as it acts as a bidentate ligand

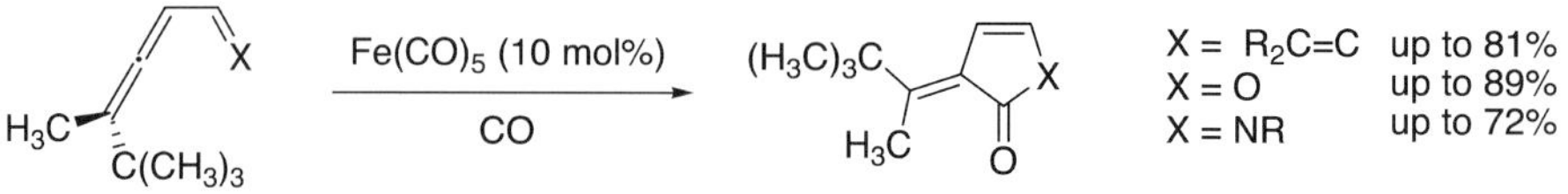

H_3C, $C(CH_3)_3$, X; $Fe(CO)_5$ (10 mol%), CO; $(H_3C)_3C$, H_3C, X, O

$X = R_2C{=}C$ up to 81%
X = O up to 89%
X = NR up to 72%

Scheme 9.24 Synthesis of five-membered carbo- and heterocyclic compounds.

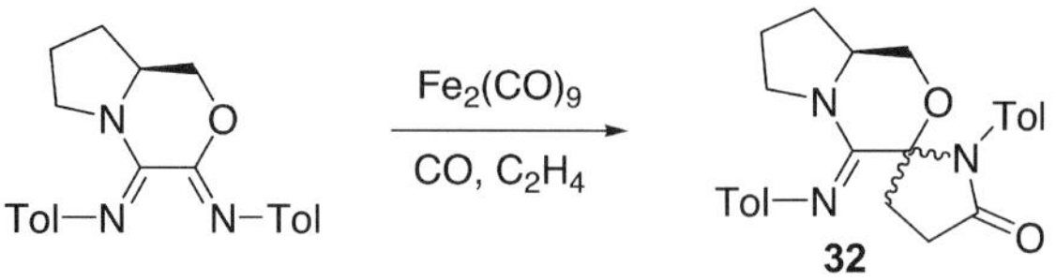

Tol–N, N–Tol, N, O; $Fe_2(CO)_9$, CO, C_2H_4; Tol–N, N, O, Tol, O, **32**

Scheme 9.25 Synthesis of complex spiro-heterocycles.

for the initial coordination of the low-valent iron complex and thereby the limitation with respect to this special functional group was rationalized [58].

9.6 [4 + 2]-Cycloadditions

The generation of six-membered ring systems by means of cycloaddition reactions can be divided into two main approaches. The first is the cyclotrimerization of alkynes utilizing low-valent iron catalyst systems, whereas the second approach is the Diels–Alder (DA) reaction of a diene and a dienophile. The latter reaction can itself be divided into three subclasses: DA reactions with normal, neutral and inverse electron demand are known. The electronic structure of the educts dictates the oxidation state of the catalyst system required to perform the diverse classes of DA reactions. Nevertheless, for each subclass examples can be found.

9.6.1 Diels–Alder Reactions with Normal Electron Demand

The rate of the DA reaction with normal electron demand can generally be accelerated by Lewis acids, which coordinate to the dienophile and lower the HOMO energy level of the reactant. The achiral versions of these DA reactions have been investigated by applying the $CpFe(CO)_2^+$ fragment in homogeneous [59, 60] and in polymer-modified catalyst systems [61]. Ferrocenium hexafluorophosphate was also reported to catalyze the DA reaction with normal electron demand; however, the extended reaction times (up to 48 h) at ambient or lower temperatures (0–20 °C) indicate that the Lewis-acidity of Cp_2Fe^+ is only moderate at best [62]. The simple iron trichloride as a stronger Lewis-acid can promote the DA reaction of α,β-unsaturated acetals. This observation can be rationalized by the Lewis acid-initiated acetal ring opening generating an intermediate allylic cation **33**, which is a good dienophile (Scheme 9.26). However, the yields are only moderate to good and the scope of this reaction is only marginally broader than that of the corresponding carbonyl compounds [63].

On the other hand, many research groups have focused their attention on the application of chiral iron-based Lewis acids for the generation of chiral products. The stereodifferentiation of these Lewis acid-based catalysts is in many cases moderate or negligible at ambient temperatures. To obtain good selectivity, only such reactants that still exhibit good reactivity at low temperatures can be employed. Typical

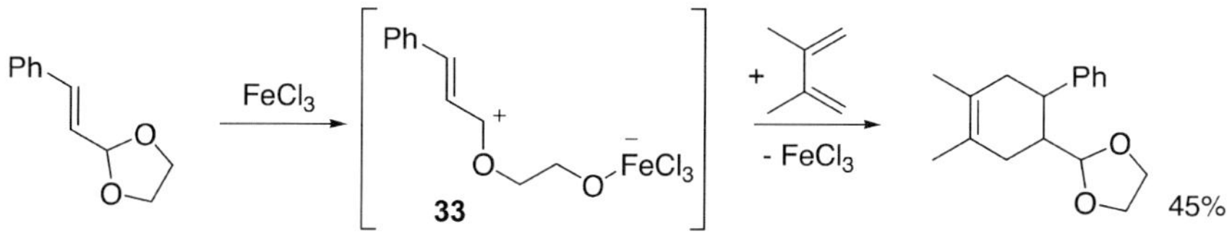

Ph O O FeCl3 [Ph + O O FeCl3 33] + - FeCl3 Ph O O 45%

Scheme 9.26 Iron-catalyzed DA reaction with acetals.

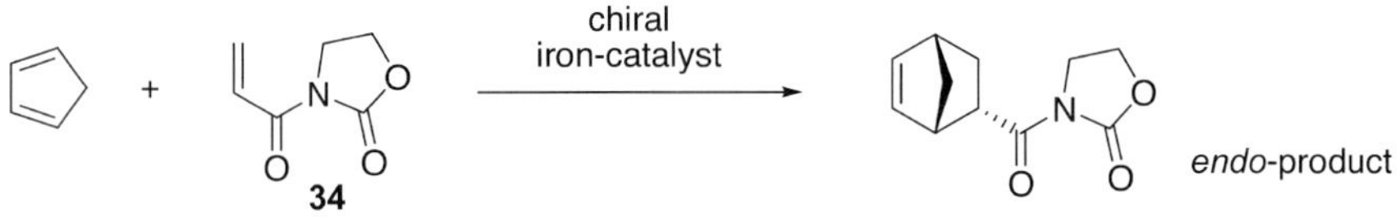

chiral iron-catalyst

34

endo-product

Scheme 9.27 Benchmark transformation for DA reaction with normal electron demand.

substrates which can be used are (a) cyclopentadiene or isoprene as 1,3-dienes and (b) acrolein, acrylates, acrylamides and benzoquinone derivatives as dienophiles. These general restrictions can be observed for almost all chiral Lewis-based catalyst systems developed so far and a solution is still not in sight.

The iron catalyst center has also been modified by chiral ligands in order to induce stereoselectivity in the DA reaction at low temperatures. As a benchmark transformation (Scheme 9.27), the conversion of cyclopentadiene and the dienophile **34** was often used to determine the effectiveness of the chiral iron catalyst system.

The ligands used in the iron-based catalyst systems and the best results for the respective catalytic system in the reaction of the acrylate derivative **34** with cyclopentadiene are summarized in Figure 9.3.

The best results were obtained with the bisoxazole class of ligands such as **35**, described by Kanemasa and coworkers [64–66] and for the derivative **36** reported by Corey and coworkers [67, 68] and Sibi *et al.* [69], who applied heterocyclic additives to the iron-catalyzed reaction. The bidentate sulfoxide ligand **37** introduced by Khiar *et al.* [70] and the phosphorus oxide ligand **38** reported by Imamoto's group [71] were less effective with respect to chiral induction.

The modification of the $CpFe(L)_2{}^+$ fragment in **39** with chiral phosphorus donor ligands based on chiral 1,2-dioles was investigated by Kündig and coworkers and tested in the reaction of acrolein derivatives (Scheme 9.28) [72–74].

In addition to these commonplace substrates, only a few extraordinary educts have been used in iron-catalyzed DA reactions, such as the naphthoquinones investigated by Brimble and McEwen [75]. Whereas the application of $FeCl_3$ and a chiral bisoxazoline ligand gave only a 25% yield and no chiral induction in the reaction of 2-acetyl-1,4-naphthoquinone with cyclopentadiene, the corresponding copper(II) triflate gave a 66% yield and moderate enantioselectivities (up to 50% *ee*). Another example was reported by Shibasaki's group in which the 2-alkoxy-1,3-butadiene **40**

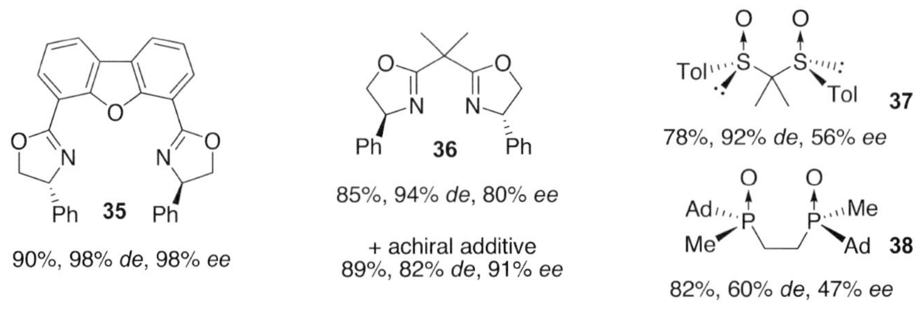

Figure 9.3 Chiral ligands utilized in asymmetric DA reactions.

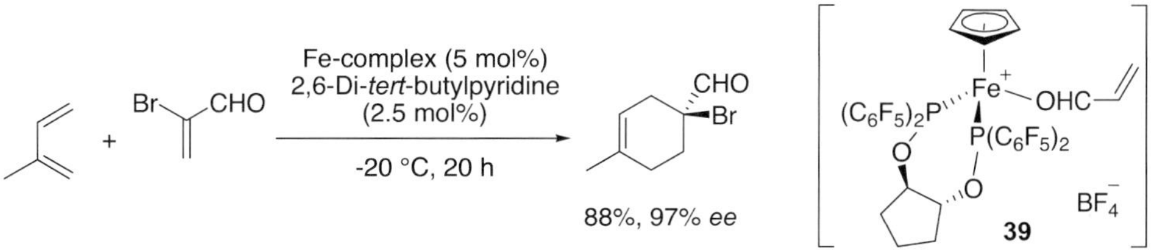

Scheme 9.28 Asymmetric DA reactions with chiral CpFe derivatives.

was used as the diene component (Scheme 9.29), discovered en route to the synthesis of polycyclic acylphlorogluranes [76].

9.6.2 Diels–Alder Reactions with Neutral Electron Demand

The DA reaction with neutral electron demand cannot be catalyzed by iron complexes in high oxidation states. Instead, low-valent iron complexes are able to coordinate both of the non-activated starting materials and induce a stepwise process incorporating formal oxidative insertion and reductive elimination steps within the ligand sphere of the iron complex. Consequently, several free coordination sites on the iron must be available for the educts. The catalysts used are either preformed as low-valent complexes with potentially labile ligands [77, 78], such as a cyclooctatetraene (COT) ligand, or the active species is generated *in situ* by reduction with organometallic reagents, such as ethyl–metal species [79, 80]. The application of diimine ligands in low-valent iron-catalyzed DA reactions investigated by tom Dieck and coworkers showed a higher level of reaction control [81] when 1,3-dienes were reacted with alkynes and allowed simple introduction of chiral side-chains to control the stereochemistry of the cycloadducts, especially when 1,3-dienes were dimerized in a formal [4 + 4]-cycloaddition process [82, 83]. A representative example for a DA reaction with neutral electron demand using the functionalized alkyne **42** as a dienophile is shown in Scheme 9.30.

This interesting and inspiring field seems to have been abandoned for some time and only recently was the pioneering work of Petit's group using an electrochemically

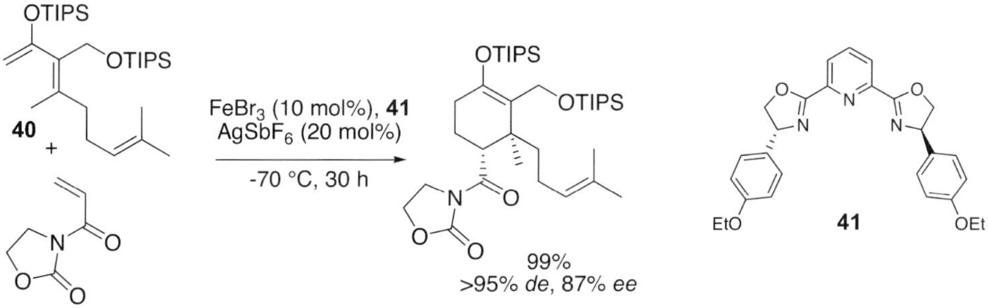

Scheme 9.29 Iron-catalyzed generation of a natural product synthesis intermediate.

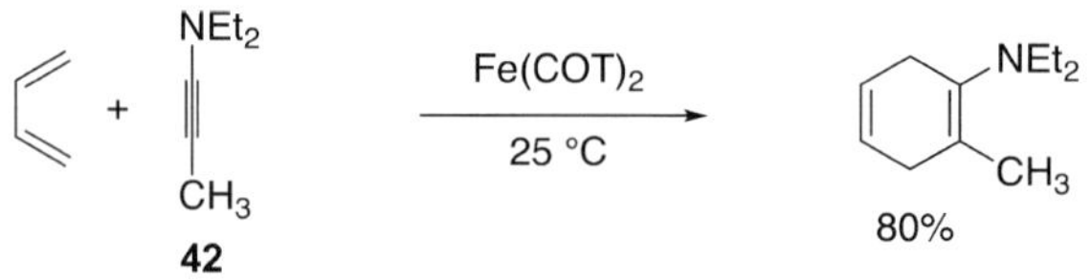

Scheme 9.30 Iron-catalyzed DA reaction with neutral electron demand.

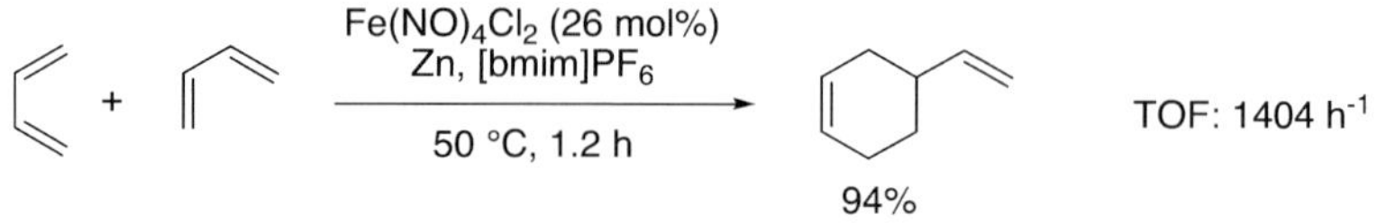

Scheme 9.31 Iron-catalyzed dimerization of 1,3-butadiene.

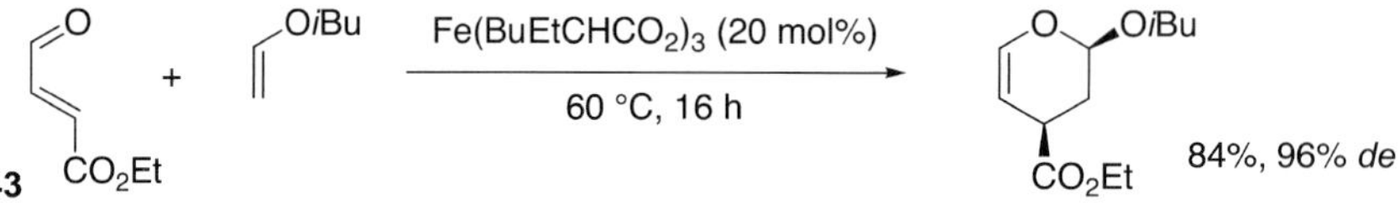

Scheme 9.32 Iron-catalyzed DA reaction with inverse electron demand.

generated $Fe(NO)_2$ complex for cycloaddition reactions [84] picked up by de Souza and coworkers in the application of *in situ*-generated "$Fe(NO)_2$" in ionic liquids ([bmim]PF_6 = 1-butyl-3-methylimidazolium hexafluorophosphate) [85]. The active species was generated through addition of reducing agents, such as zinc powder, to catalyze the dimerization of 1,3-butadiene very efficiently (Scheme 9.31).

When isoprene was used, the reactivity was somewhat diminished (TOF: up to $707\,h^{-1}$) and also mixtures of isomers were encountered.

9.6.3 Diels–Alder Reactions with Inverse Electron Demand

So far, only a single report, by Gorman and Tomlinson of an iron-catalyzed DA reaction with inverse electron demand, has appeared [86]. The transformation of a 4-oxobutenoate (**43**) as a rather electron-poor hetero-1,3-diene and an enol ether as the electron-rich dienophile can be seen as an extreme example of a diastereoselective hetero-DA reaction controlled by an iron catalyst (Scheme 9.32).

The yields for this transformation are good and the diastereoselectivities are generally excellent; however, the scope of the iron-catalyzed DA reaction with inverse electron demand seems to be limited.

9.7 Cyclotrimerization

While the DA reaction is a very elegant way to generate cyclohexene and cyclohexadiene derivatives, the cyclotrimerization of alkynes is an atom-economic and efficient route to the synthesis of benzene and other aromatic compounds. The reaction is very similar to the DA reaction with neutral electron demand, since the starting materials must coordinate within the ligand sphere of the transition metal and insertion processes and also reductive eliminations are involved in the reaction mechanism. In addition to many other metals, of which cobalt catalysts are probably the most important, cyclotrimerization can also be performed with low-valent iron complexes. In pioneering work by Hübel and Hoogzand [87] and later by Carbonaro *et al.* [88] and Usieli *et al.* [89] concerning the cyclotrimerization of alkynes with cobalt complexes and diverse iron–carbonyl complexes, it could be shown that the activity of iron–carbonyl complexes (reaction temperatures mostly >250 °C) is much lower than that of similar cobalt complexes (which require temperatures of approximately 150 °C). Nevertheless, later work by Zenneck's group demonstrated that with the appropriate labile ethylene ligand bearing iron complexes, such as bis($H_2C{=}CH_2$)(toluene)iron (0) [90], cyclotrimerizations of alkynes can be performed at ambient temperature and below. However, the instability of the iron complex inhibits a broad application of this methodology.

An improvement was reported by Pertici's group by applying an iron sandwich complex which is able to cyclotrimerize terminal and internal alkynes in good to excellent yields (Scheme 9.33) [91].

However, functional groups, other than a trimethylsilyl group, were not used and complex molecular architectures were not addressed. The regioselectivity for many terminal alkynes is only moderate ($R^1 = Ph$, $R^2 = H$; **44** : **45** = 65 : 35), whereas the regioselectivity for trimethylsilylacetylene is excellent ($R^1 = SiMe_3$, $R^2 = H$: **44** : **45** = 5 : 95).

Okamoto and coworkers recently described the iron-catalyzed cyclotrimerization of alkynes utilizing a low-valent iron–diimine complex that was generated *in situ* upon reduction with zinc dust (Scheme 9.34) [92].

The problem of regioisomerism was avoided by utilizing intramolecular triynes such as **46** as substrates for the cyclotrimerization process. However, many diimines and pyridinimines seem to give very reactive complexes so that further developments can be envisaged.

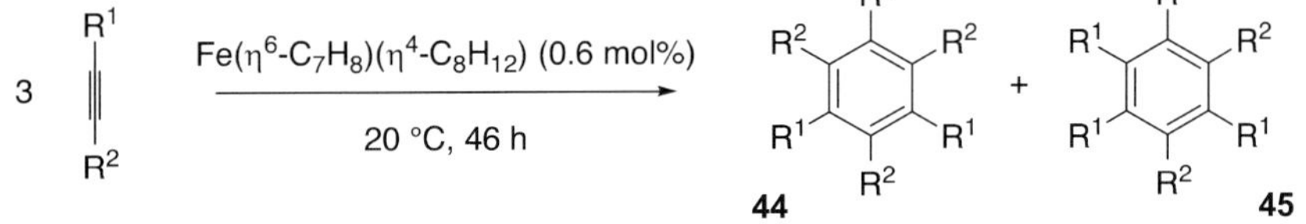

3 R^1–≡–R^2 → $Fe(\eta^6\text{-}C_7H_8)(\eta^4\text{-}C_8H_{12})$ (0.6 mol%), 20 °C, 46 h → **44** + **45**

Scheme 9.33 Synthesis of regioisomers by an iron-catalyzed trimerization of alkynes.

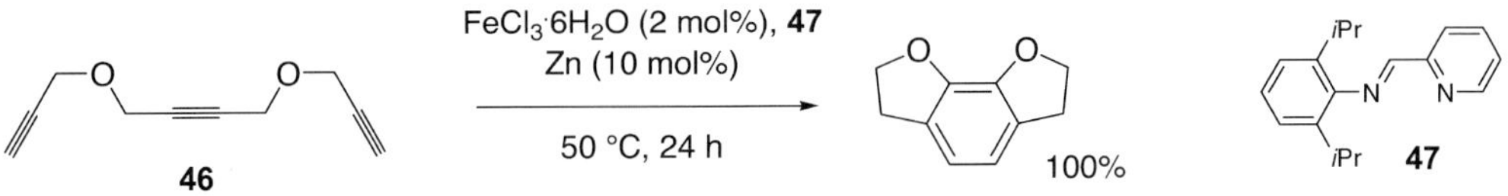

46 → $FeCl_3 \cdot 6H_2O$ (2 mol%), **47**, Zn (10 mol%), 50 °C, 24 h → 100%; *i*Pr, *i*Pr, N, N **47**

Scheme 9.34 Intramolecular iron-catalyzed alkyne trimerization.

9.8
[3 + 2]-Cycloadditions

The iron-catalyzed [3 + 2]-cycloaddition (Huisgen reaction) of nitriles and carbonyl compounds as reported by Itoh *et al.* is one of the rare examples reported where an iron reagent can be utilized for the synthesis of 1,2,4-oxadiazoles (Scheme 9.35) [93]. In this reaction, methyl ketones are nitrated at the α-position by $Fe(NO_3)_3$ to generate an α-nitro ketone. This intermediate rearranges to an acyl cyanate, which reacts further with the nitrile to give the heterocyclic product **48** in good to excellent yields ($R^1 = Ph$, $R^2 = CH_3$; 95% yield).

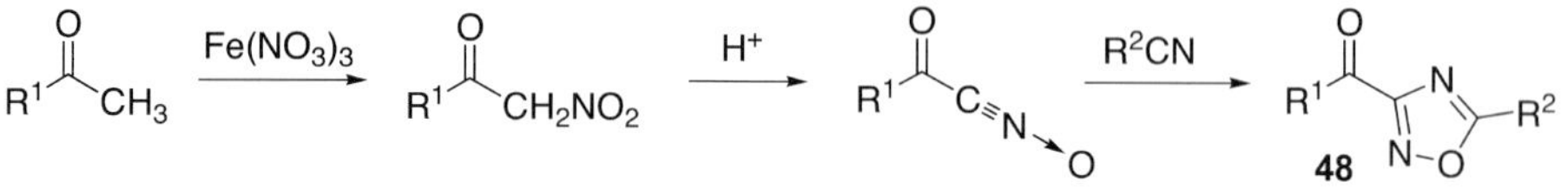

Scheme 9.35 Iron-initiated Huisgen reaction.

[3 + 2]-Cycloadditions can also be initiated by the $CpFe(CO)_2^+$ fragment, as demonstrated by Rosenblum's group in an early report where a $CpFe(CO)_2{-}CH_2CH{=}CH_2$ complex reacts as the C_3-component with the $CpFe(CO)_2^+$ adduct **49** of cyclohexenone to give the bicyclic product **50** (Scheme 9.36) in very moderate yield of 10% [94].

However, in the presence of Lewis acids such as $AlBr_3$, without the presence of the $CpFe(CO)_2^+$ Lewis acid the [3 + 2]-cycloaddition process was more effective, giving the product in 45% yield. The application of a chiral cationic CpFe(diphosphine) complex as the catalyst (Scheme 9.37) for the asymmetric [3 + 2]-cycloaddition of nitrones to acrolein derivatives was described by Kündig and coworkers [95].

The products are generated in good to excellent yields and with a high level of chiral induction. The reaction can be applied to cyclic and acyclic nitrones whereas only α,β-unsaturated aldehydes are applicable as the dipolarophile.

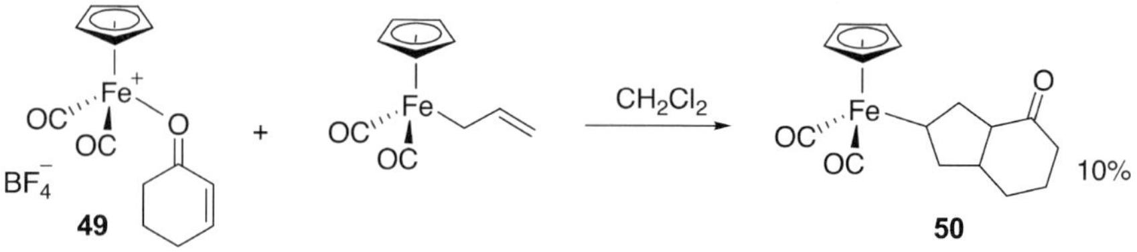

Scheme 9.36 Iron-mediated [3 + 2]-cycloaddition.

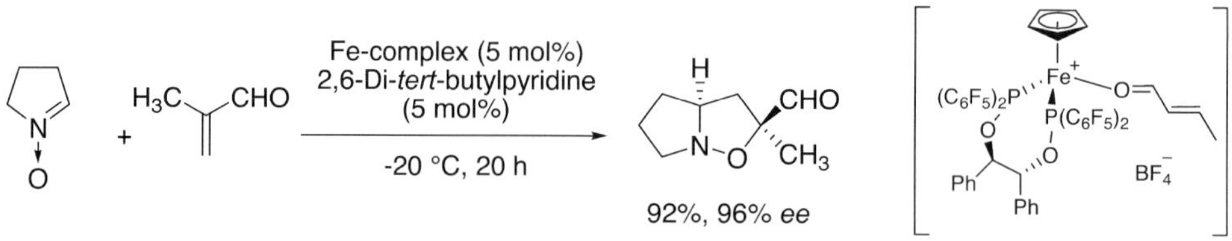

Scheme 9.37 Asymmetric iron-catalyzed Huisgen reaction.

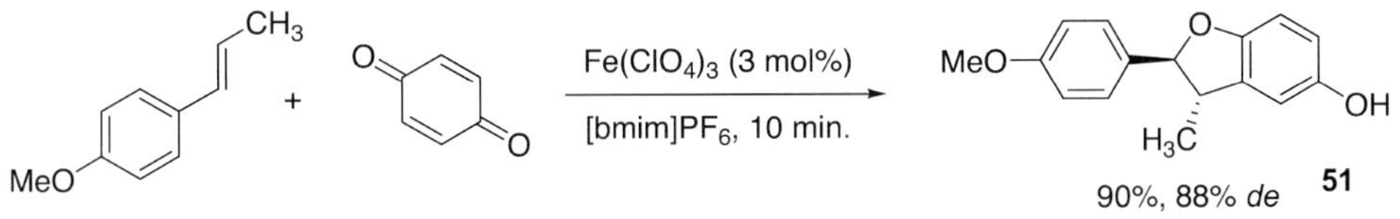

Fe(ClO$_4$)$_3$ (3 mol%)
[bmim]PF$_6$, 10 min.
90%, 88% *de*
51

Scheme 9.38 Formal [3 + 2]-cycloaddition to benzofurans.

A formal iron-catalyzed [3 + 2]-cycloaddition of styrene derivatives with benzoquinone was reported by Itoh's group [96]. The process is believed to proceed via electron-transfer reactions mediated by a proposed Fe^{3+}/Fe^{2+} couple, which generates a styrene radical cation and a semiquinone. These intermediates undergo stepwise addition to yield the benzofuran product **51** (Scheme 9.38). The reaction seems to be limited to electron-rich alkoxy-functionalized styrenes, as the Fe^{3+}/Fe^{2+} redox couple is otherwise unable to transfer the electrons from the styrene to the quinone.

9.9
[3 + 3]-Cycloadditions

Along with several other Lewis acids, $FeCl_3$ was also tested by Hsung and coworkers in the formal [3 + 3]-cycloaddition of enolized 1,3-diketones with α,β-unsaturated carbonyl compounds (Scheme 9.39) [97]. The iron-catalyzed reaction gave the desired bicyclic compound **52** in good yields; better results were obtained utilizing $BF_3{\cdot}OEt_2$ or $In(OTf)_3$ as Lewis acid.

9.10
Ring Expansion Reactions

One possibility for achieving ring expansion is the use of carbenium ion intermediates, which can be generated starting from a variety of functional groups. Among these reactions are transformations of alcohols with Lewis acids resulting in a migration of substituents, such as the Wagner–Meerwein rearrangement, so that the most stable carbenium cation is formed. According to this general scheme, iron-initiated ring expansion reactions are known, where iron salts act as Lewis acid reagents in dehydrative ring expansion reactions of cyclobutanol derivatives (Scheme 9.40), as described by Fadel and Salaün [98].

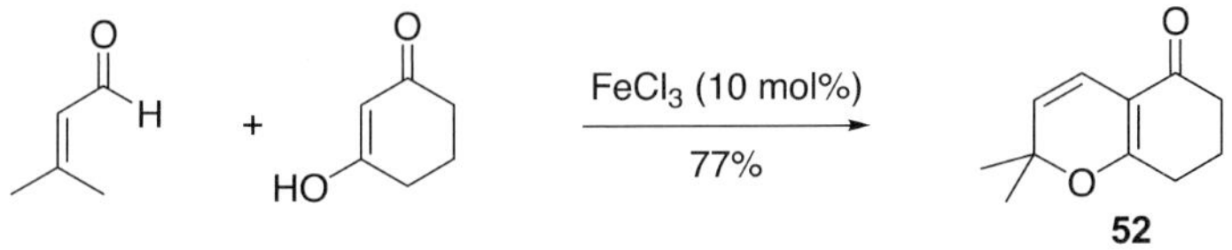

$FeCl_3$ (10 mol%)
77%
52

Scheme 9.39 Formal [3 + 3]-cycloaddition to bicyclic products.

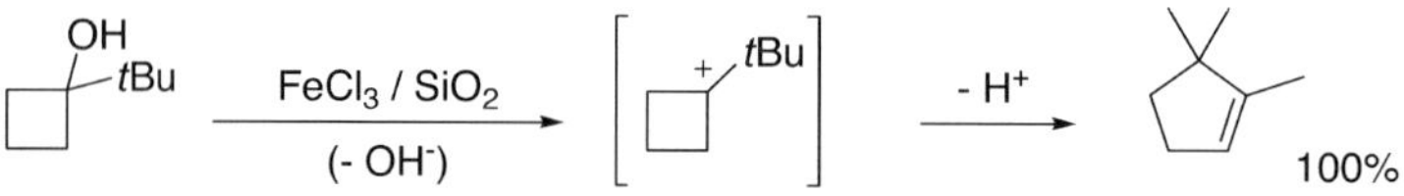

OH tBu $FeCl_3$ / SiO_2 (- OH^-) + tBu - H^+ 100%

Scheme 9.40 Iron-initiated ring expansion reaction via carbenium ions.

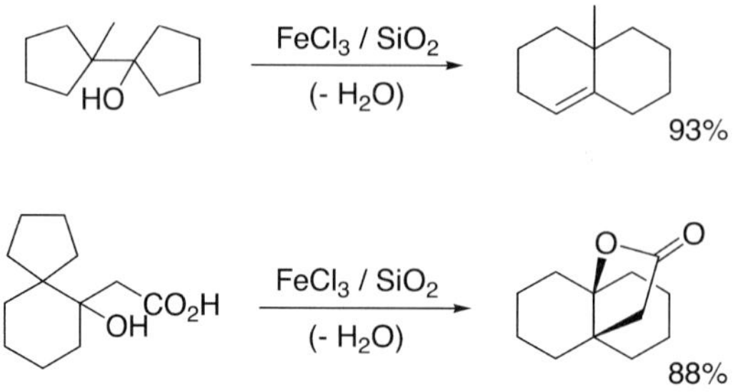

HO $FeCl_3$ / SiO_2 (- H_2O) 93%

CO_2H OH $FeCl_3$ / SiO_2 (- H_2O) O O 88%

Scheme 9.41 Ring expansion of dicyclopentane and spiro derivatives.

In addition to the ring expansion of cyclobutane derivatives, 1-cyclopentylcyclopentanol derivatives and spiro compounds (Scheme 9.41) could be utilized in the iron-initiated transformation utilizing anhydrous iron(III) chloride on silica, with good yields [99].

The ring expansion of 1-alkoxy-functionalized cyclohexane carbaldehydes (Scheme 9.42) was investigated by Kuwajima's group with 50 mol% of $FeCl_3$ to yield the ring expansion products **53**/**54** in 89% yield as a 14 : 1 mixture of constitutional isomers [100].

Other methods leading to ring enlargement utilize low-valent iron complexes such as $Fe(CO)_5$, as was shown by Taber and coworkers in the ring expansion of vinylcyclopropane derivatives **55** under carbonylating conditions to give the cyclohexenone derivatives (Scheme 9.43) as a 5.9:1 mixture of the 2,5- (**56**) and the constitutional isomer, the 2,6-substituted cyclohexenones [101, 102].

Most remarkable is the fact that this method allows the synthesis of interesting cyclohexenone derivatives which are functionalized in the 5-position. Direct functionalization of the 5-position starting from cyclohexenones is unknown, to the best

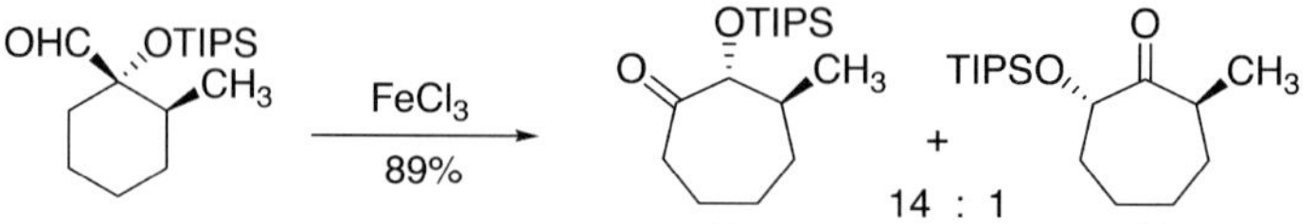

OHC OTIPS CH_3 $FeCl_3$ 89% O OTIPS CH_3 + TIPSO O CH_3 14 : 1 **53** **54**

Scheme 9.42 Ring expansion of cyclohexane carbaldehyde derivatives.

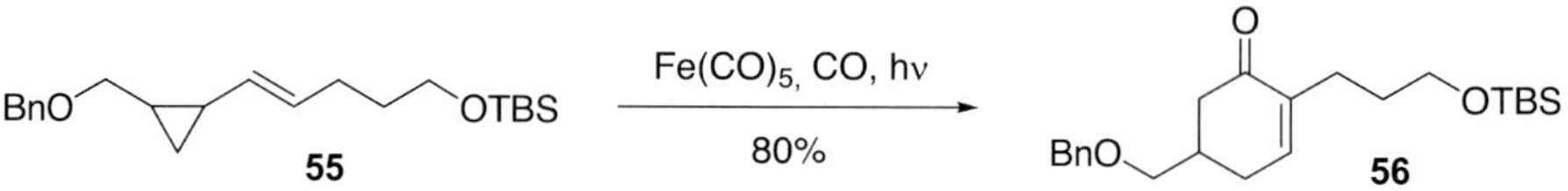

BnO OTBS **55** $Fe(CO)_5$, CO, hν 80% O OTBS BnO **56**

Scheme 9.43 Iron-catalyzed ring expansion of vinylcyclopropanes.

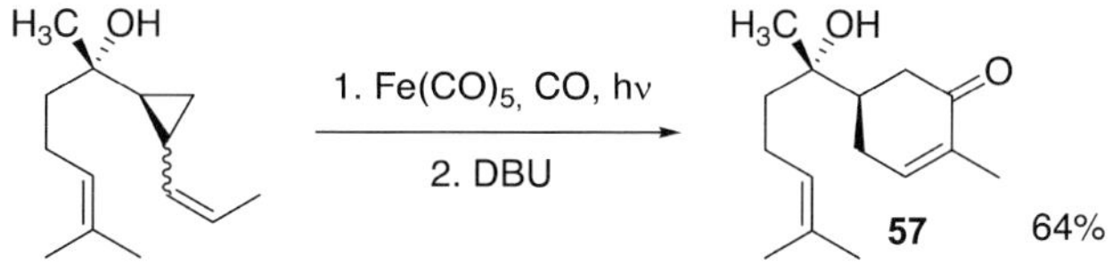

H_3C OH → 1. $Fe(CO)_5$, CO, hν; 2. DBU → **57** 64%

Scheme 9.44 Iron-catalyzed key step in the synthesis of (−)-delobanone.

Ph-epoxide + CO_2 → $FeBr_{2/3}$, Bu_4NI → 5-6%

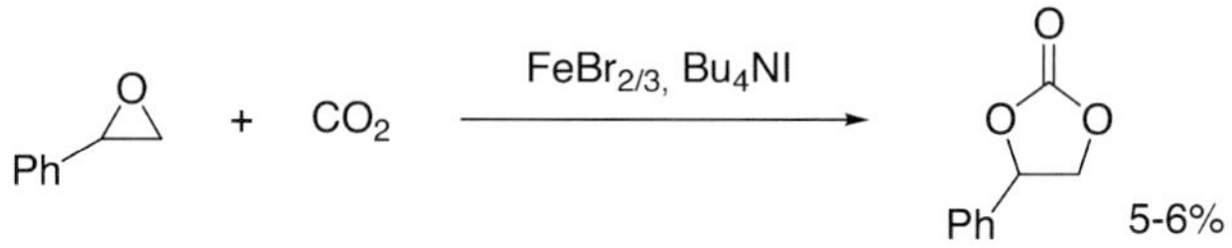

Scheme 9.45 Carbonylative iron-catalyzed ring expansion of epoxides.

of our knowledge, and only a few other methods are able to address, indirectly at best, the synthesis of such 5-substituted compounds [103]. An application of such a functionalization is exemplified by the synthesis of (−)-delobanone (**57** in Scheme 9.44) utilizing an iron-initiated carbonylative ring expansion reaction of a complex vinylcyclopropane in the key step [104].

Ring expansion reactions have also been achieved starting from epoxides under formal CO_2 insertion applying Lewis acids in supercritical CO_2, as reported by Fujita and coworkers [105]. Whereas iron salts such as $FeBr_2$ and $FeBr_3$ are only moderately active (Scheme 9.45), acceptable results (54%) were obtained using a $ZnBr_2$–Bu_4NI catalyst system for the synthesis of styrene carbonate.

Another approach for the ring expansion of epoxides uses low-valent iron complexes which open epoxides under reductive conditions, as reported by Hilt *et al.* [106]. The iron complexes are reduced and after coordination of the epoxide to the iron center an electron transfer initiates the radical-type ring opening of the epoxide. Under formal insertion of an alkene, regioselective formation of tetrahydrofurans was observed (Scheme 9.46). The reaction is applicable to a broad range of acceptor-substituted alkenes bearing another double or triple bond system in conjugation with the inserted carbon–carbon double bond.

The intermolecular epoxide ring expansion does not rely on the Thorpe–Ingold effect and is highly regioselective, with moderate to good diastereoselectivities depending on the ligands used. The intramolecular ring expansion has the advantage of less radical polymerization side-products and exhibits good to excellent diastereoselectivities, especially when iron–salen-type complexes are used [107, 108]. The intramolecular ring expansion of epoxyalkenes leads to bicyclic structures

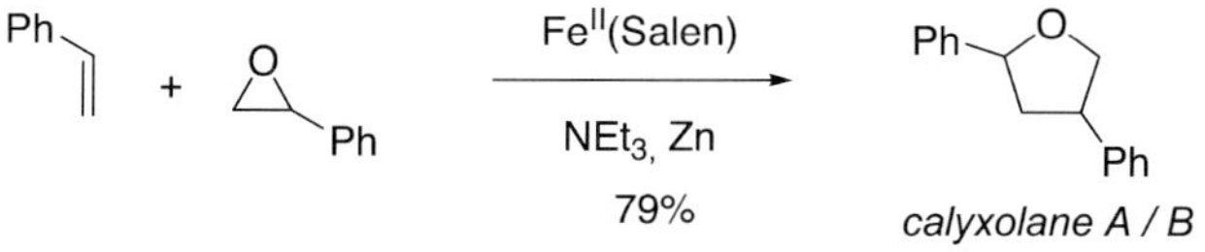

Ph-alkene + Ph-epoxide → Fe^{II}(Salen), NEt_3, Zn, 79% → *calyxolane A / B*

Scheme 9.46 Intermolecular ring expansion of epoxides to tetrahydrofurans.

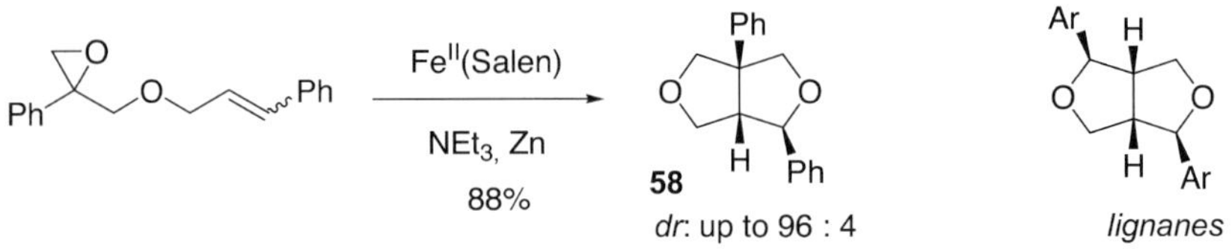

Scheme 9.47 Intramolecular ring expansion to bicyclic products.

(Scheme 9.47). Interestingly, if epoxyalkene ethers are used, structures such as **58** are generated independently of the *E*/*Z*-configuration of the alkene and, most strikingly, the products have close similarity to lignanes [109].

The iron-catalyzed ring expansion reaction is a complementary alternative to Ti(III) chemistry for the ring expansion of epoxides [110]. However, so far the reaction is limited to styrene oxide derivatives, while the alkene can be broadly varied.

9.11 Conclusion

The use of iron catalysts in various types of cycloaddition and ring expansion processes has a long tradition in methodology and recent applications in the synthesis of complex molecules and natural products show promising signs in a prospering field. Generally, iron catalysts are relatively inexpensive and non-toxic alternatives to many other transition metal catalyst systems and in the future these aspects will attract more scientists to explore the possibilities of using iron catalysts in organic synthesis.

References

1 L. Eberson, B. Olofsson, J.-O. Svensson, *Acta Chem. Scand.* 1992, **46**, 1005.

2 M. D. Redlich, M. F. Mayer, M. M. Hossain, *Aldrichim. Acta* 2003, **36**, 3.

3 J. A. Varela, C. Saá, *Chem. Rev.* 2003, **103**, 3787.

4 C. Bolm, J. Legros, J. Le Paih, L. Zani, *Chem. Rev.* 2004, **104**, 6217.

5 For a review see: K. Mikami, M. Shimizu, *Chem. Rev.* 1992, **92**, 1021.

6 J. M. Takacs, P. W. Newsome, C. Kuehn, *Tetrahedron* 1990, **46**, 5507.

7 J. M. Takacs, L. G. Anderson, *J. Am. Chem. Soc.* 1987, **109**, 2200.

8 L. F. Tietze, U. Beifuss, *Synthesis* 1988, 359.

9 B. E. Takacs, J. M. Takacs, *Tetrahedron Lett.* 1990, **31**, 2865.

10 J. M. Takacs, L. G. Anderson, M. W. Creswell, B. E. Takacs, *Tetrahedron Lett.* 1987, **28**, 5627.

11 J. M. Takacs, L. G. Anderson, G. V. B. Madhavan, M. W. Creswell, F. L. Seely, W. F. Devroy, *Organometallics* 1986, **5**, 2395.

12 J. M. Takacs, Y. C. Myoung, L. G. Anderson, *J. Org. Chem.* 1994, **59**, 6928.

13 J. M. Takacs, L. G. Anderson, *J. Am. Chem. Soc.* 1987, **109**, 2200.

14 J. M. Takacs, P. W. Newsome, C. Kuehn, *Tetrahedron* 1990, **46**, 5507.

15 J. M. Takacs, Y. C. Myoung, *Tetrahedron Lett.* 1992, **33**, 317.

16 J. M. Takacs, S. C. Boito, *Tetrahedron Lett.* 1995, **36**, 2941.

17 J. M. Takacs, S. Vayalakkada, S. J. Mehrmann, C. L. Kingsbury, *Tetrahedron Lett.* 2002, **43**, 8417.

18 J. M. Takacs, J. J. Weidner, B. E. Takacs, *Tetrahedron Lett.* 1993, **34**, 6219.

19 J. M. Takacs, J. J. Weidner, P. W. Newcome, B. E. Takacs, R. Chidambaram, R. Shoemaker, *J. Org. Chem.* 1995, **60**, 3473.

20 A. Fürstner, R. Martin, K. Majima, *J. Am. Chem. Soc.* 2005, **127**, 12236.

21 A. J. Pearson, X. Wang, *Tetrahedron Lett.* 2005, **46**, 3123.

22 A. J. Pearson, X. Wang, *Tetrahedron Lett.* 2005, **46**, 4809.

23 R. Vyas, G.-Y. Gao, J. D. Harden, X. P. Zhang, *Org. Lett.* 2004, **6**, 1907.

24 F. Avenier, J.-M. Latour, *Chem. Commun.* 2004, 1544.

25 B. D. Heuss, M. F. Mayer, S. Dennis, M. M. Hossain, *Inorg. Chim. Acta* 2003, **342**, 301.

26 M. F. Mayer, Q. Wang, M. M. Hossain, *J. Organomet. Chem.* 2001, **630**, 78.

27 M. F. Mayer, M. M. Hossain, *J. Org. Chem.* 1998, **63**, 6839.

28 M. F. Mayer, M. M. Hossain, *J. Organomet. Chem.* 2002, **654**, 202.

29 M. Redlich, M. M. Hossain, *Tetrahedron Lett.* 2004, **45**, 8987.

30 S. J. Mahmood, A. K. Saha, M. M. Hossain, *Tetrahedron* 1998, **54**, 349.

31 S. J. Mahmood, M. M. Hossain, *J. Org. Chem.* 1998, **63**, 3333.

32 J. Picione, S. J. Mahmood, A. Gill, M. Hilliard, M. M. Hossain, *Tetrahedron Lett.* 1998, **39**, 2681.

33 C. Roger, C. Lapinte, *Chem. Commun.* 1989, 1598.

34 V. Guerchais, S. Lévêque, A. Hornfeck, C. Lapinte, S. Sinbandhit, *Organometallics* 1992, **11**, 3928.

35 W. J. Seitz, A. K. Saha, D. Casper, M. M. Hossain, *Tetrahedron Lett.* 1992, **33**, 7755.

36 W. J. Seitz, M. M. Hossain, *Tetrahedron Lett.* 1994, **35**, 7561.

37 W. J. Seitz, A. K. Saha, M. M. Hossain, *Organometallics* 1993, **12**, 2604.

38 D. J. Casper, A. V. Sklyarov, S. Hardcastle, T. L. Barr, F. H. Försterling, K. F. Surerus, M. M. Hossain, *Inorg. Chim. Acta* 2006, **359**, 3129.

39 Q. Wang, F. H. Försterling, M. M. Hossain, *J. Organomet. Chem.* 2005, **690**, 6238.

40 P. Tagliatesta, A. Pastorini, *J. Mol. Catal. A* 2003, **198**, 57.

41 T.-S. Lai, F.-Y. Chan, P.-K. So, D.-L. Ma, K.-Y. Wong, C.-M. Che, *J. Chem. Soc., Dalton Trans.* 2006, 4845.

42 C. G. Hamaker, G. A. Mirafzal, L. K. Woo, *Organometallics* 2001, **20**, 5171.

43 S. K. Edulji, S. T. Nguyen, *Pure Appl. Chem.* 2004, **76**, 645.

44 S. K. Edulji, S. T. Nguyen, *Organometallics* 2003, **22**, 3374.

45 P. Braunstein, G. Clerc, X. Morise, *New. J. Chem.* 2003, **27**, 68.

46 A. G. M. Barrett, J. Mortier, M. Sabat, M. A. Sturgess, *Organometallics* 1988, **7**, 2553.

47 A. G. M. Barrett, N. E. Carpenter, J. Mortier, M. Sabat, *Organometallics* 1990, **9**, 151.

48 M. Rosenblum, D. Scheck, *Organometallics* 1982, **1**, 397.

49 M. Periasamy, A. Mukkanti, D. S. Raj, *Organometallics* 2004, **23**, 6323.

50 M. W. Bouwkamp, A. C. Bowman, E. Lobkovsky, P. J. Chirik, *J. Am. Chem. Soc.* 2006, **128**, 13340.

51 M. S. Sigman, B. E. Eaton, *J. Am. Chem. Soc.* 1996, **118**, 11783.

52 B. E. Eaton, B. Rollman, *J. Am. Chem. Soc.* 1992, **114**, 6245.

53 M. S. Sigman, C. E. Kerr, B. E. Eaton, *J. Am. Chem. Soc.* 1993, **115**, 7545.

54 M. S. Sigman, B. E. Eaton, J. D. Heise, C. P. Kubiak, *Organometallics* 1996, **15**, 2829.

55 M. S. Sigman, B. E. Eaton, *J. Org. Chem.* 1994, **59**, 7488.

56 W. Imhof, E. Anders, *Chem. Eur. J.* 2004, **10**, 5717.

57 W. Imhof, A. Göbel, *J. Mol. Catal. A* 2003, **197**, 15.

58 W. Imhof, E. Anders, A. Göbel, H. Görls, *Chem. Eur. J.* 2003, **9**, 1166.

59 A. S. Olson, W. J. Seitz, M. M. Hossain, *Tetrahedron Lett.* 1991, **32**, 5299.

60 P. V. Bonnesen, C. L. Puckett, R. V. Honeychuck, W. H. Hersh, *J. Am. Chem. Soc.* 1989, **111**, 6070.

61 A. K. Saha, M. M. Hossain, *Tetrahedron Lett.* 1993, **34**, 3833.

62 T. R. Kelly, S. K. Maity, P. Meghani, N. S. Chandrakumar, *Tetrahedron Lett.* 1989, **30**, 1357.

63 S. P. Chavan, A. K. Sharma, *Synlett* 2001, 667.

64 S. Kanemasa, Y. Oderaotoshi, H. Yamamoto, J. Tanaka, E. Wada, *J. Org. Chem.* 1997, **62**, 6454.

65 S. Kanemasa, Y. Oderaotoshi, S.-i. Sakaguchi, H. Yamamoto, J. Tanaka, E. Wada, D. P. Curran, *J. Am. Chem. Soc.* 1998, **120**, 3074.

66 In this case a chiral bisoxazoline ligand was used: S. Kanemasa, K. Adachi, H. Yamamoto, E. Wada, *Bull. Chem. Soc. Jpn.* 2000, **73**, 681.

67 E. J. Corey, K. Ishihara, *Tetrahedron Lett.* 1992, **33**, 6807.

68 E. J. Corey, N. Imai, H.-Y. Zhang, *J. Am. Chem. Soc.* 1991, **113**, 728.

69 M. P. Sibi, S. Manyem, H. Palencia, *J. Am. Chem. Soc.* 2006, **128**, 13660.

70 N. Khiar, J. Fernández, F. Alcudia, *Tetrahedron Lett.* 1993, **34**, 123.

71 S. Matsukawa, H. Sugama, T. Imamoto, *Tetrahedron Lett.* 2000, **41**, 6461.

72 E. P. Kündig, C. M. Saudan, F. Viton, *Adv. Synth. Catal.* 2001, **343**, 51.

73 E. P. Kündig, B. Bourdin, G. Bernardinelli, *Angew. Chem.* 1994, **106**, 1931; *Angew. Chem. Int. Ed. Engl.*1994, **33**, 1856.

74 M. E. Bruin, E. P. Kündig, *Chem. Commun.* 1998, 2635.

75 M. A. Brimble, J. F. McEwan, *Tetrahedron: Asymmetry* 1997, **8**, 4069.

76 H. Usuda, A. Kuramochi, M. Kanai, M. Shibasaki, *Org. Lett.* 2004, **6**, 4387.

77 J. P. Genet, J. Ficini, *Tetrahedron Lett.* 1979, **17**, 1499.

78 A. Carbonaro, A. Greco, G. Dall'Asta, *J. Org. Chem.* 1968, **33**, 3948.

79 H. tom Dieck, R. Diercks, *Angew. Chem.* 1983, **95**, 801; *Angew. Chem. Int. Ed. Engl.*1983, **22**, 778.

80 A. Greco, A. Carbonaro, G. Dall'Asta, *J. Org. Chem.* 1970, **35**, 271.

81 H. tom Dieck, J. Dietrich, *Angew. Chem.* 1985, **97**, *795; Angew. Chem. Int. Ed. Engl.* 1985, **24**, 781.

82 H. tom Dieck, J. Dietrich, *Chem. Ber.* 1984, **117**, 694.

83 K.-U. Baldenius, H. tom Dieck, W. A. König, D. Icheln, T. Runge, *Angew. Chem.* 1992, **104**, 338; *Angew. Chem. Int. Ed. Engl.*1992, **31**, 305.

84 E. Le Roy, F. Petit, J. Hennion, J. Nicole, *Tetrahedron Lett.* 1978, **27**, 2403.

85 R. A. Ligabue, J. Dupont, R. F. de Souza, *J. Mol. Catal. A* 2001, **169**, 11.

86 D. B. Gorman, I. A. Tomlinson, *Chem. Commun.* 1998, 25.

87 W. Hübel, C. Hoogzand, *Chem. Ber.* 1960, **93**, 103.

88 A. Carbonaro, A. Greco, G. Dall'Asta, *J. Organomet. Chem.* 1969, **20**, 177.

89 V. Usieli, R. Victor, S. Sarel, *Tetrahedron Lett.* 1976, **31**, 2705.

90 A. Funhoff, H. Schäufele, U. Zenneck, *J. Organomet. Chem.* 1988, **345**, 331.

91 C. Breschi, L. Piparo, P. Pertici, A. M. Caporusso, G. Vitulli, *J. Organomet. Chem.* 2000, **607**, 57.

92 N. Saino, D. Kogure, K. Kase, S. Okamoto, *J. Organomet. Chem.* 2006, **691**, 3129.

93 K.-i. Itoh, H. Sakamaki, C. A. Horiuchi, *Synthesis* 2005, 1935.

94 A. Bucheister, P. Klemarczyk, M. Rosenblum, *Organometallics* 1982, **1**, 1679.

95 F. Viton, G. Bernerdinelli, E. P. Kündig, *J. Am. Chem. Soc.* 2002, **124**, 4968.

96 H. Ohara, H. Kiyokane, T. Itoh, *Tetrahedron Lett.* 2002, **43**, 3041.

97 A. V. Kurdyumov, N. Lin, R. P. Hsung, G. C. Gullickson, K. P. Cole, N. Sydorenko, J. J. Swidorski, *Org. Lett.* 2006, **8**, 191.

98 A. Fadel, J. Salaün, *Tetrahedron* 1985, **41**, 413.

99 A. Fadel, J. Salaün, *Tetrahedron* 1985, **41**, 1267.

100 T. Matsuda, K. Tanino, I. Kuwajima, *Tetrahedron Lett.* 1989, **30**, 4267.

101 D. F. Taber, K. Kanai, Q. Jiang, G. Bui, *J. Am. Chem. Soc.* 2000, **122**, 6807.

102 D. F. Taber, P. V. Joshi, K. Kanai, *J. Org. Chem.* 2004, **69**, 2268.

103 G. Hilt, F. Galbiati, *Synthesis* 2006, 3589; see also M. T. Valahovic, J. M. Keane, W. D. Harman, in D. Astruc (ed.), *Modern Arene Chemistry*, Wiley, New York, 2002, p. 297; A. R. Pape, K. P. Kaliappan, E. P. Kündig, *Chem. Rev.* 2000, **100**, 2917.

104 D. F. Taber, G. Bui, B. Chen, *J. Org. Chem.* 2001, **66**, 3423.

105 J. Sun, S.-I. Fujita, F. Zhao, M. Arai, *Appl. Catal. A* 2005, **287**, 221.

106 G. Hilt, P. Bolze, I. Kieltsch, *Chem. Commun.* 2005, 1996.

107 G. Hilt, C. Walter, P. Bolze, *Adv. Synth. Catal.* 2006, **348**, 1241.

108 G. Hilt, P. Bolze, K. Harms, *Chem. Eur. J.* 2007, **13**, 4312.

109 G. Hilt, P. Bolze, M. Heitbaum, K. Hasse, K. Harms,W. Massa, *Adv. Synth. Catal.* 2007, **349**, 2018.

110 A. Gansäuer, J. Justicia, C.-A. Fan, D. Worgull, F. Piestert, *Top. Curr. Chem.* 2007, **279**, 25.

Index

Iron Catalysis in Organic Chemistry. Edited by Bernd Plietker

ISBN: 978-3-527-31927-5

d

k

l

m